A New Introduction to Organic Chemistry

SI Edition

G. I. Brown B.A., B.Sc.

Formerly Assistant Master, Eton College

Longman

Longman Group UK Limited
Longman, House,
Burnt Mill, Harlow, Essex CM20 2JE, England
and Associated Companies throughout the world.

First published 1978
Twelfth impression 1994

ISBN 0 582 35128 6

Produced by Longman Singapore Publishers Pte Ltd
Printed in Singapore

The publisher's policy is to use paper manufactured from
sustainable forests.

Contents

List of Summarising Charts.

Preface

This book is a replacement for my earlier Introduction to Organic Chemistry which is now almost twenty years old. Like the earlier book it is intended to provide a thorough introduction to organic chemistry and to arouse a student's interest in the subject. Its contents are based mainly on the requirements of such examinations as the G.C.E. (A and S levels) and University scholarship examinations.

The book has been completely rewritten. The preparation, properties and uses of all simple organic compounds are fully described in relation to the functional group or groups which they contain and the book has been rearranged by treating, for the most part, similar aliphatic and aromatic compounds together. The factual content has been limited so that much greater stress can be put on reaction mechanisms.

The presentation is in numbered sections so that particular topics can be omitted or rearranged to suit individual requirements and the theoretical parts have been presented in such a way that they can, if necessary, be omitted at a first reading.

Summarising charts and tables are included in most chapters for those who find them useful.

SI units have been used throughout and the nomenclature is based on the recommendations of the ASE report on Chemical Nomenclature, Symbols and Terminology (1972).

Acknowledgements

We are grateful to the following examining boards for permission to use past examination questions from various examinations:

The Associated Examining Board, University of Cambridge Local Examinations Syndicate and St. John's College and Trinity College Cambridge, Joint Matriculation Board, University of London School Examinations Department, Oxford Delegacy of Local Examinations, Oxford and Cambridge Schools Examination Board, Southern Universities Joint Board and the Welsh Joint Education Committee.

1

Introduction

1 The carbon atom

Organic chemistry is the study of carbon compounds. It owes its name to the fact that the original compounds studied came from plants or animals. Such compounds are still studied, particularly in biochemistry, but many so-called organic compounds, e.g. plastics, have no relation to living matter.

Carbon is unique amongst the elements in that its known compounds are much more numerous than the known compounds of all the other elements put together. This is mainly due to the fact that carbon can form a variety of strong covalent bonds.

The carbon atom has 6 electrons with 4 in the outermost orbital. The energy changes involved in gaining or losing four electrons are very high so that simple C^{4+} and C^{4-} ions do not exist, but the four electrons are available for sharing in covalent bond formation and carbon forms such bonds in three important ways (Figs. 1 and 2).

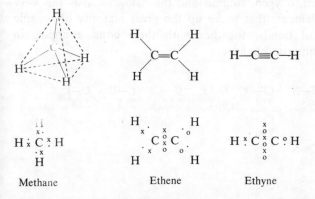

Methane   Ethene   Ethyne

Fig. 1. The arrangement of bonds (top) and electrons (bottom) in molecules of methane, ethene and ethyne.

a. Formation of four single covalent bonds. Four other monovalent atoms or radicals can be linked to a carbon atom by four single covalent bonds, as in methane, CH_4. The carbon atom will be surrounded by four shared pairs of electrons and as these will repel each other the four hydrogen atoms will be arranged at the corners of a tetrahedron. The H—C—H bond angle is 109.5 °.

b. Formation of two single and one double bond. This arrangement of bonds occurs in ethene, C_2H_4. Each carbon atom is surrounded by three lots of shared pairs of electrons and as they repel each other the molecule is planar with bond angles of 120 °.

c. Formation of one single and one triple bond. This occurs in ethyne, C_2H_2. The shared pairs, once again, repel each other so that the molecule is collinear, with bond angles of 180 °.

This simple way of describing the bonding of carbon will have to be extended later but it is adequate for many purposes. It illustrates one of the reasons why carbon is unique in forming so many stable compounds. In all three modes of bonding the carbon atom attains a noble gas arrangement of electrons (Fig. 1). The octet of electrons in the atom is both a fully shared octet and an octet which cannot expand because the maximum covalency of carbon is 4. Carbon atoms in carbon compounds cannot, therefore, act as donors or acceptors.

Other reasons are the great facility with which carbon atoms form strong covalent bonds between themselves in long chains or in rings, and the strength of the covalent bonds which carbon can form with hydrogen, nitrogen, oxygen, sulphur and the halogens. It is this very small group of elements that make up the great majority of organic molecules. Typical bonds, together with their bond energies, in $kJ\,mol^{-1}$, are summarised below:

C—H	C—C	C—N	C—O	C—F	C—Cl	C—Br	C—I
413	346	305	358	485	327	285	213
	C=C		C=O				
	610		736				
	C≡C	C≡N					
	835	890					

It will be seen that, as expected, more energy is required to break multiple bonds.

2 Representation of organic molecules

Covalent bonds are spatially directed as summarised in Fig. 2 so that molecules containing them may be three-dimensional, as for methane (Fig. 1). It is not easy to represent such spatial arrangements on paper and models must be used. Many of these give, too, a scaled representation of the relative sizes of the atoms concerned (Fig. 3) so that inter-atomic distances can be measured from the models.

On paper, either the molecular formula can be used, e.g. CH_4 for methane or CH_3OH for methanol, or a planar structural formula,

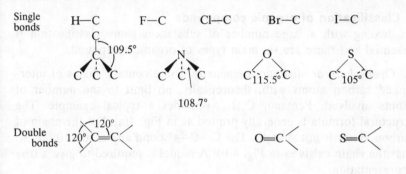

Fig. 2. The directional nature of bonds formed by H, C, N, O, S and the halogens with carbon.

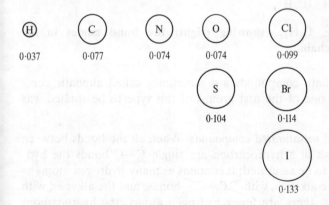

Fig. 3. The relative sizes of the atoms when covalently bonded. The figures give the covalent radii (in nm) which, when added together, give an approximate measure of the single bond distance between the two atoms concerned.

which may attempt to indicate the geometry, can be used, e.g.

H
|
H—C—H
|
H
Methane

H   H
| |
H—C—O
|
H
Methanol

It is easy when reading a book to come to regard all organic molecules as being flat; it is important to remember that most of them are not.

3 Classification of organic compounds
In dealing with a large number of substances some classification is essential and there are six main types of organic compound.

a. Open-chain or aliphatic compounds. These contain chains of inter-linked carbon atoms with, theoretically, no limit to the number of atoms involved. Pentane, C_5H_{12}, provides a typical example. The structural formula is generally printed as in Fig. 4(a), but the chain of carbon atoms is not straight. The C—C—C bond angles are 109.5° so that the chain exists as in Fig. 4(b). A model is required to give a true representation.

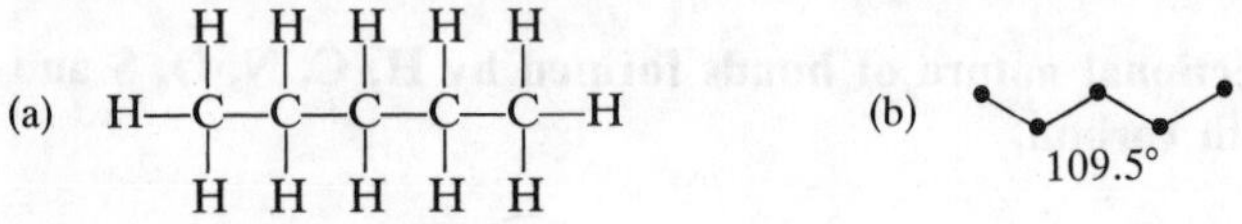

(a)
H H H H H
| | | | |
H—C—C—C—C—C—H
| | | | |
H H H H H

(b)   109.5°

Fig. 4. Pentane, C_5H_{12}, showing (right) the bond angles in the 5-carbon atom chain.

These open-chain compounds are sometimes called aliphatic compounds because one of the first groups of this type to be studied was the fats.

b. Saturated and unsaturated compounds. When all the bonds between the carbon atoms in a hydrocarbon are single C—C bonds the hydrocarbon is said to be saturated; it contains as many hydrogen atoms as is possible. In the alkenes, with >C=C< bonds, and the alkynes, with —C≡C— bonds, there are fewer hydrogen atoms; the hydrocarbons are said to be unsaturated.

The term unsaturation is also used, sometimes, to describe a compound containing multiple bonds between other pairs of atoms, e.g. >C=O, —C≡N.

Unsaturated compounds have significantly different properties from saturated ones.

c. Aromatic compounds. These are carbocyclic compounds, i.e. they contain rings of carbon atoms, based on benzene, C_6H_6. All the atoms in the benzene molecule lie in one plane and it may be represented as in Fig. 5. (a) is an over-simplification as it suggests the existence of both C—C and C=C bonds which do not, in fact, occur (p. 26); (b) is a conventional form of writing the molecule which attempts to indicate the intermediate nature of the bonds between the carbon atoms; (c) is a conventional short-hand version of (b).

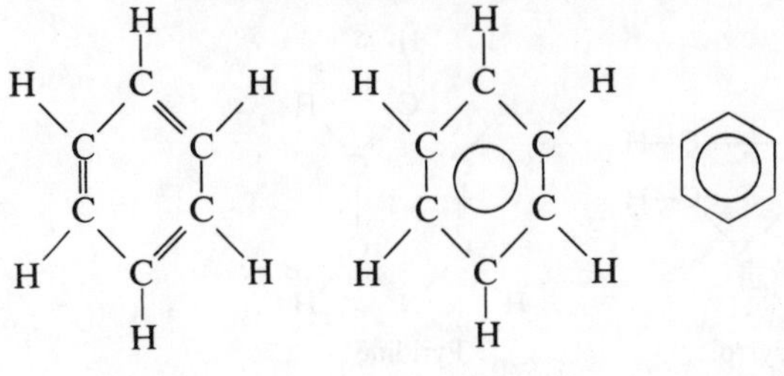

(a)   (b)   (c)

Fig. 5. Benzene, C_6H_6.

The term aromatic is due to the fact that one of the early groups of this type to be studied was the aromatic oils and spices; benzenoid is sometimes used as an alternative term.

The presence of a benzene ring in a compound endows the compound with what is referred to as *aromatic character* (p. 139) and there are important differences in properties between aliphatic and aromatic compounds, e.g. between hexane, C_6H_{14} and benzene, C_6H_6.

d. Alicyclic compounds. These are carbocyclic compounds with C—C and/or C=C bonds between the carbon atoms in the ring, e.g.

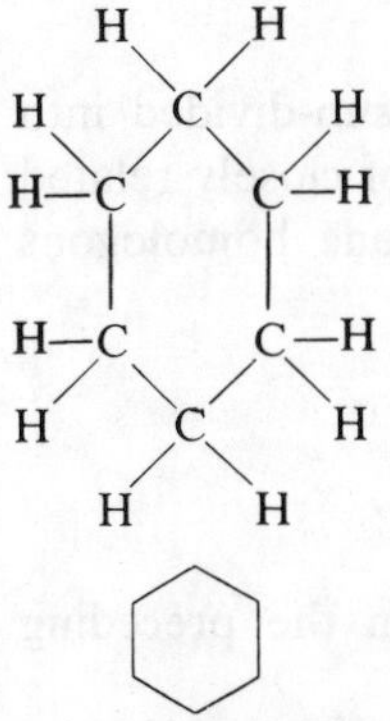

Cyclohexane 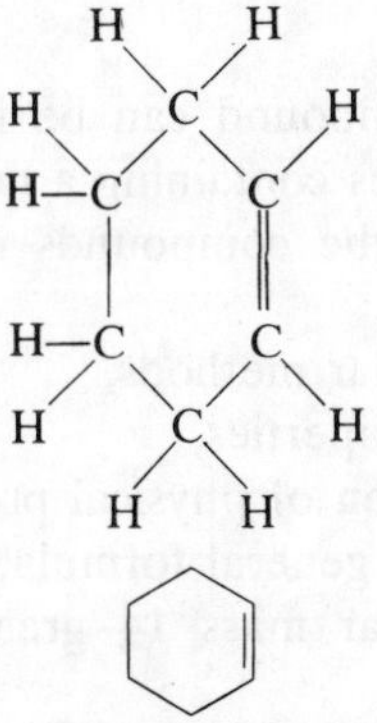   Cyclohexene 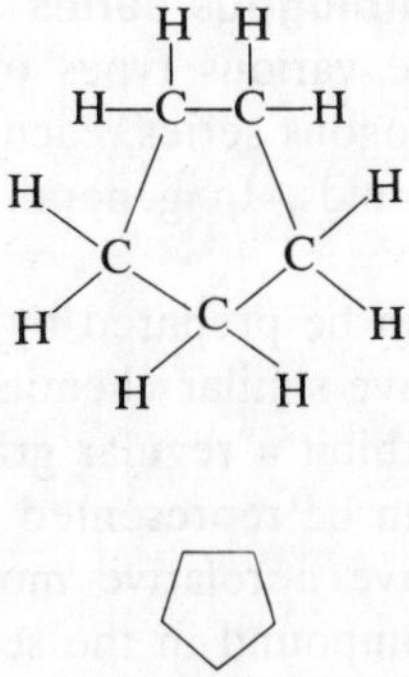   Cyclopentane

They do not contain a benzene ring so that they do not have aromatic character. They are, in general, not unlike comparable aliphatic compounds; hence the name alicyclic.

Care must be taken not to mistake the symbolic way of representing cyclohexane with the benzene ring formulation.

e. Heterocyclic compounds. These are cyclic compounds containing elements other than carbon, particularly oxygen, nitrogen or sulphur, in the ring, e.g.

Furan   Pyrrole   Pyridine

f. Polycyclic compounds. These contain two or more rings linked together, the rings being aromatic or heterocyclic, e.g.

Napthalene, $C_{10}H_8$ Quinoline, C_9H_7N

4 Homologous series

All the various types of compound can be further sub-divided into homologous series, each series containing a number of closely related compounds. In general, all the compounds in any one homologous series

(i) can be prepared by similar methods,
(ii) have similar chemical properties,
(iii) exhibit a regular gradation of physical properties,
(iv) can be represented by a general formula,
(v) have a relative molecular mass 14 greater than the preceding compound in the series.

a. Aliphatic series. The simplest aliphatic homologous series is that of the alkanes. The five simplest (or lowest) members are listed below:

		m.p./K	b.p./K	Density of liquid/g cm^{-3}	Relative molecular mass
Methane	CH$_4$	90.67	111.66	0.424	16
Ethane	C$_2$H$_6$	89.88	184.52	0.546	30
Propane	C$_3$H$_8$	85.46	231.08	0.582	44
Butane	C$_4$H$_{10}$	134.80	272.65	0.579	58
Pentane	C$_5$H$_{12}$	143.43	309.22	0.626	72

The general formula for any alkane is C_nH_{2n+2}.

Many other common homologous series can be regarded as derived from the alkanes by replacing one hydrogen atom in an alkane molecule by other monovalent atoms or radicals. Some of the commoner aliphatic homologous series are listed on p. 8; a fuller list is given in Appendix I (p. 368).

It will be seen that these homologous series are made up of molecules containing a C_nH_{2n+1} group together with some other group. The C_nH_{2n+1} groups are extremely common in organic chemistry; they are called *alkyl groups* and they contain one hydrogen atom less than the corresponding alkane. Thus

Methyl	Ethyl	Propyl	Butyl	Pentyl
—CH$_3$	—C$_2$H$_5$	—C$_3$H$_7$	—C$_4$H$_9$	—C$_5$H$_{11}$

They are not individual substances; they are monovalent groups which link with other monovalent groups to give individual chemicals.

The alkyl groups are so common that it is convenient to represent them by the letter R. Alkanes can then be written as R—H, amines as R—NH$_2$, halogenated alkanes (alkyl chlorides) as R—Cl, and so on. Me, Et, Pr etc, are also used as abbreviations for methyl, ethyl, propyl etc.

b. Aromatic series. Most of the aliphatic homologous series have a parallel aromatic series. These are derived by substitution of a hydrogen atom in aromatic hydrocarbons (*arenes*). The *aryl groups*, notably —C$_6$H$_5$ (phenyl), which are generalised as Ar correspond in aromatic chemistry to the alkyl groups in aliphatic chemistry. Thus Ar—H represents any arene, Ar—NH$_2$ any aromatic amine, and Ar—Cl any halogenoarene.

Name of series	General* formula	Suffix and/or prefix used	First two members of series*
Alkanes	C_nH_{2n+2}	-ane	Methane, CH_4 Ethane, C_2H_6
Alkenes	C_nH_{2n}	-ene	Ethene, C_2H_4 Propene, C_3H_6
Alkynes	C_nH_{2n-2}	-yne	Ethyne, C_2H_2 Propyne, C_3H_4
Chloroalkanes	R—Cl	chloro-ane	Chloromethane, CH_3—Cl Chloroethane C_2H_5—Cl
Alcohols	R—OH	-ol	Methanol, CH_3—OH Ethanol, C_2H_5—OH
Aldehydes	$*R-\overset{\displaystyle O}{\underset{\displaystyle H}{C}}$	-al	Methanal, H—CHO Ethanal, CH_3—CHO
Ketones	$\overset{\displaystyle R}{\underset{\displaystyle R}{C}}=O$	-one	Propanone, $(CH_3)_2CO$ Butanone, C_3H_7CO
Carboxylic acids	*R—COOH	-oic acid	Methanoic acid, H—COOH Ethanoic acid, CH_3—COOH
Primary amines	$R-\overset{\displaystyle H}{\underset{\displaystyle H}{N}}$	-amine	Methylamine CH_3—NH_2 Ethylamine C_2H_5—NH_2
Nitroalkanes	$R-NO_2$	nitro-ane	Nitromethane, CH_3—NO_2 Nitroethane, C_2H_5—NO_2

* It will be seen that the name of any individual compound is based on the *total* number of carbon atoms in its molecule. Meth- for 1; eth- for 2, prop- for 3 etc. In those general formula marked with * the R can stand for an H atom as well as an alkyl group.

5 Functional groups

The various groups which make up organic molecules are known as functional groups and the commonest are listed below (see Appendix II, p. 370; for a fuller list). Their names are denoted by suffixes and/or prefixes as shown.

—X	—Cl	—Br	—I
halogeno-	chloro-	bromo-	iodo-

$$\diagdown C = C \diagup$$
-ene

$$-C \equiv C-$$
-yne

$$-OH$$ hydroxy- or -ol

$$-C \diagup \substack{O \\ H}$$ -al

$$\diagdown C = O$$ -one

$$-C \diagup \substack{O \\ O-H}$$ -oic acid

$$-N \substack{H \\ H}$$ amino- or -amine

$$- NO_2$$ nitro-

$$-C \equiv N$$ cyano- or -nitrile or cyanide

They all have characteristic properties which are evident in any compound containing them. The number of common functional groups is small in comparison with the total number of organic compounds. The facts of organic chemistry are, therefore, systematised if the emphasis is put on the characteristics of individual functional groups. Once these characteristics have been understood it is possible, with reasonable accuracy, to predict the properties of any compounds containing such groups.

A completely accurate prediction is not always possible because the characteristics of a particular group depend, to some extent, on the other groups to which it is linked in a molecule. In particular, a functional group attached directly to a benzene ring does not exhibit all the same properties as it does when it is attached to an open-chain of carbon atoms. Ethanol, phenol and ethanoic acid all contain the

Ethanol   Phenol

—OH group, but they are significantly different. Similarly the carbonyl, $\ce{>C=O}$, group functions differently in propanone and in ethanoic acid,

$$\underset{\text{Propanone}}{\ce{H-\overset{\overset{\displaystyle H}{|}}{\underset{\underset{\displaystyle H}{|}}{C}}-\overset{\overset{\displaystyle O}{||}}{C}-\overset{\overset{\displaystyle H}{|}}{\underset{\underset{\displaystyle H}{|}}{C}}-H}}$$

$$\underset{\text{Ethanoic acid}}{\ce{H-\overset{\overset{\displaystyle H}{|}}{\underset{\underset{\displaystyle H}{|}}{C}}-C{\overset{\displaystyle O}{\diagdown \text{OH}}}}}$$

This effect of the environment on a functional group is one of the major points of interest in studying organic chemistry.

6 Compounds with more than one major functional group

The majority of compounds dealt with in this book will contain only one major functional group. Many other compounds are known, however, with two or more such groups.

Urea (carbamide), aminoethanoic acid (glycine) and ethanedioic (oxalic) acid provide well-known examples

$$\underset{\substack{\text{Urea} \\ \text{(Carbamide)}}}{\ce{O=C{\overset{\displaystyle NH_2}{\underset{\displaystyle NH_2}{<}}}}}
\qquad
\underset{\substack{\text{Aminoethanoic} \\ \text{acid (Glycine)}}}{\ce{H-\overset{\overset{\displaystyle H}{|}}{\underset{\underset{\displaystyle NH_2}{|}}{C}}-C{\overset{\displaystyle O}{\diagdown \text{OH}}}}}
\qquad
\underset{\substack{\text{Ethanedioic} \\ \text{(oxalic) acid}}}{\ce{C{\overset{\displaystyle O}{\diagup}}{\underset{\displaystyle OH}{}}}}$$

A particularly interesting situation arises when a carbon atom is linked to four different groups as in alanine, lactic acid and butan-2-ol,

$$\underset{\substack{\text{Alanine} \\ \text{(2-amino propanoic} \\ \text{acid)}}}{\ce{H-\overset{\overset{\displaystyle CH_3}{|}}{\underset{\underset{\displaystyle NH_2}{|}}{C}}-COOH}}
\qquad
\underset{\substack{\text{Lactic acid} \\ \text{(2-hydroxypropanoic} \\ \text{acid)}}}{\ce{H-\overset{\overset{\displaystyle CH_3}{|}}{\underset{\underset{\displaystyle OH}{|}}{C}}-COOH}}
\qquad
\underset{\text{Butan-2-ol}}{\ce{H-\overset{\overset{\displaystyle CH_3}{|}}{\underset{\underset{\displaystyle OH}{|}}{C}}-C_2H_5}}$$

The carbon atom is said to be asymmetric and the tetrahedral shape of the molecules means that they are, as a whole, asymmetric. Because of this they exhibit an important type of isomerism (p. 276).

Many naturally occurring substances, e.g. vitamin A and nicotine, and many synthetic ones, e.g. aspirin and benzedrine have large molecules made up of a wide range of functional groups, rings and chains. They illustrate the infinite number of possible ways in which the atoms concerned can be linked. Though they contain complicated molecules their chemistry is best understood by considering the contribution made by each functional group and by the possible effect of one group on another.

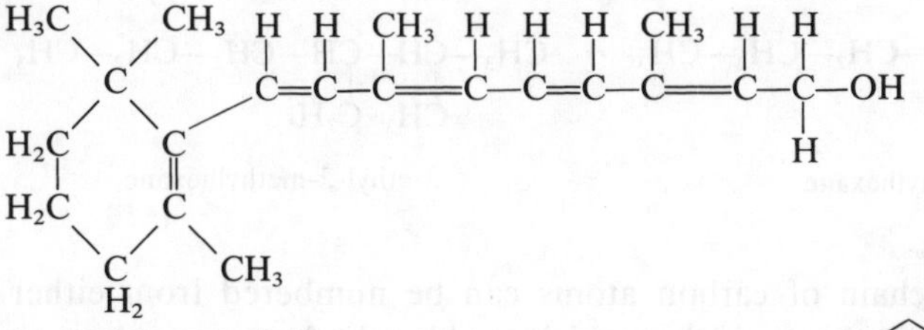

Vitamin A

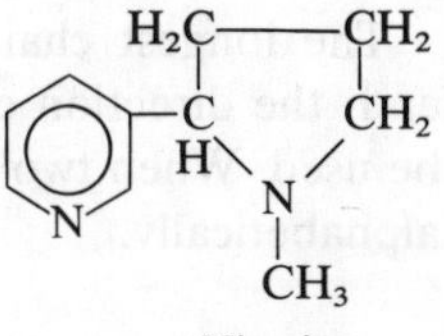

Nicotine

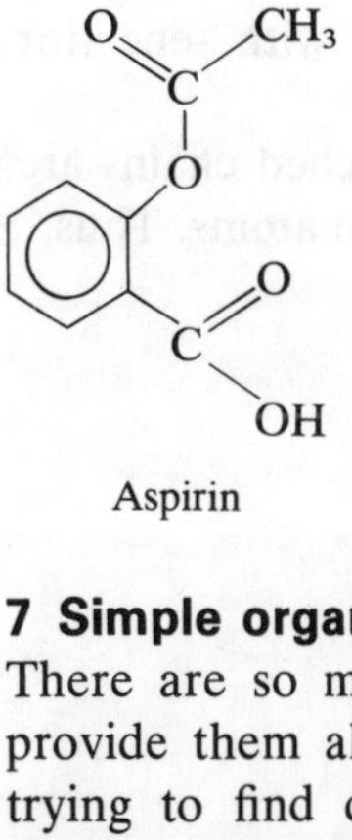

Aspirin

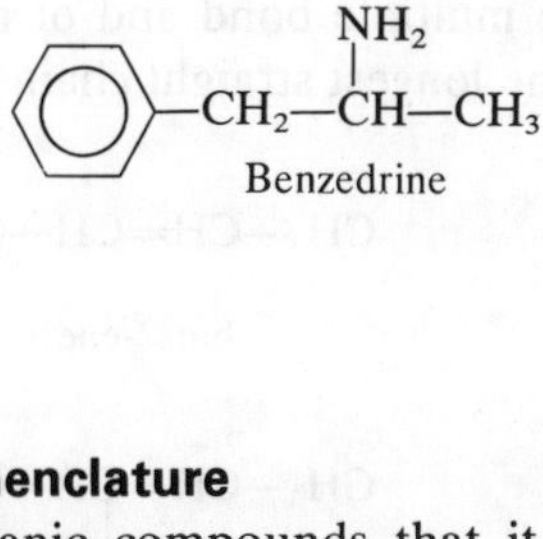

Benzedrine

7 Simple organic nomenclature

There are so many organic compounds that it is a real problem to provide them all with different, yet sensible, names. It is rather like trying to find different, yet related, surnames for everyone in the world. A set of rules (the IUPAC rules) has been agreed internationally by the International Union of Pure and Applied Chemistry and their recommendations are slowly being put into effect. The full set of rules to cover every eventuality is complicated; a brief outline is given here and extended throughout the book.

a. Alkanes. The alkanes have an -ane ending (suffix) with meth-, eth-, prop- and but- prefixes for the first four members. Thereafter, the prefix is based on the number of carbon atoms involved, e.g. pent-, hex-, etc. These names are used for the alkane with a straight chain of carbon atoms. The corresponding alkyl groups are named by replacing the -ane ending by -yl.

In alkanes with branched chains the name is based on the number of carbon atoms in the longest straight chain. This longest chain is then numbered from one end so that the position of the branched chains can be indicated. Thus,

$$CH_3-\underset{\underset{CH_3}{|}}{CH}-CH_2-CH_2-CH_2-CH_3 \qquad CH_3-\underset{\underset{CH_3}{|}}{CH}-\underset{\underset{C_2H_5}{|}}{CH}-CH_2-CH_2-CH_3$$

2-methylhexane   3-ethyl-2-methylhexane

The longest chain of carbon atoms can be numbered from either end; the direction chosen is that which enables the lowest numbers to be used. When two or more branched chains are present they are listed alphabetically.

b. Alkenes and alkynes. The number of carbon atoms is indicated by using the meth-, eth-, prop-, but-, pent- etc. prefixes with -ene (for alkenes) or -yne (for alkynes) suffixes.

The positions of the multiple bond and of any branched chains are shown by numbering the longest straight chain of carbon atoms. Thus,

$$CH_2{=}CH-CH_2-CH_3 \qquad CH_3-CH{=}CH-CH_3$$

but-1-ene   but-2-ene

$$CH_2{=}\underset{\underset{CH_3}{|}}{C}-CH_2-CH_3 \qquad CH_3-CH{=}\underset{\underset{CH_3}{|}}{CH}-CH_2-CH_3$$

2-methylbut-1-ene   3-methylpent-2-ene

c. Substitution of hydrogen atoms by functional groups. When a hydrogen atom in a hydrocarbon is replaced by a functional group the name of the group (p. 370) is indicated by using a prefix or a suffix and the position is indicated numerically.

In halogenoalkanes, for example, the particular halogen group is denoted by a prefix; in alcohols the hydroxy (OH) group is denoted by

Introduction 13

an -ol suffix. Thus

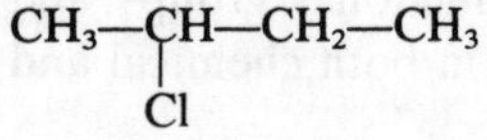

CH₃—CH—CH₂—CH₃
    |
    Cl

2-chlorobutane

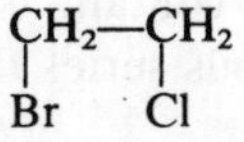

CH₂—CH₂
| |
Br Cl

1-bromo-2-chloroethane

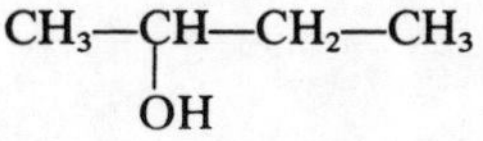

CH₃—CH—CH₂—CH₃
    |
    OH

Butan-2-ol

CH₂—CH₂
| |
OH OH

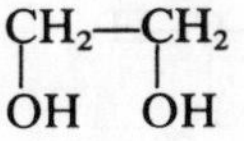

Ethan-1,2-diol

d. Trivial names. Some traditional, trivial names are still in use, e.g.

Trivial name	Formula	IUPAC name
Chloroform	CHCl₃	Trichloromethane
Acetic acid	CH₃COOH	Ethanoic acid

Glycine	CH₂COOH
	   |
	  NH₂
		Aminoethanoic acid

The use of trivial names is dying out but some may continue to be used as they are so much more concise than the systematic name and many of them are in common use in non-chemical activities.

8 Isomerism
It is very common in organic chemistry to find that many different compounds, known as isomers, can all be represented by the same molecular formula.

a. Structural isomerism (p. 269). In structural isomers the molecules contain the same number of each kind of atom but differ in regard to which atom is linked to which. They are like anagrams in which the same numbers of the same letters can be written as different words, e.g. Ronald, Roland, Arnold.

The isomers are all separate, distinct substances. If they contain the same functional groups and are members of the same homologous series, e.g.

CH₃—CH₂—CH₂—CH₃

Butane

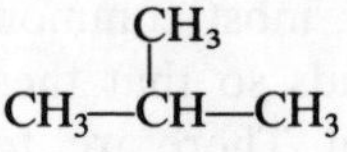

        CH₃
        |
CH₃—CH—CH₃

2-methylpropane

they have many chemical properties in common but have different physical properties. If they contain different functional groups and belong to different homologous series they differ in both chemical and physical properties, e.g.

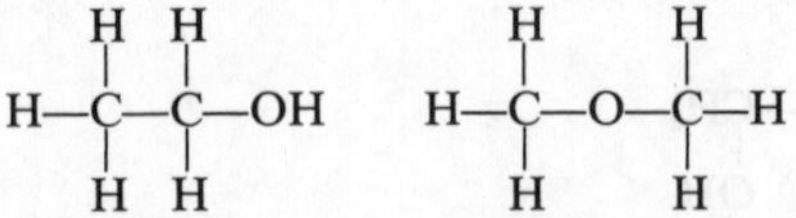

$$
\begin{array}{cc}
\mathrm{H} \quad \mathrm{H} & \mathrm{H} \quad \mathrm{H} \\
\mid \quad \mid & \mid \quad \mid \\
\mathrm{H-C-C-OH} & \mathrm{H-C-O-C-H} \\
\mid \quad \mid & \mid \quad \mid \\
\mathrm{H} \quad \mathrm{H} & \mathrm{H} \quad \mathrm{H}
\end{array}
$$

  Ethanol   Methoxymethane

The greater the number of atoms in a molecule the higher the number of possible isomers. There are, for example, no isomers of methane, ethane or propane; $C_{13}H_{28}$ is the molecular formulae for 802 isomeric alkanes; $C_{40}H_{82}$ has 625×10^{11} isomers. Further details about structural isomerism are given on p. 269.

b. Stereoisomerism (p. 271). Stereoisomers contain the same numbers of each kind of atom and do not differ in regard to which atom is linked to which. They do, however, differ in the way in which the atoms are arranged in space. There are two main types; *optical isomerism* and *geometric isomerism*.

Optical isomers occur in substances containing an asymmetric carbon atom, i.e. a carbon atom linked to four different groups (p. 276). The molecules, Cwxyz, are asymmetric so that they can exist in two forms one being the mirror image of the other; the two isomers are different in the same way as a right- and a left-hand are different (Fig. 6). The chemical properties of the two isomers do not differ greatly for the same functional groups are involved but the optical properties differ as explained on p. 276.

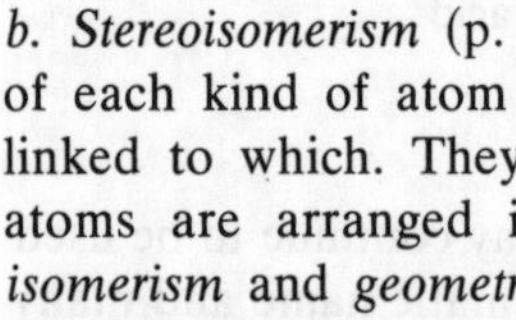

$$
\begin{array}{ccc}
\mathrm{w} & & \mathrm{w} \\
\mid & & \mid \\
\mathrm{C} & & \mathrm{C} \\
y \swarrow_{x}^{} \searrow z & & z \swarrow_{x}^{} \searrow y
\end{array}
$$

                    Mirror

Fig. 6. The two stereoisomers of a Cwxyz molecule.

Geometric isomers arise, most commonly, because there is no free rotation about double bonds so that these bonds can 'lock' the two groups they link in position. There are, for example, three isomers of dichloroethene, CH_2Cl_2:

H   Cl   H   Cl   Cl   Cl
  \\   //   \\   //   \\   //
   C=C   C=C   C=C
  /   \\   /   \\   /   \\
H   Cl   Cl   H   H   H

                             trans-form   *cis*-form
    (i)   (ii)   (iii)

1,1-dichloroethene   1,2-dichloroethene

(i) is a structural isomer of (ii) and (iii), but (ii) and (iii) are geometric isomers (the *cis*- and the *trans*-isomers). If there were free rotation about the double bond these geometric isomers would not exist. Further details and examples are given on p. 271.

2

Bonding in Organic Molecules

1 The carbon atom

The dot-and-cross method of representing covalent bonds as shared pairs (p. 1) is adequate for some simple purpose but a more accurate model can be built up; this involves the spin of an electron, the distribution of electrons in subsidiary orbits around the nucleus, and the wave nature of electrons.

Some electrons are regarded as spinning in one direction and others in the opposite direction. Two electrons with opposite spins are said to be paired; a single electron is unpaired. When two electrons occupy the same orbit they must be paired (the *Pauli* principle) and as electrons in the same orbit can only differ in having different spins it follows that no orbit can contain more than two electrons. The innermost shell within an atom, with principal quantum number of 1, can only hold two electrons; these are referred to as $1s$ electrons. The shell with principal quantum number of 2 can hold eight electrons. Of these, two, with opposite spins will be in the $2s$ subsidiary orbit. The remaining six will be in $2p$ subsidiary orbits, but, one orbit cannot hold six electrons so that it is necessary to subdivide the $2p$ orbit into three, each containing two electrons with opposed spins. These three $2p$ orbits are envisaged as in different planes (Fig. 7) and are labelled as $2p_x$, $2p_y$ and $2p_z$ orbits.

Of the six electrons in the carbon atom, two are in the $1s$ orbit and two of the remaining four are in the $2s$ orbit, which is of lower energy level than the $2p$ orbits. The remaining two electrons occupy two of the three $2p$ orbits as unpaired electrons.

The simple representation of the atomic structure of carbon (Fig. 8a) can, then, be replaced by the more detailed structure shown in Fig. 8b, each arrow representing one electron and its relative spin.

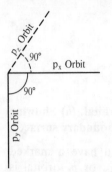

Fig. 7. Illustration of the directional arrangement of three *p* orbits.

2 The shapes of *s*- and *p*-orbitals

A beam of electrons can be diffracted which indicates that the electron has a wave-like nature, and this wave-like nature can be treated mathematically by a specialised technique known as wave-mechanics or quantum-mechanics. The idea of an electron in an atom existing as a tiny, negatively charged particle within a fixed orbit around the nucleus is replaced by the idea of a charge-cloud of varying electron

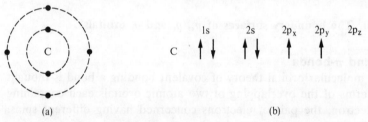

(a)   (b)

Fig. 8. (a) The simple atomic structure of carbon and (b) a more detailed structure indicating the electron spin.

density around the nucleus. The region in which the electron may be said to exist is known as an orbital and wave-mechanical calculations enable the shapes of *s*- and *p*-orbitals to be calculated.

The atomic orbital is accurately represented by a complex mathematical equation. Diagrammatically, it can be represented as a charge-cloud of varying electron density or, more simply, by mapping out a boundary surface within which a particular electron may be said to occur.

On this basis *s*-orbitals are spherical in shape (Fig. 9). This means that an *s*-electron exists somewhere within a sphere with the atomic nucleus as its centre. *s*-orbitals are said to be spherically symmetrical; the electron density is not concentrated in any particular direction.

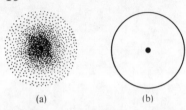

(a)   (b)

Fig. 9. Two methods of representing a 1s atomic orbital. (a) Showing the variation of charge density by shading. (b) The boundary surface.

p-orbitals, however, are dumb-bell in shape and have a marked directional character depending on whether a p_x, p_y or p_z orbital is being considered. The boundary surfaces are shown in Fig. 10; it will be seen that the p_x, p_y and p_z orbitals are concentrated in directions at right-angles to each other.

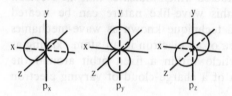

Fig. 10. The boundary surfaces of p_x, p_y and p_z orbitals.

3 σ- and π-bonds

In the molecular orbital theory of covalent bonding a bond is thought of in terms of the overlapping of two atomic orbitals each containing one electron, the pair of electrons concerned having different spins. The two electrons represented by atomic orbitals in the separate atoms are represented as a molecular orbital in the molecule formed when the two atoms combine. Because the two electrons have different spins they become paired in the molecular orbital.

Overlap of s- and p-orbitals can take place in different ways; as a general rule the larger the possible overlap the stronger the bond formed.

a. Overlap of two s-orbitals. Two spherically-symmetrical s-orbitals overlap to form a plum-shaped molecular orbital (Fig. 11a). This is known as a σ-orbital and the resulting bond is a σ-bond. It is the accumulation of charge between the two atomic nuclei that is responsible for holding the two nuclei together.

b. Overlap of s- and p-orbitals. An s- and a p-orbital can overlap to form a σ-bond as shown in Fig. 11b.

(a)

(b)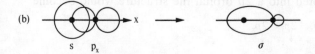

(c)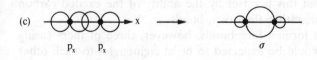

(d)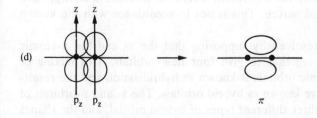

Fig. 11. Showing various ways in which atomic orbitals can overlap to form molecular orbitals. In (a), (b) and (c) there is a build-up of electron density between the two atoms; σ-bonds are formed. In (d) the build-up is alongside the two atoms; a π-bond is formed.

c. Overlap of p-orbitals. Two $2p_x$ (or $2p_y$ or $2p_z$) orbitals can overlap 'end-on' to form a σ-bond as shown in Fig. 11c.

p-orbitals can also overlap 'broadside-on'. If two $2p_z$ atomic orbitals are concerned they will form a molecular orbital as shown in Fig. 11d. This orbital has a distinctly different shape from those involved in σ-bonds. It is called a π-orbital and forms what is known as a π-bond. The accumulation of negative charge holding the two nuclei together is *alongside* the molecular axis.

4 Hybridisation
The arrangement of electrons in the carbon atom is

C $\underset{\upharpoonleft\downharpoonright}{1s}$ $\underset{\upharpoonleft\downharpoonright}{2s}$ $\underset{\downharpoonright}{2p_x}$ $\underset{\downharpoonright}{2p_y}$ $2p_z$

and only two of the electrons are unpaired and, therefore, available for bond formation. It is, however, well known that carbon can form four bonds. To account for this it must be assumed that the electrons rearrange prior to, or in the course of, chemical combination. If one $2s$ electron is promoted into a $2p$ orbital the structure would become

C $\underset{\uparrow\downarrow}{1s}$ $\underset{\uparrow}{2s}$ $\underset{\downarrow}{2p_x}$ $\underset{\downarrow}{2p_y}$ $\underset{\downarrow}{2p_z}$

providing four unpaired electrons. The promotion of the $2s$ electron requires energy but this is offset by the ability of the excited carbon atom to form four, rather than two, bonds.

If the electrons formed four bonds, however, three of them (using the $2p$ orbitals) would be expected to be at rightangles to each other and the fourth (using the $2s$ orbital) would be different in strength and have no directional nature. This is not in accordance with the known facts (p. 1).

The matter is resolved by supposing that the s- and the p-atomic orbitals combine together to give four new orbitals. This mixing or combining of atomic orbitals is known as hybridisation and the resulting new orbitals are known as hybrid orbitals. The s- and p-orbitals of carbon can form three different types of hybrid orbital, and the shapes of these can be calculated by combining the necessary atomic orbitals.

a. sp^3 hybridisation. Hybridisation of one $2s$- and three $2p$-orbitals forms four equivalent hybrid orbitals, known as sp^3 orbitals, directed towards the corners of a tetrahedron (Fig. 12). It is these four orbitals that are concerned when carbon forms four single bonds. In methane, CH_4, for example, the four sp^3 hybrid orbitals of the carbon atom each overlap with a $1s$ orbital of a hydrogen atom to form four σ-bonds arranged tetrahedrally.

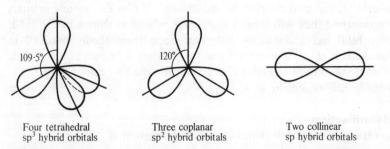

Four tetrahedral sp^3 hybrid orbitals

Three coplanar sp^2 hybrid orbitals

Two collinear sp hybrid orbitals

Fig. 12. The ways in which a carbon atom can form hybrid orbitals. Only the general, approximate shape of the orbitals is shown.

b. sp² hybridisation. Hybridisation of one $2s$ and two $2p$ atomic orbitals gives three sp^2 hybrid orbitals which are coplanar and directed at angles of $120°$ to each other (Fig. 12); one $2p$ atomic orbital remains unchanged and is directed at rightangles to the plane of the three hybrid orbitals (Fig. 13). It is these orbitals that are concerned when carbon forms two single and one double bond, as in ethene, C_2H_4.

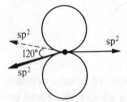

Fig. 13. The $2p$ orbital remaining after the formation of three coplanar sp^2 orbitals.

One of the sp^2 hybrid orbitals from each carbon atom forms a σ-bond between the two carbon atoms. The remaining sp^2 orbitals of each carbon atom form σ-bonds with the hydrogen atoms. The wo carbon, and four hydrogen, atoms all lie in the same plane with bond angles of $120°$. At rightangles to this plane there remains the unchanged $2p$ orbital of each carbon atom and these two orbitals interact to form a π-bond between the two carbon atoms (Fig. 14). A double bond between carbon atoms consists, then, of a σ-bond and a π-bond.

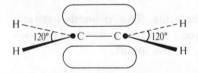

Fig. 14. The ethene molecule. σ-bonds between carbon and hydrogen atoms. One σ-bond and one π-bond between carbon atoms.

The two bonds are not of equal strength for the σ-bond is stronger than the π-bond as the sp^2 orbitals can overlap more fully than the $2p$ orbitals.

Two atoms linked by a σ-bond can rotate freely about the bond (p. 270), unless there is some steric hindrance, but the π-bonding prevents free rotation about a C=C bond (p. 271). Free rotation does not affect the overlap of orbitals forming σ-bonds but it would interfere with the p-orbital overlap involved in π-bonding.

c. sp hybridisation. Hybridisation of one $2s$ and one $2p$ orbital leads to two equivalent collinear sp orbitals (Fig. 12); two $2p$ atomic orbitals remain unchanged. It is these orbitals that are used when carbon forms one single and one triple bond, as in ethyne, C_2H_2.

One of the sp orbitals from each carbon atom overlap to form a σ-bond between the two carbon atoms; the remaining sp orbitals form

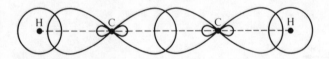

Fig. 15. σ-bonds in ethyne formed by overlap of two sp hybrid orbitals (between the two carbon atoms) and by overlap of an sp hybrid and an s orbital (between the carbon and hydrogen atoms).

σ-bonds with hydrogen atoms. The two carbon and two hydrogen atoms are all collinear (Fig. 15). The two remaining $2p$ orbitals of each carbon atom interact to form two π-bonds between the carbon atoms, these two bonds being in planes at rightangles to each other (Fig. 16).

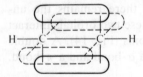

Fig. 16. The bonding in an ethyne molecule. Two σ-bonds between carbon and hydrogen atoms. One σ-bond and two π-bonds (in planes at rightangles to each other) between carbon atoms.

The $C\equiv C$ bond consists, then, of a σ-bond and two π-bonds. As in ethene, the σ-bond is stronger than the π-bonds, and there is no free rotation about the triple bond.

d. The strength of σ- and π-bonds. σ-bonds are generally stronger than π-bonds and this is well illustrated by the bond strengths and bond lengths of the carbon–carbon bonds in ethane, ethene and ethyne,

	C—C	C=C	C≡C
Bond energy/kJ mol^{-1}	346	598	813
Bond length/nm	0.154	0.134	0.121

The bond energies rise and the bond lengths decrease, in passing from C—C to C≡C, but the values are not in the ratio 1:2:3 or 3:2:1.

e. Summary. The three ways in which a carbon atom can form covalent bonds (p. 2) use different hybrid orbitals:

Nature of bonding	Orbitals used
Four single tetra-hedral bonds	All four bonds are σ-bonds using sp^3 hybrid orbitals
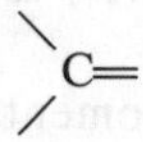 Two single and one double bond, all in the same plane	The single bonds are σ-bonds using sp^2 hybrid orbitals. The double bond consists of a σ-bond using an sp^2 orbital and a π-bond using a $2p$ atomic orbital.
One single and one triple bond; collinear.	The single bond is a σ-bond using an sp orbital. The triple bond consists of a σ-bond using an sp orbital and two π-bonds using $2p$ atomic orbitals.

5 Polar nature of covalent bonds. The inductive effect

The pair of electrons forming a single bond between two like atoms, X—X, is equally shared between the two atoms, but in a bond A—X this is not necessarily so. If A has a stronger attraction for electrons, i.e. a higher electronegativity, than X, the shared pair will be attracted towards A and away from X. This will cause a permanent displacement of electrical charge, which is known as an inductive effect.

In the C—Cl bond, with chlorine more electronegative than carbon, the inductive effect is represented as C→Cl; in the C—H bond, with hydrogen less electronegative than carbon, by C←H. The charges built up on the atoms may also be shown, or the bond may be represented by a distorted molecular orbital (Fig. 17).

C→Cl $C^{\delta+}$—$Cl^{\delta-}$ C Cl

Fig. 17. Three ways of showing the inductive effect in the C—Cl bond.

Polarity in a covalent bond renders the bond more liable to attack by other charged reagents, causes dipole moments, affects bond angles and may influence the acidity or alkalinity of a compound containing the bond. Many organic facts can be accounted for by considering the bond polarities in the chemicals concerned and, as a general rule, the higher the polarity within a molecule the more reactive it is.

a. Dipole moments. The two equal and opposite charges associated with a polar bond constitute an electrical dipole moment just as separated magnetic poles constitute a magnetic moment. Dipole moments are defined as the electrical charge (in C) multiplied by the bond distance, i.e. the charge separation, (in m). The Debye unit (D) is commonly used; it is equal to $3.335\,640 \times 10^{-30}$ C m.

Compounds containing polar bonds may have a zero dipole moment for the moment of a molecule as a whole is the *vector* sum of the individual bond moments. The C—Cl bond is definitely polar but the dipole moment of the symmetrical CCl_4 molecule is zero. Some other values* are summarised below:

$CHCl_3$	CH_2Cl_2	CH_3Cl	CH_3Br	CH_3I
1.02 D	1.54 D	1.86 D	1.79 D	1.64 D

b. +I and −I groups. It is most useful, in organic chemistry, to consider the electron-attraction of an atom or group *in relation to* the hydrogen atom. Any atom or group which attracts electrons more strongly than hydrogen is said to have a negative inductive effect (−I); any atom or group which attracts electrons less strongly than hydrogen is a +I group. It may be convenient to remember that a **plus** I group re**pels** electrons.

Dipole moment measurements enable various groups to be compared on this basis, as indicated below:

$C \twoheadrightarrow NO_2$

$\qquad C \twoheadrightarrow Cl$ $\qquad\qquad\qquad\qquad\qquad\qquad C \twoheadleftarrow C(CH_3)_3$

$\qquad\quad C \twoheadrightarrow Br$ $\qquad\qquad\qquad\qquad\qquad C \twoheadleftarrow CH(CH_3)_2$

$\qquad\qquad C \twoheadrightarrow I$ $\qquad\qquad\qquad C \twoheadleftarrow C_2H_5$

increasing $\qquad\qquad\qquad\qquad\qquad\qquad\qquad\qquad$ increasing

−I effect $\qquad\quad C \twoheadrightarrow C_6H_5 \qquad C \twoheadleftarrow CH_3 \qquad$ +I effect

$\qquad\qquad\qquad\qquad C—H$
$\qquad\qquad\qquad\quad$ (Arbitrary
$\qquad\qquad\qquad$ reference point)

* Dipole moments can be obtained from measurements on (a) the relative permittivity (dielectric constant) and density of vapours at different temperatures, (b) the relative permittivity of a vapour at one temperature together with its refractive index, (c) the relative permittivity of a dilute solution together with the density of the solvent and the solution.

c. Effect of polarity on bond angles. The C—Cl bonds in a CCl_4 molecule are polar but the molecule is symmetrical and the repulsive forces between the negatively charged chlorine atoms are 'evened-out' so that the bond angles are all $109.5°$. In CH_3Cl, however, the C—Cl and C—H bonds have different polarity so that the repulsive forces between the hydrogen and chlorine atoms surrounding the carbon atom are uneven. The bond angles are given, together with some related values, below:

	CH_4	CCl_4	CH_3Cl	CH_2Cl_2	$CHCl_3$
H—C—H angle	109.5		110.5	113.0	
Cl—C—Cl angle		109.5		111.8	110.9

6 Delocalised bonds

In a σ-bond, the concentration of charge is between the bonded atoms even though the molecular orbital may be distorted if the bond is polar (Fig. 17). The bond is said to be localised (between the atoms) and it is reasonably well represented, at least positionally, by the traditional A—X or A$^{\times}_{\cdot}$X symbolism.

In a π-bond between two atoms the concentration of charge is alongside the two atoms, though a single π-bond is still localised between the two atoms. When more than two adjacent atoms have available p-orbitals, however, the overlap can take place across *all* the orbitals concerned to form what is known as a delocalised bond.

a. Chloroethene, C_2H_3Cl. In this molecule, the two carbon atoms and the chlorine atom all have available p-orbitals which all overlap to give a delocalised π-bond spread out over the C—C—Cl chain length. The bonding in the molecule consists, then, of localised σ-bonds together with delocalised π-bonds (Fig. 18).

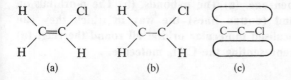

(a)   (b)   (c)

Fig. 18. Bonding in chloroethene, C_2H_3Cl. (a) Traditional representation. (b) The σ-bonds. (c) The delocalised π-bonding.

The effect of this delocalised bonding is to make the carbon–carbon bond rather less of a double bond than it is in ethene (where delocalisation is not possible as the H atoms have no available p-orbitals), and to give some double bond character to the carbon–chlorine bond as compared with the bond in C_2H_5Cl (where delocalisation is not possible as the carbon atom has no available p-orbitals). The relevant bond lengths bear this out:

	Carbon–carbon bond	Carbon–chlorine bond
C_2H_3Cl	0.138 nm	
C_2H_4	0.134 nm	
C_2H_5Cl		0.169 nm
		0.177 nm

The delocalised bonds also affect the nature of the Cl as a functional group, rendering it less reactive than in chloroethane. Similar delocalisation occurs in chlorobenzene (p. 141).

b. Benzene, C_6H_6. The formation of delocalised bonds is particularly important in benzene. Each of the six carbon atoms form three σ-bonds (two to adjacent carbon atoms and one to a hydrogen atom) using sp^2 hybrid orbitals (Fig. 19): because the sp^2 orbitals are planar all the six carbon and six hydrogen atoms are in the same plane. Each carbon atom still has an available p-orbital (with an axis at rightangles to the plane of carbon atoms) and the six p-orbitals all overlap with each other to form delocalised bonds all round the ring.

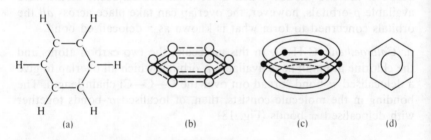

(a)   (b)   (c)   (d)

Fig. 19. Bonding in benzene. (a) The σ-bonds. (b) The p-orbitals on each carbon atom and (dotted lines) the way in which they can overlap. (c) The delocalised molecular orbital all round the ring. (d) The modern way of representing the C_6H_6 molecule.

The bonds between carbon atoms are, therefore, neither C—C nor C=C bonds; they consist of a σ-bond together with some part of delocalised π-bonding. Measurement shows that all the carbon–carbon bonds are of equal length (0.1397 nm), shorter than the C—C bond in ethane (0.154 nm) and longer than the C=C bond in ethene (0.134 nm).

It is the lack of C—C and C=C bonds which make the properties of benzene different from those of the alkanes and alkenes, and the delocalisation in the benzene molecule is the cause of aromatic character (p. 139). It would be inconvenient to represent the benzene molecule as in Fig. 19(c) every time it had to be drawn; the arrangement shown in Fig. 19(d) is traditionally used.

c. The ethanoate ion, CH_3COO^-. This ion is traditionally represented as in Fig. 20, which shows the presence of both a C=O and a C—O bond. Measurement shows, however, that the carbon–oxygen bonds are both of equal length (0.127 nm), shorter than the C—O bond in methanol (0.143 nm) and longer than the C=O bond in ethanal (0.122 nm) (Fig. 20).

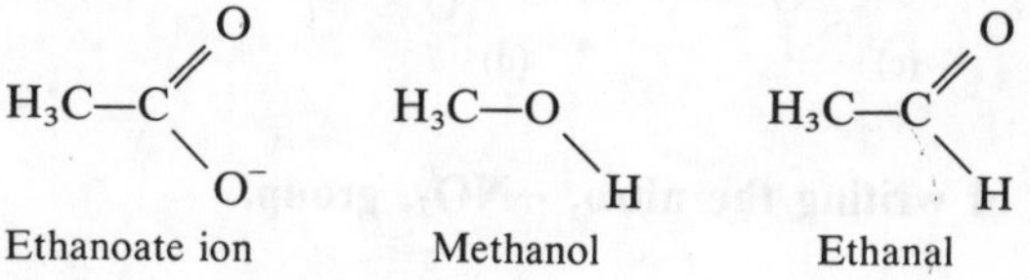

Ethanoate ion   Methanol   Ethanal

Fig. 20. Traditional ways of writing structural formulae for the ethanoate ion, methanol and ethanal.

This comes about because delocalised π-bonds can be formed between the carbon atom in the ethanoate ion and its two adjacent oxygen atoms. The carbon oxygen bonds consist, then, of a σ-bond and some part of the π-bonding. The drawing of the delocalised bonds is not very easy, though Fig. 21(a) shows their axes. The situation is conveniently represented as in Fig. 21(b).

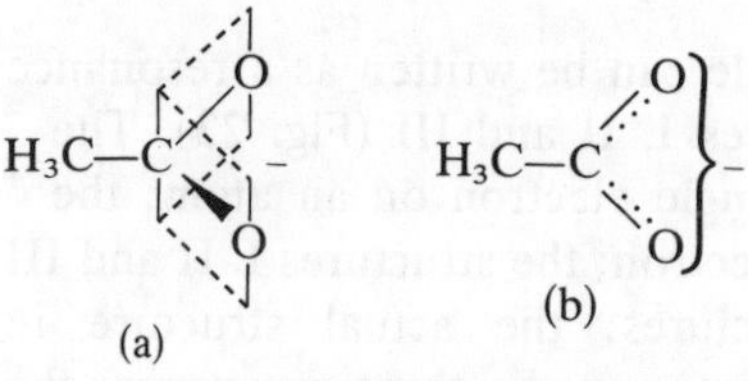

(a)   (b)

Fig. 21. Delocalised bonding in the ethanoate ion. (a) The axes of the *p*-orbitals on the C and O atoms and (dotted lines) the way in which they can overlap. (b) Symbolic representation of the delocalised bonds. See, also, Fig. 48 p. 235.

Similar delocalisation cannot occur in methanol or in ethanal for there are not two oxygen atoms adjacent to the carbon atom.

d. The nitro group. The nitro, $—NO_2$, group is traditionally written as in Fig. 22(a) or (b), but, as in the ethanoate ion, delocalised bonds can be formed between the nitrogen atom and the two oxygen atoms so

that Fig. 22(c) or (d) gives a better representation. Both nitrogen–
oxygen bonds in the nitro group are of equal length (0.122 nm); a
single bond would have a length of 0.136 nm and a double bond a
length of 0.115 nm.

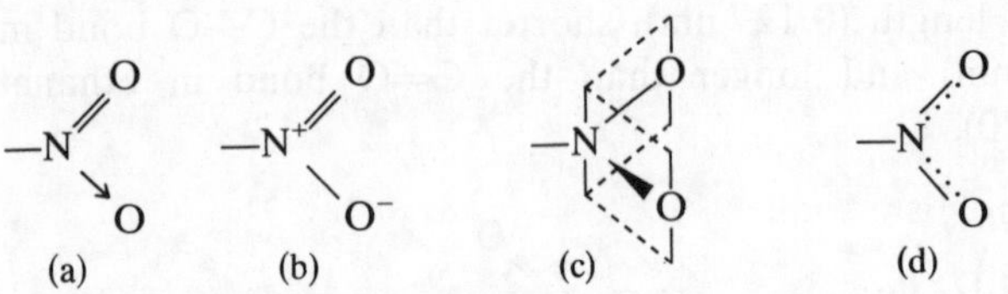

Fig. 22. Different ways of writing the nitro, $-NO_2$, group.

7 Resonance

Because a covalent bond may be polar and because of the possibility of
forming delocalised bonds the simple —, = and ≡ symbolism for
denoting covalent bonds is inadequate for many purposes, though still
useful.

An attempt to overcome the problem was made, in the theory of
resonance due to Pauling, by representing the actual bonding in a
molecule as a resonance hybrid between various possible (but non-
existing) structures. The theory has played a big part in the development of chemistry but its use is becoming less and less common.

a. Chloroethene. The C_2H_3Cl molecule can be written as a resonance
hybrid between the possible structures I, II and III (Fig. 23). The ⁻
symbol denotes the presence of a single electron on an atom; the ⁺
symbol denotes the lack of a single electron; the structures I, II and III
are referred to as canonical structures; the actual structure in
chloroethene is a resonance hybrid between the three structures; the
↔ symbol is used to represent this. Alternatively, the term *mesomeric
structures* is used and the conception is known as *mesomerism* ('between the parts').

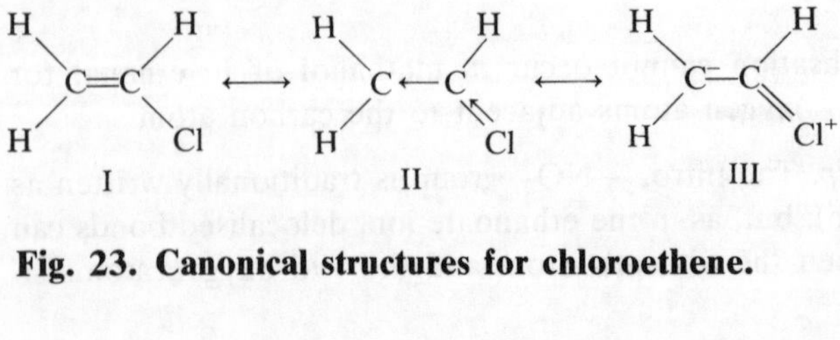

Fig. 23. Canonical structures for chloroethene.

The actual structure in chloroethene is closely related to the canonical structures, yet it differs from all of them. Chloroethene does not consist of a mixture of three different kinds of molecule. The canonical structures simply use the classical symbolism to indicate the actual, single structure in chloroethene, which cannot be written as a single structure using the classical symbols. Structure I shows double bonding between carbon atoms and single bonding between carbon and chlorine; structures II and III show single bonding between carbon atoms and double bonding between carbon and chlorine. In chloroethene the bonds are of intermediate character.

b. Benzene. The C_6H_6 molecule can be regarded as a resonance hybrid between structures I–V (Fig. 24). Structures I and II were originally

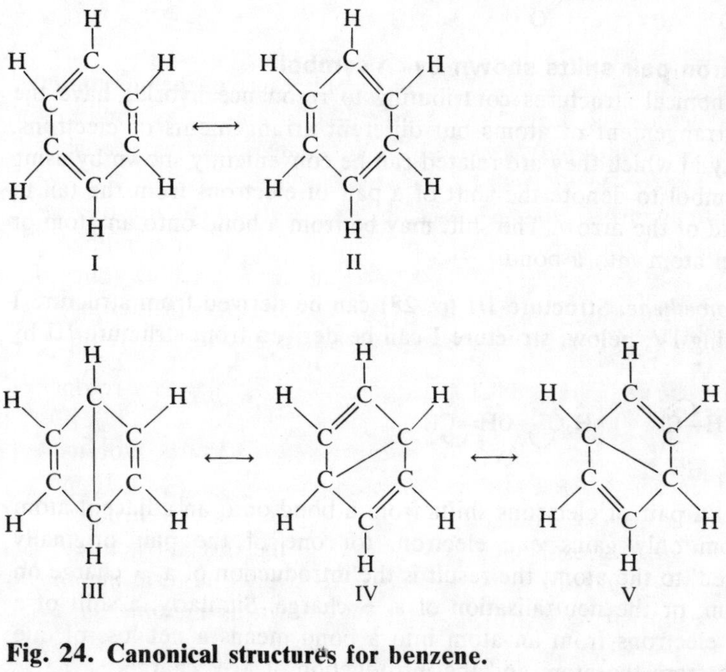

Fig. 24. Canonical structures for benzene.

suggested by Kekulé (1865); structures III–V are known as Dewar structures. The Dewar structures would be expected to be fairly unstable for they contain such long carbon–carbon bonds; they contribute to the resonance hybrid less than the Kekulé structures.

c. The ethanoate ion. This ion can be represented as a resonance hybrid between structures I and II (Fig. 25).

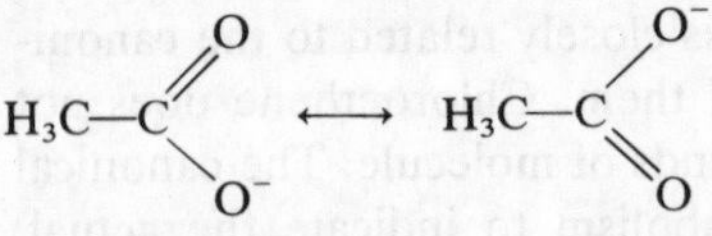

$$H_3C-C \overset{\displaystyle O}{\underset{\displaystyle O^-}{\big\langle}} \longleftrightarrow H_3C-C \overset{\displaystyle O^-}{\underset{\displaystyle O}{\big\langle}}$$

Fig. 25. Canonical structures for the ethanoate ion.

d. The nitro group. The resonance hybrids are written as

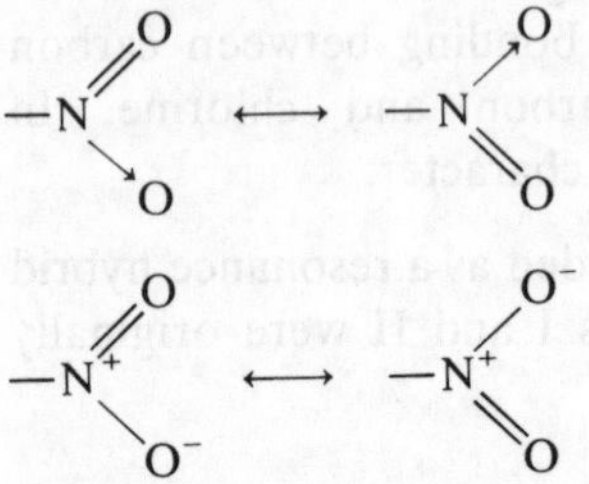

$$-N \overset{\displaystyle O}{\underset{\displaystyle O}{\big\langle}} \longleftrightarrow -N \overset{\displaystyle O}{\underset{\displaystyle O}{\big\langle}}$$

$$-N^+ \overset{\displaystyle O}{\underset{\displaystyle O^-}{\big\langle}} \longleftrightarrow -N^+ \overset{\displaystyle O^-}{\underset{\displaystyle O}{\big\langle}}$$

8 Electron pair shifts shown by ⌢ symbol

The canonical structures contributing to resonance hybrids have the same arrangement of atoms but different arrangements of electrons. The way in which they are related can be conveniently shown by using a ⌢ symbol to denote the shift of a pair of electrons from the tail to the head of the arrow. The shift may be from a bond onto an atom or from an atom into a bond.

a. Chloroethene. Structure III (p. 28) can be derived from structure I by writing IV, below; structure I can be derived from structure III by writing V.

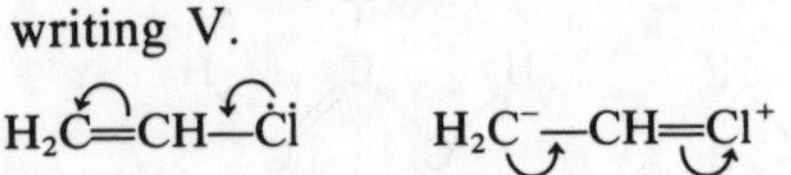

$$H_2\overset{\frown}{C}=CH\overset{\frown}{-}\overset{\frown}{Cl} \qquad H_2C^-\overset{\frown}{-}CH=\overset{\frown}{Cl}{}^+$$
$$\text{IV} \qquad\qquad \text{V}$$

When a pair of electrons shifts from a bond onto an adjacent atom the atom only gains *one* electron, for one of the pair originally 'belonged' to the atom; the result is the introduction of a − charge on the atom, or the neutralisation of a + charge. Similarly, a shift of a pair of electrons from an atom into a bond means a net loss of one electron from the atom and the introduction of a + charge.

b. The ethanoate ion. The two canonical structures (Fig. 25) can be related by writing the structure below,

$$H_3C-C \overset{\displaystyle \overset{\frown}{O}}{\underset{\displaystyle \overset{\frown}{O^-}}{\big\langle}}$$

c. The nitro group. The structures below relate the two resonance hybrids, (p. 30),

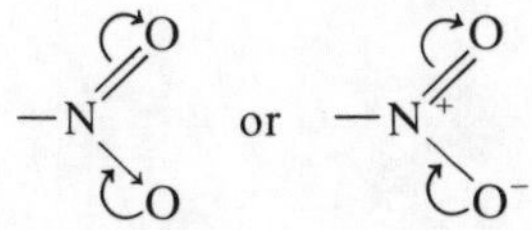

$$-N\begin{smallmatrix}\nearrow O\\ \searrow O\end{smallmatrix} \quad \text{or} \quad -N^{+}\begin{smallmatrix}\nearrow O\\ \searrow O^{-}\end{smallmatrix}$$

9 The use of straight arrows

A glance at any relatively advanced book of organic chemistry shows a lot of formulae and equations involving both straight and curved arrows; it is essential to understand their meaning.

a. →. This represents a coordinate or dative bond. Once formed the bond is very much like a covalent bond but the pair of electrons involved both originate from the same atom, as in the nitro-group (Fig. 26). The partial transfer of a pair of electrons from N to O introduces a charge separation which may also be indicated.

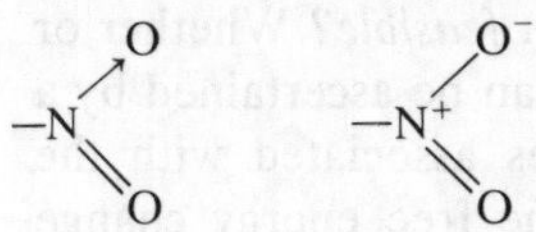

$$-N\begin{smallmatrix}\nearrow O\\ \searrow O\end{smallmatrix} \qquad -N^{+}\begin{smallmatrix}\nearrow O^{-}\\ \searrow O\end{smallmatrix}$$

Fig. 26. Coordinate or dative bonding in the nitro group.

The → symbol is also used to connect the products and reactants in an equation, e.g.

$$CH_4(g) + Cl_2(g) \rightarrow CH_3Cl(g) + HCl(g)$$

b. ⇀. This represents an inductive effect (p. 23).

c. ⇌. This is used to denote an equilibrium between two species, A and B,

$$A \rightleftharpoons B \qquad A \underset{i}{\overset{}{\rightleftharpoons}} B \qquad A \underset{ii}{\overset{}{\rightleftharpoons}} B$$

Predominance of B in the equilibrium mixture is shown by i and predominance of A by ii.

d. ↔. This is used to indicate the existence of a resonance hybrid between various canonical forms (p. 28).

3

The Nature of Organic Reactions

1 Introduction

An organic reaction is conveniently summarised by a balanced equation showing the reagents and the products, but not all equations represent a reaction which can be satisfactorily carried out in practice. A detailed understanding of any reaction involves four main issues.

a. Thermodynamic considerations. Is the reaction feasible? Whether or not a particular reaction is likely to take place can be ascertained by a theoretical consideration of the energy changes associated with the reaction. This involves the enthalpy change, the free energy change and the entropy change and is the province of thermodynamics. The free energy change is of particular importance for, from it, the equilibrium constant for the reaction can be calculated.

b. Reaction kinetics. How fast is the reaction? A reaction may be perfectly feasible from the thermodynamic point of view yet it may be of no practical value because it takes place far too slowly. A study of reaction rates is the domain of reaction kinetics.

c. Reaction mechanisms. What actually happens? The fullest understanding of a reaction involves a study of the possible mechanisms by which one set of bonds are broken whilst another set are formed. Many reactions are complex, involving a series of successive elementary steps; most reactions involve the formation of intermediates. Neither the steps nor the intermediates are shown in the simple, overall equation for the reaction.

d. Separation of the products. If a reaction is intended to make one particular product it will only be useful if it is relatively easy to separate that product from the mixture remaining at the end of the reaction. This involves purification processes as described on pp. 53-63.

2 The yield of a reaction

The theoretical yield of a reaction is the amount, usually expressed in grammes or moles, of a product which could be obtained under ideal conditions if the reaction proceeded to completion. The actual yield is the amount of pure product which can actually be obtained in the laboratory or industrially. The percentage yield is usually quoted and is obtained from the relationship,

$$\text{Percentage yield} = \frac{\text{Actual yield} \times 100}{\text{Theoretical yield}}$$

In the reaction between ethanol and ethanoic acid to form ethyl ethanoate,

$$C_2H_5OH(l) + CH_3COOH(l) \rightarrow CH_3COOC_2H_5(l) + H_2O(l)$$

the theoretical yield of ethyl ethanoate from 1 mol (46 g) of ethanol and 1 mol (60 g) of ethanoic acid would be 1 mol (88 g). A typical actual yield, from the same amounts of reagents, would be 53 g, giving a percentage yield of 5 300/88, i.e. 60 %.

This is quite a good yield for an organic reaction for the equation idealises the situation. It disregards any impurities and any side-reactions, and it presupposes that all the desired product can be extracted. In particular, it assumes that the reaction goes to completion, when, in reality, many organic reactions are reversible giving equilibrium mixtures containing both products and reagents.

3 Equilibrium constants

How fully a reaction will go is determined by the value of its equilibrium constant, which, for a simple reaction

$$A + B \rightleftharpoons C + D$$

is given by

$$K = \frac{[C][D]}{[A][B]}$$

The amount of C (x mol) in the equilibrium mixture from 1 mol of A and 1 mol of B will be given by

$$K = \frac{x^2}{(1-x)^2}$$

A high value for K means a high value for *x* and a high percentage theoretical yield as follows:

K	0	10^{-6}	10^{-4}	10^{-2}	1/9	1	9	10^2	10^4	10^6
$x \times 10^2$										
(% yield)	0	0.09	0.99	9		25	50	75	90.9	99	99.9

The equilibrium, in this simple type of reaction, can, therefore, be regarded as completely over to the right when the *K* value is larger than 10^6, and there will be no significant reaction when the *K* value is less than 10^{-6}. These figures give a rough working guide but they only apply to the simple reaction quoted and to equimolecular mixtures of reagents.

Simple Thermodynamics

4 Standard enthalpy of formation

The standard enthalpy of formation of a compound is the enthalpy change when 1 mol of the compound is formed from its elements in their standard states. Values are generally quoted at 25 °C, in $kJ\,mol^{-1}$, and symbolised by $\Delta H^{\ominus}(298\ K)$, e.g.

$$2C(s) + 3H_2(g) \rightarrow C_2H_6(g) \qquad \Delta H^{\ominus}(298\ K) = -84.6\ kJ$$
$$2C(s) + 2H_2(g) \rightarrow C_2H_4(g) \qquad \Delta H^{\ominus}(298\ K) = \ \ 52.3\ kJ$$
$$2C(s) + \ \ H_2(g) \rightarrow C_2H_2(g) \qquad \Delta H^{\ominus}(298\ K) = 226.8\ kJ$$

As the enthalpy of all elements in their standard states is arbitrarily taken as zero the standard enthalpies of formation represent the enthalpy content of the compound on this arbitrary scale. The enthalpy content of ethane is lower than that of its component elements; the enthalpy contents of ethene and ethyne are higher. In other words, energy has to be put into ethane to split it into its elements in their normal states, but energy is liberated when ethene and ethyne split. Ethane is said to be an exothermic compound; ethene and ethyne are endothermic compounds.

The situation can be summarised as in Fig. 27. The more negative the $\Delta H^{\ominus}$ value for a compound the more stable *it is likely* to be with regard to decomposition into its elements; the more positive the $\Delta H^{\ominus}$ value the more unstable the compound. It is, however, the standard free energy of formation that gives a truer indication of the stability of a compound (p. 37).

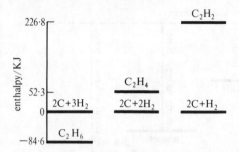

Fig. 27. Standard enthalpies of formation of C_2H_6, C_2H_4 and C_2H_2.

5 Use of standard enthalpies of formation

Chemical tables give the values of the standard enthalpies of formation of many compounds and these can be used to calculate the standard enthalpy change for a reaction for

$$
\begin{bmatrix}
\text{Standard enthalpy} \\
\text{change for a} \\
\text{reaction}
\end{bmatrix}
=
\begin{bmatrix}
\text{Sum of standard} \\
\text{enthalpies of for-} \\
\text{mation of products}
\end{bmatrix}
-
\begin{bmatrix}
\text{Sum of standard} \\
\text{enthalpies of} \\
\text{formation of reagents}
\end{bmatrix}
$$

For example,

$$
\begin{array}{llll}
C_2H_2(g) & + H_2(g) & \rightarrow C_2H_4(g) & \Delta H^{\ominus}(298\ K) = 52.3 - 226.8 \\
226.8 & 0 & 52.3 & \qquad\qquad\quad = -174.5\ kJ \\[4pt]
C_2H_4(g) & + H_2(g) & \rightarrow C_2H_6(g) & \Delta H^{\ominus}(298\ K) = -84.6 - 52.3 \\
52.3 & 0 & -84.6 & \qquad\qquad\quad = -136.9\ kJ \\[4pt]
2C_2H_6(g) & + 7O_2(g) & \rightarrow 4CO_2(g)\ + 6H_2O(1) & \Delta H^{\ominus}(298\ K) = -3\,289.4 \\
2(-84.6) & 0 & 4(-393.5)\ 6(-285.9) & \qquad\qquad\quad -(-169.2) \\
-169.2 & 0 & -1\,574.0\ \ -1\,715.4 & \qquad\qquad\quad = -3\,120.2\ kJ \\
& & \quad\ -3\,289.4 &
\end{array}
$$

These standard enthalpy changes may be described in relation to the particular type of reaction. Thus, $-136.9\ kJ\ mol^{-1}$ is the enthalpy of *hydrogenation* of ethene, and $-1\,560.1\ kJ\ mol^{-1}$ is the enthalpy of *combustion* of ethane.

6 Use of bond energies

In any reaction the overall energy change comes about because the bonds in the reagents have to be split (and this requires an input of energy) and the free atoms recombine to form different bonds (which gives out energy). If less energy is required to break the bonds of the

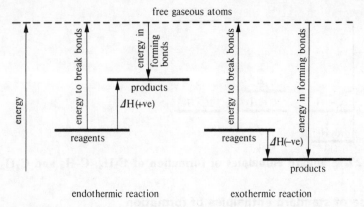

Fig. 28. Energy changes involved in breaking and forming bonds in a reaction.

reagents than is liberated in forming the bonds of the products the reaction will be exothermic; otherwise it will be endothermic (Fig. 28).

Average bond energy values can, therefore, be used to calculate the enthalpy change in a reaction. In the formation of ethane from carbon and hydrogen,

$$2C(s) + 3H_2(g) \rightarrow C_2H_6(g)$$

the bonds in solid carbon have to be split to form free gaseous atoms (this involves the enthalpy of atomisation of carbon) together with three H—H bonds; one C—C bond and six C—H bonds are formed. The energy values involved may be summarised as follows:

Energy in $\begin{cases} \text{To split bonds in C(s)} &= 2 \times 715 = 1\,430 \\ \text{To split three H—H bonds} = 3 \times 436 = 1\,308 \end{cases} = 2\,738$

Energy out $\begin{cases} \text{In forming one C—C bond} &= 346 \\ \text{In forming six C—H bonds} = 6 \times 413 = 2478 \end{cases} = 2\,824$

giving a liberation of $2824 - 2738$, i.e. 86 kJ which leads to a $\Delta H^{\ominus}$ value of -86 kJ.

In the hydrogenation of ethene to ethane,

$$C_2H_4(g) + H_2(g) \rightarrow C_2H_6(g)$$

the overall change in the bonding can be written as

$$\underset{598}{\overset{\diagdown}{/}C=C\overset{\diagup}{\diagdown}} \;+\; \underset{436}{H-H} \;\rightarrow\; \underset{346}{-\overset{|}{\underset{|}{C}}-\overset{|}{\underset{|}{C}}-} \;+\; \underset{2(413)}{2C-H}$$

$$\underset{1\,034}{} \qquad\qquad\qquad \underset{1\,172}{}$$

The figures for the average bond energies given lead to liberation of 1 172–1 034, i.e. 138 kJ, which corresponds to an enthalpy change of −138 kJ.

7 Resonance energy

The resonance energy of a substance is the extra stability of the resonance hybrid structure for the substance as compared with the most stable of the single canonical form structures. The resonance hybrid structure for benzene, for example, is more stable than structure I.

Structure I

Resonance energies can be obtained by comparing calculated and measured enthalpies. The actual measured enthalpy of formation of benzene from its free atoms is 5 501 kJ mol^{-1} but the value calculated on the basis of structure I is only 5 339 kJ mol^{-1}. The resonance energy is, therefore, 162 kJ mol^{-1}. A similar value is obtained by comparing the calculated and measured enthalpies of hydrogenation (p. 139).

The resonance energy of a molecule arises because of the delocalisation of the bonds within the molecule (p. 26); it is, therefore, sometimes referred to as delocalisation energy.

This delocalisation can only occur in conjugated systems, i.e. molecules with alternating single and double bonds (pages 114 and 141) or molecules with double bonds and adjacent atoms with lone pairs, e.g.

$$-C=C-C=C- \qquad -C=C-\ddot{O}- \qquad -C=C-\ddot{C}l:$$

8 Standard free energy changes

Standard free energy of formation can be used, in just the same way as standard enthalpies of formation, to calculate the standard free energy change in a reaction. For example,

$C_2H_2(g) + H_2(g) \rightarrow C_2H_4(g)$ $\Delta G^{\ominus}(298 \text{ K}) = 68.1 - 209.2$
209.2   0   68.1 $= -141.1 \text{ kJ}$

$C_2H_4(g) + H_2(g) \rightarrow C_2H_6(g)$ $\Delta G^{\ominus}(298 \text{ K}) = -32.8 - 68.1$
68.1   0   −32.8 $= -100.9 \text{ kJ}$

$2C_2H_6(g) + 7O_2(g) \rightarrow 4CO_2(g) + 6H_2O(l)$ $\Delta G^{\ominus}(298 \text{ K}) = -2\,936 \text{ kJ}$
2(−32.8) 0   4(−394.6) 6(−237.2)
−65.6   0   −1\,578.4 −1\,423.2

The standard free energy change for these reactions is not equal to the standard enthalpy change, the difference being accounted for by entropy changes (p. 39) in the relationship

$$\Delta G^{\ominus} = \Delta H^{\ominus} - T\,\Delta S^{\ominus}$$

It is the standard free energy change that gives the most valuable, detailed information about the feasibility of a reaction, for, under standard conditions, it is related to the equilibrium constant for the reaction;

$$\Delta G^{\ominus} = -2.303RT \lg K = -RT \ln K$$

$\Delta G^{\ominus}(298 \text{ K})$	−100	−57	−10	0	+10	+57	+100
$\lg K$	+17.53	10	1.753	0	−1.753	−10	−17.53
K	3.4×10^{17}	10^{10}	57	1	0.018	10^{-10}	2.95×10^{-18}

It follows that the more negative the value of $\Delta G^{\ominus}$ the higher the value of K, and the better the theoretical yield of the reaction (p. 33). Alternatively, high positive values of $\Delta G^{\ominus}$ mean low K values. A $\Delta G^{\ominus}$ value of 0 means a K value of 1 and relatively small changes of the $\Delta G^{\ominus}$ value, on either side of 0 give large changes in K because of the log scale relationship.

As a rough guide, if the calculated $\Delta G^{\ominus}$ value for a reaction is higher than 35 kJ ($K = 10^{-6}$) the reaction is not likely to be of any use; if the $\Delta G^{\ominus}$ value is less than −35 kJ ($K = 10^{6}$) the products will predominate in the equilibrium mixture formed and the reaction will be useful so long as conditions can be achieved in which it will take place at a reasonable rate.

9 Standard entropy changes

Differences between $\Delta H^{\ominus}$ and $\Delta G^{\ominus}$ values are due to entropy changes ($\Delta S^{\ominus}$), the relationship being

$$\Delta G^{\ominus} = \Delta H^{\ominus} - T\,\Delta S^{\ominus}$$

Entropy is best thought of, in an oversimplified way, as a measure of the disorderliness of a system. Any increase in disorderliness causes an increase in entropy ($\Delta S^{\ominus}$ is positive) and this generally comes about in an organic reaction when there is an increase in the number of gas molecules.

The standard entropy change in a reaction can be calculated, in much the same way as the standard enthalpy or free energy changes, by using standard molar entropy values obtainable from chemical tables. These values are quoted in J (not kJ) mol^{-1}, at 25 °C, and the values for the elements are not zero. For example,

$$C_2H_2(g) + H_2(g) \rightarrow C_2H_4(g) \qquad \Delta S^{\ominus}(298\ K) = 219.5 - 331.4$$
200.8   130.6   219.5 $\qquad\qquad\qquad\qquad\qquad = -111.9\ J$
    331.4

$$C_2H_4(g) + H_2(g) \rightarrow C_2H_6(g) \qquad \Delta S^{\ominus}(298\ K) = 229.5 - 350.1$$
219.5   130.6   229.5 $\qquad\qquad\qquad\qquad\qquad = -120.6\ J$
    350.1

$$2C_2H_6(g) + 7O_2(g) \rightarrow 4CO_2(g) + 6H_2O(l) \qquad \Delta S^{\ominus}(298\ K) = 1275.2 - 1893.3$$
2(229.5)   7(204.9)   4(213.8)   6(70) $\qquad\qquad\qquad\qquad = -618.1\ J$
459   1 434.3   855.2   420
       1893.3   1275.2

The entropy change in the examples given is negative because there is a decrease in the number of gas molecules. The $T\Delta S^{\ominus}$ value, at 25 °C, for the first reaction is $-(298 \times 111.9)$ J, i.e. -33.35 kJ, and this accounts for the difference in value between $\Delta H^{\ominus}(-174.5$ kJ) and $\Delta G^{\ominus}(-141.1$ kJ), i.e. -33.4 kJ.

10 Typical effects of entropy change

For many reactions the contribution of the $T\Delta S^{\ominus}$ term to the value of $\Delta G^{\ominus}$ can be disregarded in comparison to the contribution of $\Delta H^{\ominus}$. This is particularly so at low temperatures, when T is small, and for reactions in which the entropy change is low. But at higher temperatures or for reactions with a large entropy change the $T\Delta S^{\ominus}$ term may be just as important as the $\Delta H^{\ominus}$ term.

a. Effect of high temperature. Many reactions will only take place effectively at high temperatures. The high temperature may well be necessary to ensure that the rate of the reaction is high but it may also be that the reaction is only thermodynamically feasible at high temperatures. This is so for the well-known water-gas reaction,

$$C(s) + H_2O(g) \rightarrow CO(g) + H_2(g) \qquad \Delta H^{\ominus}(298\ K) = 131.3\ kJ \qquad \Delta S^{\ominus}(298\ K) = 134.1\ J$$
$$\Delta G^{\ominus}(298\ K) = 91.3\ kJ$$

The reaction is endothermic and the increase in the number of gas molecules gives a positive $\Delta S^{\ominus}$ value. The $T\Delta S^{\ominus}$ term $(298 \times 134.1\ J = 39.96$ kJ) accounts for the difference between $\Delta H^{\ominus}$ and $\Delta G^{\ominus}$ (131.3–91.3 = 40 kJ).

As the $\Delta G^{\ominus}$ value is highly positive the equilibrium constant, at 25 °C, is very low $(K = 10^{-16})$, and the reaction will not 'go'. Increase in temperature does not greatly affect the value of $\Delta H^{\ominus}$ but it markedly increases the value of the $T\Delta S^{\ominus}$ term. At 1 000 K, the $T\Delta S^{\ominus}$ term becomes 134.1 kJ which is bigger than the $\Delta H^{\ominus}$ value (131.3 kJ). The $\Delta G^{\ominus}$ value at 1 000 K is, therefore, negative $(-2.8$ kJ) so that the reaction becomes feasible $(K = 3.1)$.

b. Effect of high entropy change. The thermodynamic data for the formation of methane from its elements is as follows:

$$C(s) + 2H_2(g) \rightarrow CH_4(g) \qquad \Delta H^{\ominus}(298\ K) = -74.8\ kJ \qquad \Delta S^{\ominus}(298\ K) = -80.7\ J$$
$$\Delta G^{\ominus}(298\ K) = -50.8\ kJ$$

The difference between the $\Delta H^{\ominus}$ and $\Delta G^{\ominus}$ values (−24 kJ) is accounted for by the $T \Delta S^{\ominus}$ term (−298×80.7 J = −24 kJ). The corresponding figures for other alkanes are summarised below:

	$\Delta H^{\ominus}$/kJ	$\Delta G^{\ominus}$/kJ	$\Delta S^{\ominus}$/J	$T \Delta S$/kJ
CH_4	−74.8	−50.8	−80.7	−24.0
C_2H_6	−84.6	−32.8	−173.7	−51.8
C_3H_8	−103.8	−23.5	−269.6	−80.3
C_4H_{10}	−126.1	−17.1	−365.7	−108.9
C_5H_{12}	−173.2	−9.6	−550.9	−164.0
C_6H_{14}	−198.8	−4.4	−652.5	−194.4
C_7H_{16}	−224.4	+1.0	−756.2	−225.3
C_8H_{18}	−249.9	+6.4	−853.9	−254.4

In the formation of methane 2 mol of hydrogen are converted into 1 mol of alkane, but in the formation of octane 9 mol of hydrogen are converted into 1 mol of alkane. This change, throughout the series, causes the $\Delta S^{\ominus}$ and $T \Delta S^{\ominus}$ terms to become more and more negative and they do so more rapidly than the $\Delta H^{\ominus}$ term does. As a result the $\Delta G^{\ominus}$ value, which is negative up to and including hexane becomes positive from heptane onwards.

The higher alkanes are, therefore, unstable with respect to decomposition into their elements.

Reaction Kinetics

11 The order of a reaction

If the experimentally measured rate of reaction between A and B is found to be proportional to $[A]^x[B]^y$ the overall order of the reaction is $(x + y)$; the order with respect to A is x and with respect to B is y. The following examples are typical:

(a) $CH_3Br + OH^- \rightarrow CH_3OH + Br^-$

Rate $\propto [CH_3Br] \times [OH^-] = k[CH_3OH] \times [OH^-]$

The reaction is of the second order, and of first order with respect to both CH_3Br and OH^-. If the rate is measured in $mol\,dm^{-3}\,s^{-1}$ and the concentrations in $mol\,dm^{-3}$ the units of k (known as the rate constant) will be $dm^3\,mol^{-1}\,s^{-1}$.

(b) $(CH_3)_3CBr + OH^- \rightarrow (CH_3)_3COH + Br^-$

Rate $\propto [(CH_3)_3CBr] = k'[(CH_3)_3CBr]$

The reaction is of the first order with respect to $(CH_3)_3CBr$, the rate being independent of the OH^- ion concentration. The units of k', in this example, are s^{-1}.

The order of a reaction is determined solely by the orders of the concentration terms in the expression which best fits the rate-concentration relationship as measured experimentally. The order need not be a whole number; the order for the thermal decomposition of ethanal at 450 °C, for example, is 1.5,

$$CH_3CHO \rightarrow CH_4 + CO$$

Rate $\propto [CH_3CHO]^{1.5}$

For some reactions with a very complicated rate-concentration relationship the idea of a numerical order for the reaction may be of no value.

12 Activation energy

It might well have been assumed that two molecules would react whenever they came close enough together in a collision, but a comparison of the calculated number of collisions at any one temperature with reaction rates shows that, in general, fewer molecules react than collide. Moreover, the number of collisions goes up with increasing temperature but reaction rates go up more rapidly.

These facts are accounted for on the basis of activation energy. The general idea, first suggested by Arrhenius in 1889, is that two molecules in a collision will only react if they have, together, more than a certain amount of energy known as the activation energy, E_A.

Values of activation energies can be obtained by measuring rate constants at different temperatures, for

$$\frac{d \ln k}{dT} = \frac{E_A}{RT^2} \quad \text{or} \quad \frac{\lg k_1}{\lg k_2} = \frac{-E_A}{2.303R} \left[\frac{1}{T_1} - \frac{1}{T_2} \right]$$

A large value for E_A means a low value for k and a slow rate of reaction; a small value for E_A means a large value for k and a quick rate of reaction.

The activation energy represents a sort of energy barrier which the reagents have to overcome before the reaction can take place. The situation for exothermic and endothermic reactions can be summarised in energy profiles as in Fig. 29.

These profiles show both the activation energy and the enthalpy change. The maximum in the profile represents what is known as the *transition state* or activated complex. It is the arrangement of atoms of maximum energy (least stability) through which the reagents must pass before the products can be formed. The transition state is not a

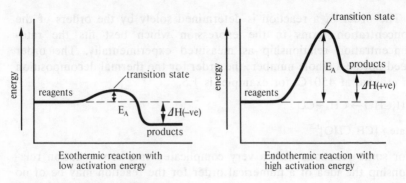

Fig. 29. Energy profiles.

discrete molecule that can be isolated; it is an activated species
containing weakened bonds due to the activation.

13 Thermodynamic or kinetic control

If a reaction can produce alternative products, e.g.

Naphthalene-1-sulphonic acid
(less stable)

$+ H_2SO_4$

Naphthalene-2-sulphonic acid
(more stable)

it may be that the more stable product is mainly formed (in which case the reaction is
said to be thermodynamically controlled) or it may be that the product which is formed
the more rapidly predominates (in which case the reaction is said to be kinetically
controlled).

In the example quoted, the less stable isomer is formed the more rapidly at 100 °C and
it predominates; the reaction is kinetically controlled. At 160 °C, or after prolonged
reaction time, the more stable isomer predominates; the reaction is thermodynamically
controlled.

14 The molecularity of a reaction

Organic reactions are not usually as simple as the overall equation
indicates for a reaction may take place in a series of steps. A reaction
mechanism attempts to define the nature of these steps by breaking
down a complex reaction into a series of elementary, one-step reactions.

a. Elementary reactions. In an elementary, one-step reaction the reagents pass through a transition state and form the products. The energy
profile is as in Fig. 29.

If only one molecule is involved in the reaction, i.e.

$$AB \rightarrow [A \cdots B] \rightarrow A + B$$
Transition state

the reaction is said to be unimolecular. If two molecules are involved, i.e.

$$AB + X \rightarrow [A \cdots B \cdots X] \rightarrow A + BX$$
Transition state

the reaction is bimolecular.

The molecularity is a theoretical postulate and not an experimentally measured quantity. It must be a whole number because fractions of molecules cannot be involved in reactions; it may or may not be equal to the order of the reaction.

b. Complex reactions. A complex reaction takes place in a series of successive elementary steps. In a two-step reaction the first step gives a product known as the intermediate. This may be stable enough to be isolated, or it may be so unstable that it only has a transient existence. The intermediate takes part in the second step of the reaction, e.g.

Overall reaction	$AB + X \rightarrow AX + B$
1st step (slow)	$AB \rightarrow [A \cdots B] \rightarrow A + B$
2nd step (fast)	$A + X \rightarrow [A \cdots X] \rightarrow AX$

The overall rate of the reaction will be fixed by the rate of the slower, rate-determining step (the first step in the example given) and the overall molecularity is taken as the molecularity of that step. The reaction as shown is, therefore, unimolecular, even though two molecules are involved in the overall reaction.

The energy profile (Fig. 30) links together the profiles of the two elementary reactions involved. It shows two transition states and one

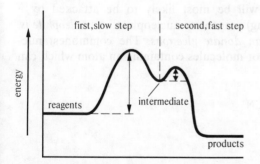

Fig. 30. Energy profile for a two-step reaction, a slow first step being followed by a fast second step.

intermediate. The activation energy for the slow, rate-determining step is larger than that for the fast step. If the intermediate is very reactive the activation energy for the second step will be very low.

General Types of Reagent and Reaction

15 Types of reagents

Many reactions can be regarded as taking place between two chemicals and, if it be necessary, one is referred to as the *reagent* and the other as the *substrate*. The reagent is thought of as the chemical which is attacking the substrate, but the use of the terms is somewhat arbitrary (see (d) below).

a. Electrophiles and nucleophiles. Many covalent bonds have some polar character so that the atoms joined by the bond carry some positive or negative charge. The initial attack of a reagent is generally on one of these charged atoms in the substrate.

Electrons will be available at a negatively charged atom within a molecule and such an atom will be most likely to be attacked by an electrophilic (electron-seeking) reagent or electrophile. *An electrophile is, therefore, a reagent which can accept electrons.* The commonest electrophiles are positively charged ions, or molecules containing an atom which can act as an acceptor, e.g.

H^+ H_3O^+ Br^+ Cl^+ NO_2^+ R^+ R_3C^+

BF_3 $AlCl_3$

Common electrophiles

There will be a shortage of electrons at a positively charged atom within a molecule and it will be most likely to be attacked by a nucleophilic (nucleus-seeking) reagent or nucleophile. *A nucleophile is, therefore, a reagent that can donate electrons.* The commonest nucleophiles are negative ions, or molecules containing an atom which can act as a donor, e.g.

HO^- RO^- Cl^- Br^- I^- CN^-

$\ddot{N}H_3$ $R\ddot{N}H_2$ $H_2\ddot{O}$ $R\ddot{O}H$

Common nucleophiles

b. Oxidising and reducing agents. Oxidation and reduction are very common processes in organic chemistry. As an oxidising agent is capable of accepting electrons it is electrophilic. A reducing agent, capable of donating electrons, is nucleophilic.

c. Lewis acids and bases. Lewis extended the older definitions of acids and bases as follows:

A Lewis acid is a substance that can accept a pair of electrons to form a covalent bond.

A Lewis base is a substance that can donate a pair of electrons to form a covalent bond.

For example,

$$H^+ + :\ddot{O}-H^- \rightleftharpoons H:\ddot{O}-H$$

$$\underset{\text{Acid}}{F \overset{\cdot x}{\underset{\cdot}{\text{x}}} B} + \underset{\text{Base}}{:N_x^x H} \rightleftharpoons F_x^x B : N_x^x H$$

A Lewis acid is, therefore, electrophilic and a Lewis base is nucleophilic.

d. Substrate and reagent. Both substrate and reagent are equally necessary for a reaction to take place. A substrate can be regarded as electrophilic when it is attacked by a nucleophile or as nucleophilic when it is attacked by an electrophile. Thus, when the C=O bond is attacked by a CN^- ion (p. 218) it can be regarded as an electrophile; when benzene is attacked by electrophilic reagents, e.g. NO_2^+, (p. 129) it can be regarded as a nucleophile.

Some substances can be either electrophilic or nucleophilic under differing circumstances.

16 Types of intermediate

The bonds in a molecule taking part in a reaction can split in a number of different ways to give different intermediates.

a. Heterolytic fission. Carbonium ions. Carbanions. When a single covalent bond splits in such a way that the pair of electrons is transferred onto *one* of the bonded atoms the process is known as heterolytic fission or heterolysis.

In a C—X bond the electron shift may be onto the C or onto the X atom. The commonest examples involve a shift onto the X atom, for, in most compounds, X is likely to be more electronegative than C. The electron pair shift is shown by the $\curvearrowright$ symbol (p. 30) and the resulting ion with a positive charge on the C atom is called a *carbonium ion*, e.g.

$$(CH_3)_3C-Br \rightarrow (CH_3)_3C^+ + Br^-$$

When the electron shift is onto the carbon atom, the ion with negative charge on the C atom is called a *carbanion*, e.g.

$$Cl_3C\!-\!H \;\rightarrow\; Cl_3C^- + H^+$$
$$\overset{\text{A}}{\underset{\text{carbanion}}{}}$$

Carbonium ions, and to a lesser extent carbanions, are involved as short-lived intermediates in many organic reactions, even though they may be present in only low concentrations. Reactions involving these ions take place most readily in polar solvents.

b. Homolytic fission. Free radicals. A single covalent bond may split in such a way that one electron from the bond becomes associated with each of the two bonded atoms. This type of bond fission, known as homolysis or homolytic fission, is shown by the ⌒ symbol. The product is an atom or group with a single unpaired electron known as a free radical. Homolysis can be brought about by ultra-violet radiation, e.g.

$$Cl\!-\!Cl \rightarrow Cl\cdot + Cl\cdot$$

$$H_3C\!-\!\overset{\overset{\text{O}}{\|}}{C}\!-\!CH_3 \longrightarrow H_3C\cdot + \cdot COCH_3$$

or by heating, e.g.

$$Pb(C_2H_5)_4 \rightarrow Pb + 4\cdot C_2H_5$$

Free radicals are involved in many reactions in the gaseous state or in non-polar solvents. They are often photochemical reactions and one free radical commonly leads to the formation of another in part of a chain reaction (p. 95), e.g.

$$Cl\cdot + CH_4 \rightarrow HCl + \cdot CH_3$$

17 Ease of formation of carbonium ions and free radicals

A carbonium ion has a positive charge (electron deficiency) on a C atom, and the more this charge can be reduced (or the electron deficiency made good) by the effect of neighbouring groups the more readily will the carbonium ion be formed. Thus, for simple alkyl carbonium ions, there is an increase in stability as H atoms are replaced by alkyl groups, i.e.

$$H-\underset{\underset{H}{|}}{\overset{\overset{H}{|}}{C^+}} \quad H-\underset{\underset{H}{|}}{\overset{\overset{R}{\uparrow}}{C^+}} \quad R\rightarrow\underset{\underset{H}{|}}{\overset{\overset{R}{\uparrow}}{C^+}} \quad R\rightarrow\underset{\underset{R}{\uparrow}}{\overset{\overset{R}{\uparrow}}{C^+}}$$

Primary   Secondary   Tertiary

$\longrightarrow$ Increase in ease of formation $\longrightarrow$

This arises because of the $+I$ effect (p. 24) of the alkyl groups. The positive charge in a C atom is reduced by an adjacent alkyl group because of the inductive, electron-release of that group as compared with hydrogen. Two alkyl groups are more effective than one, and three more effective than two. A C atom linked to three alkyl groups is called a tertiary C atom; one linked to two alkyl groups is secondary; and one linked to one alkyl group is primary.

Alkyl free radicals also contain a C atom with an electron deficiency and the same arguments apply. Thus

$$H-\underset{\underset{H}{|}}{\overset{\overset{H}{|}}{C\cdot}} \quad H-\underset{\underset{H}{|}}{\overset{\overset{R}{\uparrow}}{C\cdot}} \quad R\rightarrow\underset{\underset{H}{|}}{\overset{\overset{R}{\uparrow}}{C\cdot}} \quad R\rightarrow\underset{\underset{R}{\uparrow}}{\overset{\overset{R}{\uparrow}}{C\cdot}}$$

Primary   Secondary   Tertiary

$\longrightarrow$ Increase in ease of formation $\longrightarrow$

The nature of a number of reactions is controlled by these differences in the ease of formation of carbonium ions and free radicals (pages 87, 111, 155, 177, 178).

18 Types of reaction
There are four basic types of reaction though one type may be followed by another in any actual process.

a. Substitution reactions. These reactions involve the substitution of one atom or group by another atom or group, e.g.

$$
\begin{array}{lll}
CH_4 & +Cl_2 \rightarrow CH_3{-}Cl & +HCl & i \\
C_6H_6 & +HNO_3 \rightarrow C_6H_5{-}NO_2 & +H_2O & ii \\
C_6H_6 & +H_2SO_4 \rightarrow C_6H_5{-}SO_3H+H_2O & iii \\
CH_3{-}Cl & +OH^- \rightarrow CH_3{-}OH & +Cl^- & iv \\
(CH_3)_3CCl & +OH^- \rightarrow (CH_3)_3COH & +Cl^- & v \\
CH_3{-}Cl & +CN^- \rightarrow CH_3{-}CN & +Cl^- & vi
\end{array}
$$

When the substitution is a common process it may be given a specific name. Thus the substitution of —H by a halogen (—X) is called *halogenation;* substitution of —H by —NO$_2$ is *nitration;* substitution of —H by —SO$_3$H is *sulphonation.* If only one hydrogen atom is substituted the product is known as a monosubstituted derivative; if two atoms are involved it gives a disubstituted derivative; and so on.

Substitution reactions may involve free radicals (as in i above), and they may be electrophilic reactions (as in ii and iii) or nucleophilic (as in iv, v and vi) depending on the nature of the reagent. An S$_N$1 reaction is a unimolecular nucleophilic substitution; an S$_N$2 reaction is a bimolecular nucleophilic substitution.

b. Addition reactions. In these reactions two or more substances combine to give a single product, e.g.

$$C_2H_4 \quad + Cl_2 \quad \rightarrow C_2H_4Cl_2$$
$$C_2H_4 \quad + HBr \rightarrow C_2H_5—Br$$
$$C_2H_2 \quad + 2H_2 \quad \rightarrow C_2H_6$$
$$CH_3—CHO + HCN \rightarrow CH_3CH(OH)CN$$
$$CH_3—CN \quad + 2H_2 \quad \rightarrow CH_3CH_2NH_2$$

The compounds that undergo addition reactions are unsaturated (contain multiple bonds), the product of the reaction being saturated, or, at least, less unsaturated. The reactions can involve electrophilic or nucleophilic reagents or free radicals.

Some unsaturated compounds can, so to speak, undergo self-addition, the product having the same empirical formula but higher relative molecular mass than the reagent, e.g.

$$3C_2H_2 \rightarrow C_6H_6$$
$$nC_2H_4 \rightarrow (C_2H_4)_n$$
$$nC_2H_3Cl \rightarrow (C_2H_3Cl)_n$$

When a large number of molecules (n) are involved the product is known as an addition polymer and the process as *addition polymerisation* (p. 322). For two and three molecules the terms dimerisation and trimerisation are used.

c. Elimination reactions. These reactions are the opposite of addition reactions, a small molecule being eliminated from a larger one to give two products, e.g.

$$C_2H_5—OH \rightarrow C_2H_4 \quad + H_2O$$
$$C_2H_5—Br \rightarrow C_2H_4 \quad + HBr$$
$$C_2H_5—OH \rightarrow CH_3—CHO + H_2$$
$$(COOH)_2 \rightarrow H—COOH + CO_2$$

Specific names are sometimes used. Elimination of H_2O is *dehydration;* elimination of HX is *dehydrogenhalogenation;* elimination of H_2 is *dehydrogenation;* elimination of CO_2 is *decarboxylation.*

When the elimination occurs between two different molecules the term *condensation* is sometimes used and if many molecules are involved the product is a *condensation polymer* (p. 322).

E1 and E2 are used to denote unimolecular and bimolecular elimination reactions, respectively.

d. Rearrangement reactions. These involve the internal rearrangement of atoms within a molecule, e.g.

$$NH_4CNO \rightarrow CO(NH_2)_2$$

The rearranged molecule may split up into other products, as in the preparation of phenol from cumene (p. 198), or it may undergo some other type of reaction following the rearrangement.

Questions on Chapter 3

1 Classify the reactions below as nucleophilic substitution, nucleophilic addition, electrophilic substitution, electrophilic addition as you judge to be appropriate: (a) bromine and ethene (ethylene), (b) bromine and benzene (with iron initiator), (c) hydrogen cyanide and propanone (acetone), (d) sodium hydroxide solution and bromoethane (ethyl bromide).

   In each case, justify the classification in terms of the reaction mechanism. Explain, in terms of electronic theory, why (e) hydrogen cyanide will not react additively with ethene, (f) bromine will not react additively with propanone. (S)

2 Outline and discuss evidence for the proposition that 'Reagents which attack carbon–carbon double bonds do not attack carbon–oxygen double bonds, and vice versa'. (Camb. Schol.)

3 Organic compounds are sometimes described as 'reactive' or 'non-reactive', as the case may be. Give two examples of each of the

following: (a) a 'very reactive' compound; (b) a 'non-reactive' compound; (c) a 'moderately reactive' compound. Discuss whether or not you think that this kind of description of an organic compound serves any useful purpose. (Camb. Schol.)

4 Organic reactions are often divided into four main types: substitution, addition, elimination and rearrangement. Give examples to illustrate these types. Discuss in more detail any one of the types, indicating the mechanisms of the reactions of this type. (Oxf. Schol.)

5 Complete the following reaction schemes, stating the products and any essential conditions:
   (a) $C_2H_5COOCH_3 + OH^- \rightarrow$
   (b) $C_6H_6 + NO_2^+ \rightarrow$
   (c) $CH_3CHO + HSO_3^- \rightarrow$
   (d) $CH_4 + Cl \cdot \rightarrow$
   Explain which of the reagents you would classify as either an electrophile or a nucleophile. Give the detailed mechanism for reaction (d) and state why it is classified as a homolytic reaction. (W)

6 The direction of change of a chemical system at constant temperature and pressure may be said to be governed by two factors; the enthalpy change and the entropy change for the given reaction. (a) Explain what you understand by the term 'entropy'. (b) Discuss, with suitable examples of your own choice, the way in which the enthalpy and entropy of a reaction determine the direction of change. (L)

7 Draw up a list showing how organic reagents are classified and apply this classification to a mechanistic consideration of reactions of similar type. (SUJB)

8 List the following carbonium ions in order of increasing stability: Cl_3C^+, H_3C^+, $CH_3—C^+H_2$, $C_2H_5—C^+H_2$, $(CH_3)_2C^+H$.

9*Calculate (a) the enthalpy change, (b) the free energy change and (c) the entropy change in the following reactions under standard

_____
* Throughout the book questions marked with * require the use of a data book.

conditions:
  (i) $CH_4(g) + Cl_2(g) \rightarrow CH_3Cl(g) + HCl(g)$
  (ii) $2C_2H_5OH(l) + O_2(g) \rightarrow 2CH_3CHO(l) + 2H_2O(l)$
  (iii) $C_6H_6(l) + 3H_2(g) \rightarrow C_6H_{12}(l)$
Comment on any significant points.

10*Use bond energy values to estimate the enthalpy changes in the
   following reactions:
   (a) $C_2H_6(g) + Br_2(g) \rightarrow C_2H_5Br(g) + HBr(g)$
   (b) $C_2H_4(g) + H_2(g) \rightarrow C_2H_6(g)$
   (c) $2C_2H_2(g) + 5O_2(g) \rightarrow 4CO_2(g) + 2H_2O(g)$

4

Investigation of Organic Compounds

1 General procedure

Once a particular organic substance has been prepared in the laboratory or isolated from some natural source it is usually investigated further in the following important stages:

a. *Purification* (p. 54). Analysis must be preceded by careful purification for it is of no use analysing an impure substance unless it is to find whether it is pure or not.

b. *Qualitative analysis* (p. 64). Once purified, an organic substance is analysed qualitatively to find what elements it contains. A substance, X, for example, might be found to contain carbon, hydrogen and oxygen.

c. *Quantitative analysis* (p. 67). This follows the successful qualitative analysis of a substance and enables the percentage composition to be determined. Substance X might have a percentage composition of C = 37.5 per cent, H = 12.5 per cent, and O = 50 per cent.

d. *Empirical formula*. The empirical formula of a compound is the simplest formula expressing the *relative* numbers of each atom present in one molecule of the compound. It does not necessarily express the *actual* numbers of atoms. The empirical formula is easily calculated from the percentage composition as follows:

Elements present	C	H	O
$\dfrac{\text{Per cent of element}}{\text{Relative atomic mass}}$	$\dfrac{37.5}{12} =$	$\dfrac{12.5}{1} =$	$\dfrac{50}{16} =$
i.e. relative number of atoms present	3.125	12.5	3.125
Simplest whole number ratio of atoms	1	4	1

The empirical formula of X is, therefore, CH_4O.

e. Molecular formula (p. 69). The molecular formula of a compound expresses both the *relative* and the *actual* number of atoms present in one molecule of the compound. Substance X, with empirical formula of CH_4O, might have a molecular formula CH_4O, $C_2H_8O_2$, $C_3H_{12}O_3$, etc., for each of these expresses the same relative number of atoms.

To determine which of the many possible molecular formulae does actually represent the substance it is necessary to measure the relative molecular mass. For X it was found to be 32 which showed that the molecular formula was, in fact, CH_4O, i.e. the same as the empirical formula. Had the relative molecular mass been 64 the molecular formula would have been $C_2H_8O_2$.

f. Structural formula (p. 71). The molecular formula of a substance expresses the actual number of atoms in one molecule but it gives no information regarding the arrangement of atoms within the molecule. In a simple substance, such as CH_4O, there may be only one possible way of arranging the atoms; X must, in fact, be methanol. In more complicated substances, however, there may be very many ways in which the atoms could be linked together. It takes a lot of chemical skill and ingenuity, for example, to decide the structural formula of something like insulin (p. 305) with a molecular formula of $C_{254}H_{377}N_{65}O_{75}S_6$ and a relative molecular mass of 5727. It is rather like doing a jig-saw puzzle in the dark, though modern physical methods of investigation have enabled the problems to be solved.

g. Synthesis. Once the structural formula of a naturally occurring compound has been established it may be possible to synthesise it. If the substance is useful, but not easily obtainable in large quantities, e.g. vitamin C or penicillin, its synthesis will generally lead to a much wider availability. Moreover, it may eventually be possible to synthesis substances with modified structural formulae which may be superior in some respects to the original substance.

Cocaine, for example, was originally extracted from coca (the dried leaves of shrubs

Cocaine

which grow in S. America) and used as a local anaesthetic and pain killer. It was eventually replaced by procaine or novocaine (the hydrochloride of procaine) which were cheaper, less toxic and less habit-forming drugs. Nowadays, lignocaine, still better, is mainly used.

$$H_5C_2\diagdown N-CH_2-CH_2-O-\overset{O}{\overset{\|}{C}}-\langle\!\!\!\!\!\!\!\bigcirc\!\!\!\!\!\!\!\rangle-NH_2$$
$$H_5C_2\diagup$$

Procaine

$$H_5C_2\diagdown N-CH_2-\overset{O}{\overset{\|}{C}}-\overset{H}{\overset{|}{N}}-\langle\!\!\!\!\!\!\!\bigcirc\!\!\!\!\!\!\!\rangle$$
$$H_5C_2\diagup$$

Lignocaine

Purification Methods

In order to purify a substance it is necessary to separate the pure substance from the impurities and in order to separate any two substances it is necessary to find some differences between them. Those generally used are differences in solubility or vapour pressure (boiling point). The greater the difference between two substances the easier it is to separate them.

2 Recrystallisation

This is the commonest method for purifying a solid. A solvent is found, by experimental trial, in which the solid is more soluble when the solvent is hot than when it is cold. The impure solid is then dissolved in the minimum amount of hot solvent (see next section), the aim being to get a solution which is nearly saturated at the boiling point of the solvent. This hot, nearly saturated solution is then filtered so that any insoluble impurities can be removed. The filtration must be rapid to avoid crystallisation during the process. Funnels with cut-off stems or Buchner funnels are most convenient. If necessary, they can be pre-heated with hot solvent or jacketed (Fig. 31).

Soluble impurities remain, together with the solid required, in the filtrate. This is cooled to facilitate complete crystallisation and the crystals formed are filtered off through a Buchner funnel. They are then washed in the funnel with a little of the cold solvent and pressed down onto the filter paper. The suction is maintained until the crystals are as dry as possible. Further drying may be carried out by pressing between filter papers or in a desiccator or oven.

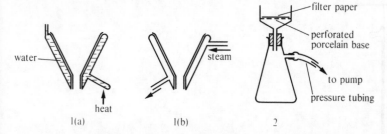

Fig. 31. Apparatus used in recrystallisation. 1. A funnel with a cut-off stem surrounded by (a) a hot-water jacket, and (b) a steam-jacket. 2. A Buchner funnel and Buchner flask.

Purification by recrystallisation depends on differences in solubility and concentration. Consider, for example, a substance, A, contaminated with 5 per cent of an impurity, B. Suppose, further, that the volume of a solvent used in the recrystallisation process can, when cold, dissolve 6 g of B and 20 g of A, i.e. the impurity is the less soluble of the two.

The state of affairs at various stages in the recrystallisation of 100 g of impure A is summarised as follows:

	Mass present in hot solution	Mass present in cold solution	Mass of crystals deposited
A	95 g	20 g	75 g
B	5 g	6 g	None

It will be seen that 75 g of pure crystals of A result. If the original percentage of B had been greater than 6 per cent then some B would have crystallised out with A, and two or more recrystallisations would be required to obtain pure A.

3 Use of a reflux condenser

In carrying out a recrystallisation a hot saturated solution must be prepared. If a volatile, and possibly inflammable, solvent has to be used the solution is best made in a flask fitted with a condenser arranged vertically (Fig. 32). Such a condenser is referred to as a reflux condenser.

The mixture in the flask can be heated as required without loss for any vapour from the solvent is condensed in the reflux condenser and the resulting liquid drops back into the flask. If the solvent being used is inflammable the flask is best heated on a sand- or water-bath.

Most organic reactions involving any volatile or inflammable compounds are carried out in a container fitted with a reflux condenser. The process is known as refluxing or heating under reflux.

4 Sublimation

It is easy to purify a substance which sublimes, e.g. naphthalene, from substances which do not sublime, simply by heating the mixture and

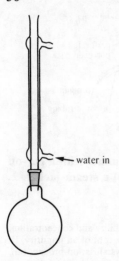

Fig. 32. A reflux condenser.

allowing the vapour formed to come into contact with a cold surface, but the method is of very limited application because so few substances do sublime.

5 Simple distillation

If a volatile liquid contains impurities which are not volatile, e.g. solids or liquids with high boiling points, purification can be achieved by a simple distillation. The mixture is boiled in a flask and the vapour from it is passed through a condenser. Pure liquid collects in the receiver and is referred to as the *distillate*; the impurities remain in the distillation flask. A typical simple arrangement is shown in Fig. 33.

6 Fractional distillation

This process is used in separating two or more volatile, miscible liquids with different boiling points, but it cannot always be carried out effectively. In its simplest aspect, a mixture of two liquids e.g. A (b.p. = 50 °C) and B (b.p. = 100 °C), is heated in a flask. The vapour coming from the mixture is richer in A (the more volatile component) than the original mixture. If this vapour is condensed, the resulting liquid will also be richer in A and may, under favourable conditions, be pure A.

The separation of two miscible liquids by fractional distillation is not, however, always straightforward. The mixture of liquids is in equilibrium with a mixture of vapours, but it is the way in which the

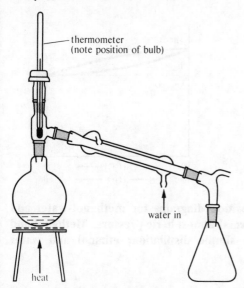

thermometer
(note position of bulb)

water in

heat

Fig. 33. Simple distillation.

compositions of these two mixtures are related at different temperatures which controls what happens during fractional distillation.

It is only when the temperature-composition curves are of the type found with methanol and water mixtures (Fig. 34) that complete separation is possible. If a mixture of methanol and water containing 50 per cent of each is boiled, it will boil at temperature T. The vapour coming from it will have a composition represented by A, and, on condensation, it will give a liquid of the same composition. This liquid will contain more than 50 per cent of methanol.

By repeating the process, pure methanol could be obtained but the process would be tedious and the same result can be achieved in one operation by using a fractionating column. If a 50–50 mixture of methanol and water is distilled using such a column the mixture boils, first, at temperature T, giving a vapour of composition A. This vapour condenses in the lower part of the column giving a liquid of composition A. As more hot vapour passes up the column this liquid is boiled giving a vapour of composition B; this condenses further up the column giving a liquid of composition B with a boiling point of T_1. This liquid, in its turn, is boiled by ascending vapour, and so on. The vapour issuing from the top of the column is pure methanol vapour and pure, liquid methanol is collected in the receiver. Pure water is eventually obtained in the distillation flask.

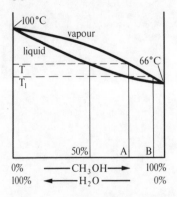
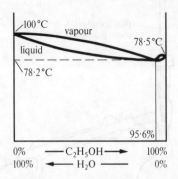

Fig. 34. Temperature-composition diagrams for methanol-water and ethanol-water mixtures at normal atmospheric pressure. Methanol and water can be separated by simple distillation; ethanol and water cannot.

There are many types of fractionating column and rather complex ones are needed to get a good separation of liquids with very similar boiling points. But the methanol-water type of temperature-composition curve is essential for getting a complete separation.

With ethanol and water the temperature-composition curve (Fig. 34) has a minimum. With initial mixtures containing less than 95.6 per cent of ethanol in the distillation flask pure water can be obtained as the residue in the flask and a mixture of 95.6 per cent ethanol and 4.4 per cent water can be collected in the receiver. With an initial mixture containing more than 95.6 per cent of ethanol, pure ethanol can be obtained as the residue and the same mixture of 95.6 per cent ethanol and 4.4 per cent water can be collected in the receiver.

Other mixtures give temperature-composition curves with maxima.

7 Distillation under reduced pressure

If a liquid decomposes at a temperature below its boiling point it cannot be distilled in the normal way under atmospheric pressure. It can often be distilled, however, at a lower temperature, under reduced pressure. The pressure is reduced by a pump and the danger of bumping is overcome by allowing a fine stream of bubbles to pass through the liquid being boiled via a fine capillary tube.

8 Steam distillation

This method is used for separating a high-boiling point liquid (immiscible with water) from non-volatile impurities such as inorganic solids.

A stream of steam is passed into the mixture in an apparatus as shown in Fig. 35. The distillate consists of the liquid required mixed with water. The liquid can be removed from the water by using a separating funnel or by extraction with a solvent (see Section 9).

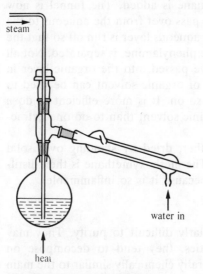

Fig. 35. Steam distillation.

The method depends on the fact that the vapour pressure of a mixture of two immiscible liquids at any temperature is equal to the sum of the vapour pressures of the individual liquids at the same temperature. The mixture boils when the total vapour pressure becomes equal to atmospheric pressure.

A mixture of phenylamine (b.p. = 184 °C) and water for example will boil at 98 °C under an atmospheric pressure of 101.325 kNm^{-2} or kPa. At this temperature, the vapour pressure of phenylamine is 7.066 kNm^{-2} and that of water is 94.259 kNm^{-2}, the total pressure being 101.325 kNm^{-2}. The vapour coming from the boiling mixture will contain both phenylamine vapour and water vapour the volume ratio of the two being equal to the ratio of their vapour pressures at 98 °C.

9 Solvent extraction

Methoxymethane, and other similar solvents, can be used to separate a solid which is soluble in them from solids insoluble in them.

They can also be used to extract an organic solid or liquid from an aqueous solution or suspension, for organic compounds are usually more soluble in organic solvents than in water. The organic solvents are, moreover, usually immiscible with water.

Pure phenylamine, for example, can be extracted from a mixture with water by the following procedure. The mixture is placed in a separating funnel and methoxymethane is added. The funnel is now shaken to allow the phenylamine to pass over from the aqueous to the organic layer. On settling, the lower aqueous layer is run off so that the organic layer, containing most of the phenylamine, is separated. Not all the phenylamine will, however, have passed into the organic layer in this one extraction, but a fresh lot of organic solvent can be used to carry out a second extraction, and so on. It is more efficient to do n extractions each with V cm^3 of organic solvent than to do one extraction with nV cm^3.

All the extracts are mixed together, dried by standing over solid potassium hydroxide, and filtered. The methoxymethane is then distilled off, taking special precautions because it is so inflammable.

10 Chromatography

Many natural products are particularly difficult to purify. They may only be available in small quantities, they tend to decompose on heating, and the impurities are generally chemically similar to the main product. The traditional methods of separation cannot, therefore, be applied very successfully. In recent years the study of many compounds has been greatly facilitated by the development of chromatography. The technique was first used by Tswett in 1906 to separate the various pigments present in leaves (hence the term chromatography) but it is now used, in a variety of ways, not only for coloured substances.

a. Column chromatography. If a solution containing various solutes is passed through a glass tube packed with a powdered material it is commonly found that the solutes concentrate at different levels of the column giving what is known as a *chromatogram*.

By pushing the whole column out of the glass tube and cutting it into separate bands each solute can be extracted separately. Alternatively, the various solutes can be washed through the column, in turn, by using more solvent or a different solvent. This is known as *elution* and the solvents as *eluants*.

If coloured solutes are involved the separation can be achieved visually. For colourless substances ultra-violet illumination may be necessary, or the column may have to be treated with some chemical to distinguish one solute from another.

Many different packing materials, e.g. aluminium oxide, calcium carbonate, magnesium oxide, charcoal, starch, calcium phosphate(V), ion-exchange materials, and many proprietary mixtures, can be used.

Typical solvents, which may be mixed together, include water, propanone, ethanol, benzene, trichloromethane and hexane. A satisfactory combination of packing material and solvent can be found for most problems.

In simple *adsorption column chromatography* the solutes are separated because they have different solubilities in the original solvent and because they are adsorbed differently on the column. It is unlikely that any two solutes will be identical in both respects so that they can be separated. In *partition column chromatography* another factor enters in. The packing material is itself wet with one liquid, e.g. water, which is held stationary, and the solutes partition between this liquid and the original, moving solvent. In practice both adsorption and partitioning may take place together.

b. Paper chromatography. Column chromatography requires quite large amounts of material and the same type of separation can be achieved with much smaller amounts by using specially made grades of absorbent paper instead of the packed column.

A very small spot of solution is placed near the bottom of a strip of paper which is then dipped into a suitable solvent. As the solvent front rises up the paper by capillary action the various solutes present are carried forward to differing extents. When the solvent front reaches almost to the top of the paper, the paper is removed and dried. If the solutes are coloured their positions on the paper can be seen. For colourless solutes the paper can be treated with some chemical to convert the solutes into coloured compounds. The paper can then be cut up and each solute extracted separately.

A mixture of adsorption and partitioning is generally involved. The cellulose of the paper acts as an adsorbent but it is also wet with water which allows for partitioning if a non-aqueous solvent is used. For any particular set of conditions, each individual substance has a fixed R_f value given by

$$R_f = \frac{\text{Distance moved by substance}}{\text{Distance moved by solvent front}}$$

The identity of any unknown substance can, therefore, be established by comparison with known materials under the same conditions.

Two substances with equal R_f values cannot be separated by one-way chromatography but a two-way process may be possible. The original mixture is spotted onto one corner of a square piece of paper and a one-way separation is carried out using one solvent. The paper is then dried and turned through 90° before carrying out a second separation with a different solvent.

c. Thin layer chromatography. Paper chromatrography is limited to separations that can be achieved on a cellulose base but the same principle can be applied by using a thin layer of material deposited evenly on a glass plate. The plate is then used in the same way as a sheet of paper. Many materials can be used and there is the normal wide choice of solvent, and even smaller amounts of material can be dealt with than in paper chromatography.

d. Gas chromatography. Gas chromatography is like column chromatography except that a gas is used instead of a liquid solvent. It is used for separating gases or volatile liquids and solids in the gaseous state.

A column is packed with a material which may simply act as an adsorbent or which may be wetted with an involatile oil so as to facilitate partitioning. A steady flow of a carrier gas, such as hydrogen, nitrogen or carbon dioxide, is then passed through the column, heated if necessary, and the mixture of gases or vapours to be analysed is fed into the stream of carrier gas, via a microsyringe inserted through a rubber cap, before it enters the column (Fig. 36).

Some constituents of the mixture are soon carried right through the column, whereas others pass through only very slowly. Each constituent will, in fact, pass through the column at a different rate and will emerge at the end of the column at different times. A detector at

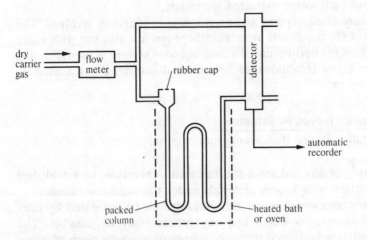

Fig. 36. Gas chromatography.

the end shows the emergence of the various constituents of the original mixture, which is therefore analysed.

The commonest type of detector is a thermal conductivity gauge. The gas emerging from the end of the column is passed round a heated filament. The heat loss from this filament depends on the thermal conductivity of the gas mixture surrounding it, and its electrical resistance, which is measured, changes as the gas mixture changes. The presence of different gases can in this way, be detected.

If the carrier gas is hydrogen or if some hydrogen is added to the carrier gas at the end of the column a flame ionisation detector can also be used. The emergent gas is burnt at a metal jet which acts as a negative electrode. The flame impinges on a metal plate above it which acts as a positive electrode; a potential difference is maintained between the two electrodes. The current passing between the electrodes depends on the concentration of ions within the flame and this varies with the composition of the gas being burnt.

Both methods of detection can be connected to automatic recording devices.

11 Criteria of purity

It is not possible to tell whether a substance is pure or not simply by looking at it and some experimental method of deciding is essential. Theoretically, many of the physical properties of the substance concerned could be measured. In practice, it is the melting point of a solid or the boiling point or refractive index of a liquid that are most commonly used.

Determination of boiling point is not so simple or quick as determination of melting point and the purity of a liquid is sometimes established by first converting it into a solid derivative.

a. Determination of melting point. If a solid has a sharp melting point, i.e. melts completely within a range of about 1 °C, it is almost certainly pure. Impure solids, e.g. butter or chocolate, or solids which decompose before they melt, give indefinite melting points. The purity of a solid can, then, be established by measuring its melting point. For a known substance, the measured melting point can be compared with the recorded value. For an unknown substance purification processes must be continued until a constant melting point is reached.

With a large amount of solid the melting point can be measured by plotting a cooling curve but the commoner procedure requires only a small amount of material. This is first dried and powdered and then placed in a melting-point tube about 7 cm long and 1 mm in diameter made by drawing out a hot soft-glass tube, cutting off the required length and sealing one end. The bottom 1 cm of the tube should be filled with solid.

The tube is then heated in a liquid bath (Fig. 37), or in an electrically heated apparatus. The temperature at which the solid is seen to melt is recorded as the melting point.

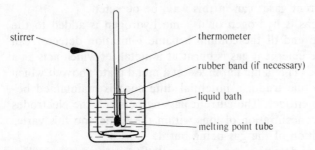

stirrer

thermometer

rubber band (if necessary)

liquid bath

melting point tube

Fig. 37. Determination of melting point.

b. The method of mixed melting points. The recorded melting point of, say, benzoic acid is 121.85 °C, so that it might be presumed that an unknown substance, X, with an equal melting point must be benzoic acid. It might be so, but it could also be that there was some other substance with a melting point of 121.85 °C. The matter can be resolved by the method of mixed melting points. If the melting point of a mixture of X with some benzoic acid is 121.85 °C then X must be benzoic acid. If the melting point of the mixture is not sharp and is not 121.85 °C then X cannot be benzoic acid.

c. Determination of boiling point. The boiling point of a liquid can be measured in a distillation apparatus (Fig. 33 p. 57). A pure liquid will have a single, sharp boiling point, but a single, sharp boiling point does not firmly establish the purity of a liquid, for the liquid may be a constant boiling mixture. If, however, the liquid gives a sharp boiling point at two different pressures the possibility of its being a constant boiling mixture is ruled out.

Qualitative Analysis

Qualitative analysis of an organic compound is carried out to find what elements are present in the compound. Since the number of different elements which do occur in organic compounds is small (p. 3) a specific test is carried out for each likely element, but there is no very satisfactory test for oxygen.

12 Test for carbon and hydrogen

If a compound is known to be organic it is usually assumed that it contains carbon and hydrogen. The presence of the elements can be proved, however, by heating the compound under test (it must be dried and powdered) with about eight times its mass of dry, powdered copper(II) oxide in a hard-glass test tube arranged as in Fig. 38. Hydrogen is indicated by the condensation of water at the top of the test tube; carbon by the lime-water turning milky. If necessary, the condensed liquid can be treated with anhydrous copper(II) sulphate(VI).

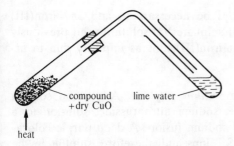

compound + dry CuO   lime water

heat

Fig. 38. Test for carbon and hydrogen.

13 Lassaigne's test for nitrogen, sulphur and halogens

The organic substance is heated in a dry ignition tube with a small pellet of sodium. The heating is gentle until the sodium melts but is then stronger until the tube is a dull red-heat. At this stage, and whilst the tube is still very hot (so that it will readily crack) it is plunged into some cold water in a beaker or evaporating dish. The operation must be carried out firmly but carefully. The mixture, containing the broken glass and the aqueous extract is boiled for a few minutes and then filtered.

The filtrate might contain sodium cyanide (if the original compound contained nitrogen), sodium sulphide (if the original compound contained sulphur) and sodium halides (if the original compound contained halogens). It is, therefore, a matter of dividing up the filtrate and doing individual tests for these three substances.

a. Test for nitrogen, present as sodium cyanide. The filtrate will be alkaline due to the presence of sodium hydroxide originating from the sodium and water. On addition of some iron(II) sulphate(VI) solution a precipitate of iron(II) hydroxide forms and this reacts with any sodium cyanide, on boiling, to form sodium hexacyanoferrate(II). On cooling, adding 2–3 drops of iron(III) chloride solution, and acidifying with concentrated hydrochloric acid (to dissolve any iron(II) or iron(III) hydroxide precipitates), a greenish-blue coloration or precipitate of iron(III) hexacyanoferrate(II) is formed and indicates the presence of nitrogen in the original compound.

$$Fe^{2+}(aq) + 2OH^-(aq) \rightarrow Fe(OH)_2(s)$$

$$Fe(OH)_2(s) + 6CN^-(aq) \rightarrow Fe(CN)_6^{4-}(aq) + 2OH^-(aq)$$

$$3Fe(CN)_6^{4-}(aq) + 4Fe^{3+}(aq) \rightarrow Fe_4\{Fe(CN)_6\}_3(s).$$

In practice it may not always be necessary to add any iron(III) chloride solution as the iron(II) sulphate(VI) solution used previously will generally contain sufficient iron(III) ions as impurity due to atmospheric oxidation.

b. Test for sulphur, present as sodium sulphide. Add some sodium pentacyanonitrosylferrate(II) i.e. sodium nitroprusside, solution to a portion of the filtrate from the sodium fusion. A deep purple coloration indicates the presence of S^{2-} ions and, therefore, sulphur in the original compound.

c. Test for halogens, present as sodium halides. If no nitrogen or sulphur was found to be present, a portion of the filtrate from the sodium fusion is acidified with dilute nitric(V) acid and silver nitrate(V) is added. A white, cream or yellow precipitate indicates the presence of a chloride, bromide and iodide. To determine which halogen is present more firmly, a further portion of the original filtrate must be acidified with dilute sulphuric(VI) acid and benzene and chlorine water added. On shaking, an iodide gives a violet-coloured benzene layer; a bromide a yellow brown layer; and a chloride a colourless layer.

If the substance under test was found to contain nitrogen or sulphur the test for halogens must be slightly modified for the sodium cyanide or sulphide present would give precipitates on adding silver nitrate(V). Sodium cyanide or sulphide must, therefore, be removed before testing for halogens. This is most conveniently done by boiling with dilute nitric(V) acid for a few minutes to expel any hydrogen cyanide or hydrogen sulphide. The acidified filtrate, now free of any cyanide or sulphide, is tested for halogens by adding silver nitrate(V) solution.

Summary of qualitative analysis

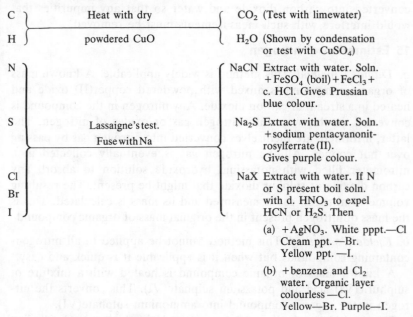

C ⎫ Heat with dry ⎧ CO₂ (Test with limewater)

H ⎭ powdered CuO ⎩ H₂O (Shown by condensation
  or test with CuSO₄)

N ⎫ NaCN Extract with water. Soln.
  + FeSO₄ (boil) + FeCl₃ +
  c. HCl. Gives Prussian
  blue colour.

S ⎬ Lassaigne's test. Na₂S Extract with water. Soln.
  Fuse with Na + sodium pentacyanonit-
  rosylferrate (II).
  Gives purple colour.

Cl ⎫ NaX Extract with water. If N
Br   or S present boil soln.
I ⎭ with d. HNO₃ to expel
  HCN or H₂S. Then

  (a) + AgNO₃. White pppt.—Cl
  Cream ppt. —Br.
  Yellow ppt. —I.

  (b) + benzene and Cl₂
  water. Organic layer
  colourless —Cl.
  Yellow—Br. Purple—I.

Quantitative Analysis

The chemical reactions on which quantitative analysis of organic compounds is based are simple but the detailed procedures required to obtain accurate results are lengthy for the reactions must be made to go to completion and all the relevant products must be measured.

Macro-analysis requires about 1 000 mg of the material being analysed; semi-micro methods need 10–50 mg; micro methods need 1–5 mg. Some modern analytical apparatus can make automatic recordings for some elements.

It is difficult to estimate the amount of oxygen in an organic compound. If it is present, its amount is generally obtained by difference.

14 Estimation of carbon and hydrogen

A known mass of the dry organic compound is heated in a combustion tube in a stream of pure, dry oxygen. The carbon in the compound is converted into carbon dioxide (measured by absorption in soda-lime tubes) and the hydrogen is converted into steam or water (measured by absorption in anhydrous magnesium chlorate(VII) tubes).

As 12/44 ths of the mass of carbon dioxide is carbon and 2/18 ths of the mass of water is hydrogen the percentage of the two elements in the original organic compound can be calculated.

The apparatus used is such that all the carbon and hydrogen are converted into carbon dioxide and water so that any impurities that would interfere with such measurements must be removed.

15 Estimation of nitrogen

a. Dumas's method. This method is widely applicable. A known mass of organic compound is mixed with powdered copper(II) oxide and heated in a stream of carbon dioxide. Any nitrogen in the compound is converted, in this way, into nitrogen gas or oxides of nitrogen. The latter, if formed, are themselves converted into nitrogen gas by passing over hot copper. All the nitrogen gas is eventually collected in a nitrometer filled with potassium hydroxide solution to absorb any carbon dioxide or sulphur dioxide that might be present. The resulting volume of nitrogen gas is measured and its mass is calculated. This is the mass of nitrogen present in the original mass of organic compound.

b. Kjeldahl's method. This method cannot be applied to all nitrogen-containing compounds but when it is applicable it is quick and easy.

A known mass of organic compound is heated with a mixture of sulphuric(VI) acid and potassium sulphate(VI). This converts the nitrogen in the organic compound into ammonium sulphate(VI).

The resulting ammonium sulphate(VI) solution is then made alkaline by adding excess sodium hydroxide. Ammonia is liberated on heating and is estimated by passing into a measured volume of standard acid.

The mass of ammonia obtained from the organic compound is calculated, and 14/17 ths of this mass is equal to the mass of nitrogen in the compound.

16 Estimation of halogens

Carius's method estimates chlorine, bromine and iodine by heating a known mass of an organic compound with fuming nitric(V) acid and silver nitrate(V) in a sealed tube at 260 °C for about 5 hours. The halogen is converted by this treatment into silver halide which is washed out of the tube, filtered off, washed, dried and weighed.

The mass of halogen in the measured mass of silver halide is easily calculated and this mass of halogen is the mass originally present in the amount of organic compound taken.

17 Estimation of sulphur

Sulphur can also be estimated by Carius's method. A known mass of organic compound is heated with fuming nitric(V) acid in a Carius tube and this converts any sulphur in the compound into sulphuric(VI) acid. When the heating is completed the acid is washed out of the tube and converted into barium sulphate(VI) by adding excess barium chloride solution. The barium sulphate(VI) is filtered off, washed, dried and weighed, and the mass of sulphur involved is calculated.

Summary of quantitative analysis

C⎱ ⎰ CO$_2$ (absorb in weighed soda-lime tubes)
    Heat in a stream
    of dry O$_2$
H⎰ ⎱ H$_2$O (absorb in weighed anhyd. Mg(ClO$_4$)$_2$ tubes)

   Heat with CuO in a current of CO$_2$ N$_2$ gas (collect in nitrometer
    (Dumas's method)   and measure volume)
N

   Heat with c. H$_2$SO$_4$+K$_2$SO$_4$ (NH$_4$)$_2$SO$_4$ $\xrightarrow[\text{and heat}]{\text{+NaOH}}$ NH$_3$ (pass into known
    (Kjeldahl's method)   volume of standard acid)

S $\xrightarrow{\text{Heat with fuming HNO}_3\text{ in}}$ H$_2$SO$_4$ $\xrightarrow{\text{+ BaCl}_2}$ BaSO$_4$ (estimate gravimetrically)
   a sealed tube (Carius's
   method)

Cl⎱
Br⎰ Heat with fuming HNO$_3$+AgNO$_3$ in sealed tube   AgCl
I ⎰ (Carius's method)   AgBr   (estimate gravimetrically)
                                                            AgI

Determination of Molecular Formulae

Determination of the molecular formula of a substance once its empirical formula is known requires a knowledge of its relative molecular mass. Really accurate methods are not always required for it may simply be a matter of distinguishing between a relative molecular mass of n, $2n$, $3n$, etc. A variety of traditional methods are still in use and mass spectroscopy provides a more modern method.

18 Traditional methods

These are described in books on Physical Chemistry but are summarised below.

a. For gases. The relative molecular mass can be obtained from the relative density which can be measured by direct weighing. Measurement of the volume changes which occur when hydrocarbons are exploded with excess oxygen can also give relative molecular mass values.

b. For volatile liquids or solids. The relative density of a volatile substance can be measured by Victor Meyer's method.

c. For non-volatile solids. The relative molecular mass of non-volatile solids is generally determined by measuring the depression of the freezing point or the elevation of the boiling point when a known mass of the solid is dissolved in a known mass of a suitable solvent. Very high relative molecular masses are obtained by measurement of osmotic pressure.

19 Mass spectroscopy

Mass spectrometers are expensive but when they are available they can be used for finding relative molecular masses and, sometimes, for giving important clues as to the structural formula (p. 79).

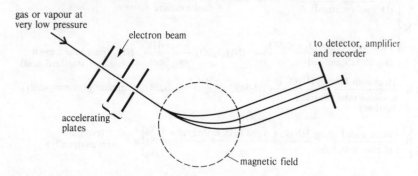

Fig. 39. The principle of a mass spectrometer.

The substance under investigation is bombarded in the gaseous state at a very low pressure by an electron beam (Fig. 39). This causes some of the molecules to lose an electron and form a positive, molecular ion, e.g.

$$M + e^- \rightarrow M^+ + 2e^-$$

Ions of higher charge may also be formed but the singly-charged ions predominate. The positive ions are then accelerated by passing between charged plates so that they emerge with equal velocities. They then pass through a magnetic field in which they are deflected according to their mass/charge ratio. For singly-charged ions the deflection depends solely on the mass.

If only one M^+ ion is present it will show up as a single peak on the recorder and the position of the peak will enable the mass of M^+ and of M to be calculated. The results obtained are very accurate and very little material is required. The presence of isotopes can be detected and positive ions differing only very slightly in mass can be separated in what is known as the mass spectrum.

With some substances the single M^+ ion is unstable, breaking down into smaller positive ions of lower mass. Bombardment of methane, for example, by electrons yields CH_4^+, CH_3^+, CH_2^+, CH^+, C^+ and H^+ ions which can all be detected by mass spectroscopy. The heights of the peaks for each ion also show the relative amounts of each ion present (see p. 79).

Determination of Structural Formulae

The elucidation of the structural formula of a substance, once its molecular formula is known, can be very simple for simple molecules or very difficult for complex ones (p. 53). Traditionally it was a matter of getting as much chemical information about the substance concerned as possible. The structure of ethanol can be decided, for example, on the basis that its reaction with phosphorus pentachloride (p. 153) indicates the presence of one —OH group, its reaction with sodium shows that one hydrogen atom differs from the other five, and its synthesis from ethane by treatment with chlorine and subsequent hydrolysis.

For much larger molecules, e.g. proteins (p. 308) it was necessary to break down the large molecule into suitable fragments whose structures could be identified. The way in which the various fragments were held together in the whole had then to be worked out.

The traditional method used large amounts of the substance concerned, was relatively slow, and could not always solve the problem. It has been largely superseded by spectroscopic methods. These require expensive apparatus but they are rapid and accurate and use only small amounts of the substance under test.

20 Spectroscopic methods

Single *atoms* can absorb energy by the promotion or excitation of electrons from lower to higher energy level orbitals. The energy is quantised corresponding to the specific energy levels involved and it is mainly ultra-violet or visible radiation that is involved. The excited atoms re-emit the energy to give atomic or line spectra with lines in the ultra-violet or visible region.

Molecules give more complicated spectra. The electrons in the component atoms can still be promoted but energy can also be absorbed by stretching or bending vibrations of bonds within the molecule or by rotation of the molecule or its parts.

These vibrational and rotational energy changes are quantised in the same way as the electronic changes but they do not involve ultra-violet or visible light because the energy changes involved are smaller.

The vibrational energy changes involve infra-red radiation and it is infra-red absorption spectra that give the most useful information about the structure of organic molecules. Rotational energy changes are very small; they are studied in micro-wave spectroscopy.

a. Infra-red spectra. Organic molecules give characteristic infra-red absorption spectra for many of the typical bonds in the molecules absorb particular frequencies of infra-red radiation as indicated in Table 1. The precise frequency of radiation absorbed by any one bond varies slightly as neighbouring bonds change but it is, generally, not difficult to tell whether bonds such as those in Table 1 are present or not by a study of infra-red absorption spectra.

The detailed interpretation of the spectra is also made easier by comparing the spectrum of an unknown substance with those of known substances with similar structures.

Table 1   Infra-red absorption wave number for typical bonds

Wave number cm^{-1}	Bond	Type of compound
1 000–1 300	C—O	Alcohol, ether, carboxylic acid
1 620–1 680	C≡C	Alkene
1 700–1 760	C=O	Aldehyde, ketone, carboxylic acid
2 000–2 300	C≡N	Nitriles
2 100–2 200	C≡C	Alkynes
2 850–3 000	C—H	Alkanes
3 020–3 100	C—H	Alkenes
3 300	C—H	Alkynes
3 300–3 500	N—H	Primary amine
3 600–3 700	O—H	Alcohols and phenols

Different bonds absorb different frequencies of radiation but the frequencies involved are generally expressed as wave numbers, i.e. the number of waves per centimetre rather than as frequencies, i.e. the number of cycles per second. The wave number (cm^{-1}), σ, is the reciprocal of the wavelength, λ, i.e. $\sigma = 1/\lambda$. Wave number, frequency, f, and velocity, c, are related by the expression,

$$\sigma = \frac{f}{c}.$$

In SI units, frequencies are expressed in hertz ($1 \, Hz = 1$ cycle per second, s^{-1}), wavelengths in metres, m, and wave numbers in m^{-1}. It follows that $1 \, cm^{-1} = 29.97925$ gigahertz (GHz).

Typical infra-red spectra are shown in Fig. 40; they are obtained on an infra-red spectrophotometer which is, nowadays, a standard piece of equipment in an advanced laboratory. An infra-red source is provided by heating a rod made of metallic oxides (a Nernst glower) and the radiation is passed through the material under investigation either after or before passing through a diffraction grating or a rock-salt prism (rock-salt is necessary as glass absorbs infra-red light). The radiation that passes through the material is detected on a sensitive thermocouple and the proportion of each frequency that is transmitted is recorded on a plot of frequency against percentage transmission. A deep trough indicates a strong absorption at the frequency concerned.

Gases are contained in rock-salt cells; liquids are studied as films on rock-salt plates or in solution; solids can be studied in solution, as mulls in liquids such as Nujol, or by compression into discs with potassium bromide.

21 Nuclear magnetic resonance (NMR) spectroscopy

NMR spectroscopy is particularly useful for giving information about the hydrogen atoms in an organic molecule. The hydrogen atom nucleus, i.e. the proton, contains an odd number of nucleons* (one) and can spin in two opposite directions. The spinning nucleus (proton) has a magnetic moment so that it can be orientated in an external magnetic field either in the same direction as the field or in the opposite direction.

Protons aligned in the same direction as the external field have lower energy than those opposed to the field, the energy difference being proportional to the field strength.

Protons in organic molecules can absorb radiation when its energy is of the right value to raise them from their lower to their higher energy level, i.e. to invert their spins. The frequency of radiation necessary to do this depends on the strength of the external magnetic field and on the nature of the hydrogen atom concerned. For a field strength of about 10 000 gauss the frequencies absorbed lie in the range of radio waves between 10 and 100 MHz.

---

* The NMR spectra associated with the ^1H nucleus are easily the most important but other nuclei with odd mass numbers, e.g. ^{19}F and ^{31}P, will also give rise to NMR spectra. Nuclei with even mass numbers and even atomic numbers, e.g. ^{12}C and ^{16}O give no NMR spectra.

Fig. 40. Typical infra-red absorption spectra (simplified). The percentage transmission, or absorption, is plotted against the wave number in cm^{-1} or the wavelength in micrometres (microns), μm. The clearest plot is obtained by using one scale for wave numbers from 2 000–4 000 cm^{-1} and another for 800–2 000, as shown.

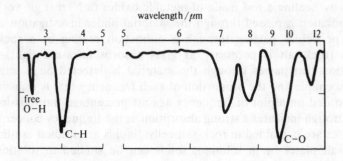

(a) *Ethanol in the vapour phase.* The sharp band at 3 700 cm^{-1} is caused by free O—H groups, i.e. ones which are not hydrogen-bonded. The band at 3 000 cm^{-1} is caused by the C—H bonds in CH$_3$ and CH$_2$ groups. Bands below 1 600 are caused by bending of O—H and C—H bonds or C—O stretching (1 050 cm^{-1}).

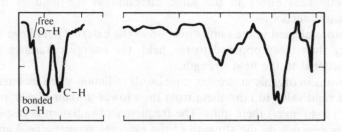

(b) *Ethanol in 10 per cent solution in CCl$_4$.* The O—H bonds in ethanol are strongly hydrogen bonded in CCl$_4$ solution so that the absorption due to free O—H bonds (3 700 cm^{-1}) is only very weak. There is strong absorption at 3 400 cm^{-1} due to the hydrogen bonded O—H bonds. The hydrogen bonding weakens the bond so that the absorption is at a lower frequency (higher wave length). C—H absorption is still strong around 3 000 cm^{-1}

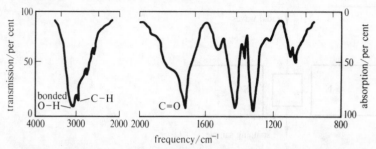

(c) *Ethanoic acid in 10 per cent solution in CCl$_4$.* The O—H groups are strongly hydrogen bonded as in (b); even more so, in fact, for their absorption is at a still lower frequency (3 200 cm^{-1}) and this tends to overlap with the C—H absorption. There is also a strong band at 1 740 cm^{-1} due to the C=O bonds, which are not present in ethanol.

For any fixed magnetic field the precise frequency of radiation absorbed depends on the nature of the hydrogen atom concerned. The hydrogen atoms in —OH, —CH$_3$, —CHO and other groups are all surrounded by slightly different arrangements of electron density so that they are shielded differently from any external magnetic field. The nature of the hydrogen atoms in organic molecules can, therefore, be determined either by measuring what frequencies of radiation the molecule will absorb when in a fixed magnetic field or what magnetic field strengths will cause the absorption of a fixed frequency of radiation. It is, in practice, easier to use the latter technique and only relatively small changes in magnetic field strength are needed. The number of any particular type of hydrogen atom in a molecule is indicated by the amount of radiation of a particular frequency that is absorbed.

a. Experimental. An NMR spectrometer is provided with an oscillator producing radio waves of a fixed frequency (typically 40, 60 or 100 MHz), a powerful electromagnet whose field can be changed slightly by altering the current, and a detector to record the level of radiation absorbed as the magnetic field is changed (Fig. 41). The poles of the electro-magnet are very close together to give a homogeneous field and the sample under test (either in liquid form or in solution) is spun between the poles.

Tetramethylsilane, TMS, (CH$_3$)$_4$Si, is commonly used as a standard. It contains only one type of hydrogen atom which absorbs a fixed frequency of radiation at one particular magnetic field strength (H) which is taken as the standard. Other protons in other organic molecules will absorb the same frequency at different field strengths

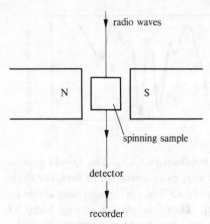

Fig. 41. Essential features of an NMR spectrometer.

(H_1) and the relative change, i.e. $(H - H_1)/H$ is a constant for any particular proton. This constant is generally multiplied by 10^6 i.e. expressed as parts per million (p.p.m.); it is then known as the *chemical shift*, δ, for the proton concerned, the value for TMS being zero. Alternatively TMS is given an arbitrary value of 10 p.p.m. and chemical shifts are expressed in Tau (τ) units, with $\tau = 10 - \delta$.

Values for the chemical shifts of hydrogen atoms in typical organic groups are given in Table 2, and typical NMR spectra are shown in Figs 42 and 43. The spectra can be plotted either at low or high resolution. A modern NMR spectrometer can also record automatically the areas under each peak thus indicating the number of protons concerned.

Table 2   Chemical shifts (δ) for protons in p.p.m. relative to TMS = 0

Type of proton	Chemical shift	Type of proton	Chemical shift
$R{-}CH_3$	0.9	$R{-}O{-}CH_3$	3.8
$R{>}CH_2$	1.3	$R{-}O{-}H$	about 5
$R{>}CH$	2.0	$R{-}\underset{\overset{\|}{O}}{C}{-}H$	9.7
$R{-}CH_2{-}I$	3.2	$R{-}COOH$	About 11

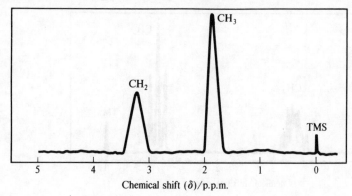

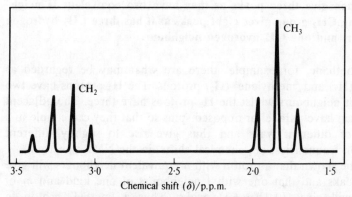

Chemical shift (δ)/p.p.m.

Fig. 42. Typical NMR spectra for iodoethane. The upper diagram shows the peaks for CH$_3$ and CH$_2$ groups at low resolution. At higher resolution (lower diagram) the CH$_3$ splits into three peaks (it has two equivalent H neighbours) and the CH$_2$ group splits into four peaks (it has three equivalent H neighbours).

b. Spin-spin coupling. A single peak in an NMR spectrum recorded at low resolution may split into more peaks at higher resolutions as shown in Figs. 42 and 43. This arises when protons are present within a molecule on adjacent carbon atoms.

$$H-\overset{\overset{\displaystyle H}{|}}{\underset{\underset{\displaystyle H}{|}}{C}}-\overset{\overset{\displaystyle H}{|}}{\underset{\underset{\displaystyle H}{|}}{C}}-I$$

Iodoethane

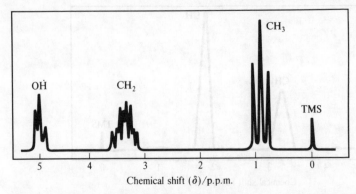

Fig. 43. High resolution NMR spectrum for ethanol. The OH and CH_3 groups each give three peaks as they have two equivalent H neighbours. The CH_2 group gives eight peaks as it has three CH_3 hydrogen neighbours and one OH hydrogen neighbour.

In iodoethane, for example, there are what may be regarded as 'methyl' (H_3) and 'methylene' (H_2) protons. The H_3 protons have two equivalent neighbours whilst the H_2 protons have three. The adjacent protons can have parallel or opposed spins so that they can couple in a number of different ways and thus give rise to slightly different absorption frequencies, i.e. chemical shifts, in the NMR spectrum.

It can be shown that a proton with n equivalent neighbours will give $(n+1)$ peaks and that one with l neighbours of one kind and m of another will give $(l+1)(m+1)$ peaks. Thus a 'methyl' proton in iodoethane will give 3 peaks (it has two equivalent neighbours) and a 'methylene' proton will give 4 peaks (it has three equivalent neighbours). Furthermore, if n is 2 the intensity of the peaks will be in the ratio $1:2:1$; if n is 3 the ratio is $1:3:3:1$ (Fig. 42).

$$H-\overset{\displaystyle H}{\underset{\displaystyle H}{C}}-\overset{\displaystyle H}{\underset{\displaystyle H}{C}}-O-H$$

Ethanol

In ethanol, at a sufficiently high resolution (Fig. 43), the 'methyl' protons give three peaks ($n=2$); the 'hydroxyl' proton gives three peaks ($n=2$); and the 'methylene' proton gives eight peaks (it has three 'methyl' protons and one 'hydroxyl' proton as neighbours).

High resolution NMR spectroscopy can, therefore, provide invaluable information about the types of hydrogen atoms in an organic molecule, about the number of each type, and about the neighbouring hydrogen atoms.

22 Use of mass spectroscopy

Some molecules can be broken up into a variety of positively charged ions by bombardment with electrons and both the masses and the amounts of each ion can be discovered using a mass spectrometer (p. 71). The so-called fragmentation pattern varies according to the energy of the bombarding electrons so that weak bonds in a molecule can be detected. A detailed study of the various fragmentation patterns obtainable from a molecule can help to elucidate its structure. One possible structure would be expected to give one fragmentation pattern; a different structure would give a different pattern.

Butane, for example, gives a strong peak at a mass of 29 ($C_2H_5^+$) because the ethyl–ethyl bond in the $C_4H_{10}^+$ molecular ion splits fairly easily,

$$C_2H_5 - C_2H_5^+ \rightarrow C_2H_5^+ + C_2H_5\cdot + e^-$$

There are no similar ethyl groups in 2-methylpropane so that it does not give any significant peak at mass 29.

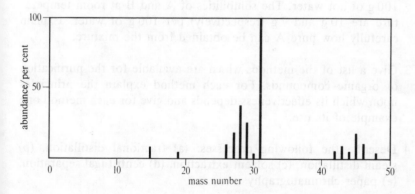

Fig. 44. The mass spectrum of ethanol showing the fragmentation pattern.

The complete fragmentation pattern obtained from ethanol is shown in Fig. 44. The abundance of each ionic species is represented by a vertical line as a percentage of the most abundant ion, which, for ethanol, is CH_2OH^+. This is formed by cleavage of the C—C bond in the original molecular ion,

$$C_2H_5OH + e^- \rightarrow C_2H_5OH^+ + 2e^-$$
$$C_2H_5OH^+ \rightarrow \cdot CH_3 + CH_2OH^+$$

The other main ions present are summarised below:

Mass number	15	26	27	28	29	31	45	46
Ion	CH_3^+	$C_2H_2^+$	$C_2H_3^+$	$C_2H_4^+$	$C_2H_5^+$	CH_2OH^+	$C_2H_5O^+$	$C_2H_5OH^+$

This particular fragmentation pattern is unique for ethanol and every other substance has its own pattern.

Questions on Chapter 4

1 Explain the varied illustrative examples the difference in meaning between the terms empirical formula, molecular formula and structural formula.

2 A mixture of 80 g of A with 20 g of B was completely dissolved in 100 g of hot water. The solubilities of A and B at room temperature are 10 g and 7 g (respectively) per 100 g of water. Explain carefully how pure A can be obtained from the mixture.

3 Give a list of the methods which are available for the purification of organic compounds. For each method explain the principles upon which its effectiveness depends and give for each method one example of its use.

4 Describe the following processes: (a) fractional distillation, (b) steam distillation, (c) solvent extraction, (d) centrifugal separation, (e) paper chromatography.

5 Explain carefully why it is not possible to obtain pure ethanol by a simple distillation process.

6 Comment on the statement that two liquids with different boiling points can be separated by fractional distillation.

7 Explain why a fractional distillation using a fractionating column gives a better separation than one without a column.

8 How would you attempt to find out whether a given organic solid was pure or not? If impure, how would you set about trying to purify it?

9 0.059 g of an organic compound having a molecular formula C_3H_9N was subjected to analysis by Dumas's method. What volume of nitrogen was obtained at s.t.p?

10 0.667 g of an organic compound treated by Kjeldahl's method gave sufficient ammonia to neutralise 22.2 cm^3 of M HCl. On combustion, 0.229 g of the compound gave 0.168 g of carbon dioxide and 0.137 g of water. Assuming that the compound contained only C, H, N and O determine its empirical formula.

11 An organic compound containing only C, H and Cl has a relative molecular mass of 119. When treated by Carius's method, 1 g of it gives 3.602 g of silver chloride. If the compound contained 10.05 % of C, what is its structural formula?

12 An organic compound containing C, H, O and S has a relative molecular mass of 238. It contains 31.4 % of C and 2.52 % of H. 0.7 g of it gives, by Carius's method, 1.37 g of barium sulphate(VI). What is (a) the molecular and (b) the structural formula of the compound?

13 A mixture of 10 cm^3 of a gaseous hydrocarbon and 200 cm^3 of oxygen on explosion gave 175 cm^3 which was reduced to 135 cm^3 on shaking with potassium hydroxide solution. What is the molecular formula of the hydrocarbon?

14 A mixture of methane, hydrogen and nitrogen occupying 10 cm^3 was made up to 90 cm^3 with air and exploded. The resulting volume was 73.75 cm^3 and, after absorption with potash, it was 69.75 cm^3. Calculate the volume composition of the mixture.

15 A hydrocarbon was exploded with excess oxygen under conditions in which the hydrocarbon was gaseous but the steam formed condensed. The contraction in volume was observed as was the further contraction when the carbon dioxide present was absorbed by potassium hydroxide. The first contraction was $\frac{3}{4}$ of the second. With what hydrocarbons will this occur?

16 When a substance is being extracted from an aqueous solution by an organic solvent it is more efficient to use a given volume of the solvent in several fractions rather than as a whole. Explain fully why this is so.

17 Discuss the usefulness and limitations, in organic preparative chemistry, of each of the following techniques: (a) fractional distillation, (b) paper partition chromatography, (c) recrystallisation. (W)

18 Give an account of the considerations involved in selecting the most appropriate solvent for purifying an organic compound by recrystallisation. Why is the method of purification by sublimation particularly suited for organic solid compounds? Describe practical details for purifying a compound by sublimation. Comment on significant features regarding melting points or boiling points in relation to molecular structure of the various classes of organic compounds you have studied. (W)

20 Both mass spectrometry and chromatography may be used to isolate a substance from a mixture whose separation would otherwise be difficult. Give an account of the principles of these two methods of separation. How may one or other of these two methods be used to separate (a) a mixture of hydrocarbons of approximately the same boiling point, and (b) the fissionable isotope of uranium from the natural material. (L)

19 Explain the meaning of the terms (a) infra-red absorption, (b) spin-spin coupling, (c) fragmentation pattern.

21 Explain, briefly, the theory governing each of the following processes, and in each case give a practical illustration: (a) ether extraction; (b) distillation under reduced pressure; (c) fractional recrystallisation to constant melting point; (d) drying in a desiccator. (L)

5

Petroleum Products

1 Fractional distillation of crude oil (Fig. 45)

Crude oil occurs in large subterranean deposits and is an extremely important raw material. It is required both to provide fuels and as a source of many organic and inorganic chemicals (p. 88). It consists, essentially, of a mixture of many different liquids containing dissolved gases and solids. The components are mainly alkanes (containing from one to about forty carbon atoms per molecule), cycloalkanes (particularly cyclopentane and cyclohexane) and aromatic hydrocarbons.

The origin of crude oil is uncertain, but the most favoured theory is that is has been formed by the bacterial decomposition, under pressure, of animal and plant remains.

Crude oil is heated to about 400 °C by passing it through coils of pipes in a gas-heated pipe-still furnace. The hot oil, which at 400 °C is a mixture of vapours and liquids, passes into a tall cylinder known as a *fractionating tower*. This tower is divided into a number of compartments by means of trays set one above the other. The trays contain holes, each of which is covered by what is known as a 'bubble-cap', and overflow pipes.

The temperature of the fractionating tower ranges from about 400 °C at the bottom to about 40 °C at the top, and each tray is at a controlled temperature. As the mixture passes into the tower, the liquids fall to the bottom whilst the vapours pass up through the trays. As each tray is at a different temperature, different mixtures of vapours condense to give different liquid fractions in each tray.

The bubble-caps force the rising vapours to pass through trays full of condensed liquid. The liquids in the trays are cooler than the rising vapours and this assists condensation of high boiling-point components. At the same time, the hot vapours re-vaporise any low boiling point compounds which may have condensed in a tray but which ought not to be there. Each tray is, too, fitted with an overflow so that liquid is constantly passing from a higher to a lower tray, there to be scrubbed, once more, by hot

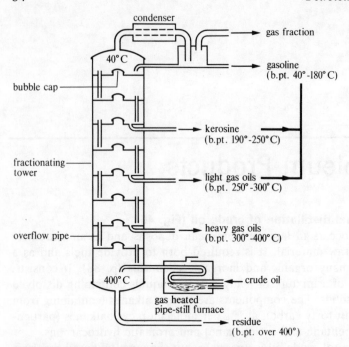

Fig. 45. Fractional distillation of crude oil.

vapours. The constant counterflow of vapour through liquid ensures a thorough separation into fractions.

The liquid mixtures in each tray are tapped off continually. Each fraction may be re-distilled, and the fractions of higher boiling point may also be re-distilled under a vacuum. This enables the distillation to be carried out at a lower temperature which minimises thermal decomposition. The fractions are also treated chemically.

The main fractions which are collected and used are as follows:

Fraction	Approx. temp. range °C	Approx. no. of C atoms	Use
Residue.	>400	25	Heavy fuel oil and bitumen.
Heavy gas-oil. Light lubricating oil.	300–400	18–25	Fuel. Lubricants. Paraffin wax. Medicinal paraffin.
Light gas-oil.	250–300	13–17	Fuel oil. Diesel engines.

Fraction	Approx. temp. range °C	Approx. no. of C atoms	Use
Kerosine. (Paraffin oil.)	190–250	11–14	Fuel oil. Tractor and jet-engines.
Naphtha.	100–200 ⎫		Solvent. Raw material.
Petrol. (Gasoline.)	40–180 ⎬ 5–10		Fuel.
Petroleum ether.	40–60 60–80 ⎭		Solvent.
Gas fraction.	<40	1–5	Gaseous fuel. Sources of alkanes.

2 Further refining processes

The fractions taken from the fractionating tower are treated and u ͻd as follows:

a. Residue. The residue can be separated into a heavy oil, used in furnaces and ship's boilers, and bitumen, used for road surfacing and roofing materials. By distilling under reduced pressure it is also possible to obtain lubricating oils and paraffin wax.

b. Heavy gas oil. This fraction is first treated with liquid propane to remove bitumen and tarry compounds. It is then cooled to crystallise out *paraffin wax*, which may also be obtained by extraction with phenol. Finally it is treated with powdered clay to improve the colour; it is this treatment which gives the oil its green, fluorescent appearance.

The waxes obtained are used in making candles, polishes and waxed papers, and *greases* are made by thickening lubricating oils with soaps. The lubricating oil fraction also provides *medicinal or liquid paraffin.* The fraction is first washed with sulphuric(VI) acid and then with sodium hydroxide solution. It is then decolourised with charcoal and further purified by distillation. The product is a colourless, odourless oil used in cosmetic creams, hair dressings and ointments, and as a mild laxative.

c. Light gas-oil. Light gas oils, of which Diesel fuel is the best known example, are used as fuels in oil-fired furnaces and in diesel engines. Some light gas-oil is also cracked (p. 87).

d. Kerosine (paraffin oil). Kerosine, long used for heating and lighting purposes, was originally the most important crude oil product. As newer methods of heating and lighting developed its use diminished but it has increased again in recent years because of the demands of tractor and turbo-jet engines. Some of the kerosine fraction is, nevertheless available for cracking (p. 87).

e. Petrol (gasoline). Part of the gasoline fraction is suitable for use in internal combustion engines after a simple sweetening process involving the conversion of evil-smelling sulphur compounds. Most of the fraction, however, requires further treatment, as it knocks easily if used directly. This means that the petrol-air mixture in the cylinders explodes prematurely and incompletely causing a metallic rattle known as knocking or pinking. The effect is particularly noticeable in high-compression engines.

The knock characteristics of a petrol is measured by its octane number. 2,2,4-trimethyl pentane (iso-octane), which knocks only under extreme conditions, and heptane, which knocks very readily, have been chosen as standards against which to measure the performance of any given petrol. Iso-octane has been given an arbitrary octane number of 100; heptane one of 0. The octane number of any fuel is the percentage of iso-octane in a mixture of iso-octane and heptane which will knock to the same extent as the fuel under test when used in a standard engine under standard conditions.

Considerable improvement in the knock qualities of a petrol can be achieved by adding anti-knock reagents such as tetrethyllead(IV) (p. 344). It is also known that straight-chain alkanes, such as heptane, cause knocking far more readily than branched-chain ones, such as iso-octane. Alkenes and aromatic hydrocarbons also have better knock qualities than straight-chain alkanes. Many conversion processes have, therefore, been developed to change the composition of the original gasoline fraction (see section 3).

f. Other fractions. The naphtha fraction is used as a solvent and as an important raw material for the chemical industry (p. 88). Petroleum ethers, with varying boiling point ranges, are used as solvents. The gas fraction serves as a source of individual C_1—C_5 alkanes (p. 90) and as a gaseous fuel.

3 Conversion processes

Fractional distillation separates crude oil into its component parts or fractions but it cannot change the proportion of the various products. These proportions vary in different samples of crude oil but they do not meet modern requirements. In particular, the yield of petrol obtained by straightforward distillation is too low, and the quality of the straight-run product is not very high. Both the yield and the quality of the petrol can be improved by various conversion processes.

a. Cracking. Cracking processes break down the larger molecules in fuel oils into the smaller molecules needed in petrol and, at the same time, they convert some of the straight-chain alkanes (which knock easily) into branched-chain ones. This can be achieved at high temperature and pressure (*thermal cracking*) but *catalytic cracking* is the commoner process. The hot, vapourised fuel oil reacts on the surface of a solid catalyst so finely divided that it flows like a liquid. Alkanes with lower relative molecular mass are produced, together with a mixture of gaseous alkenes, e.g. C_2H_4, C_3H_6, C_4H_8, known as *refinery gas*,

$$C_8H_{18} \text{ octane} \nearrow \begin{array}{l} C_6H_{14} + C_2H_4 \\ \text{hexane} \quad \text{ethene} \end{array}$$
$$\searrow \begin{array}{l} C_4H_{10} + C_4H_8 \\ \text{butane} \quad \text{butene} \end{array}$$

b. Polymerisation. In this process, gaseous and low boiling-point alkenes found in the gas fraction and occurring as by-products of cracking processes are polymerised in the presence of a catalyst. The polymer is still an unsaturated hydrocarbon but may be converted into an alkane by treatment with hydrogen, e.g.

$$2H_2C = C \begin{array}{c} CH_3 \\ | \\ CH_3 \end{array} \xrightarrow[100\,°C]{c.\,H_2SO_4} (CH_3)_3 C - \overset{\displaystyle H}{\underset{\displaystyle |}{C}} = C(CH_3)_2$$

$$\xrightarrow{H_2} 2,2,4\text{-trimethylpentane (iso-octane)}$$

c. Isomerisation. In this process, straight-chain alkanes are converted into branched chain isomers at about 100 °C and under pressure, in the presence of an aluminium chloride catalyst e.g.

$$CH_3 - CH_2 - CH_2 - CH_3 \xrightarrow{AlCl_3} H_3C - \overset{\displaystyle CH_3}{\underset{\displaystyle CH_3}{\overset{\displaystyle |}{\underset{\displaystyle |}{C}}}} - H$$

The isomerisation occurs because of the conversion of an initial secondary carbonium ion into the more stable tertiary ion (p. 47).

d. Alkylation. The process of adding a branched-chain alkane to the C=C bond of an ethene is known as alkylation, e.g.

$$H_2C = C \begin{array}{c} CH_3 \\ | \\ CH_3 \end{array} + H_3C - \overset{\displaystyle CH_3}{\underset{\displaystyle H}{\overset{\displaystyle |}{\underset{\displaystyle |}{C}}}} - CH_3 \xrightarrow[100\,°C]{c.\,H_2SO_4} 2,2,4\text{-trimethylpentane (iso-octane)}$$

e. Reforming. This involves the conversion of a straight-chain alkane into an aromatic hydrocarbon. The process involves a simultaneous *cyclisation* and *dehydrogenation*, e.g.

C_6H_{14} $\xrightarrow{-4H_2}$

hexene

benzene

A catalyst of platinum catalyses the dehydrogenation whilst one of aluminium oxide catalyses the cyclisation. The process, sometimes called platforming, was originally used to increase the amount of aromatic hydrocarbons in petrol but it is now also used for making pure benzene and methylbenzene (p. 127).

4 Natural gas

Deposits of crude oil often have natural gas associated with them and, when this is so, it is a mixture of oil and gas which comes from the well. Natural gas may also occur in places where there is no oil.

The composition of natural gas varies with its source. North sea gas is predominantly methane but in American sources there are important amounts of other alkanes together with some hydrogen, nitrogen, carbon monoxide, carbon dioxide, water-vapour and hydrogen sulphide. The natural gas from Dexter in the U.S.A. also contains about 3 per cent of helium and is an important commercial source of this gas.

Natural gas is used as a gaseous fuel and, in this country, it is rapidly replacing coal-gas. It is also an important chemical raw material.

5 Petrochemicals

Crude oil was originally used as a source of fuels but, beginning in 1930 in the U.S.A. and in 1950 in European countries, it is now a very important source of many individual chemicals.

The gas and naphtha fractions from crude oil distillation, together with refinery gas from cracking, and natural gas, provide the major starting materials. In particular they serve as a source of methane, ethane, propane and butane, ethene, propene and butenes, and aromatic hydrocarbons such as benzene, methylbenzene (toluene) and dimethylbenzenes (xylenes). These basic hydrocarbons act, in their turn, as the starting materials for a very wide range of important products.

Questions on Chapter 5

1 Explain the meaning of the following terms: (*a*) cracking, (*b*) octane number, (*c*) knocking, (*d*) sweetening, (*e*) bubble cap.

2 Describe briefly the refining of crude oil and explain the nature and importance of cracking.

3 Explain carefully how a fractionating tower enables the components of crude oil to be separated.

4 "The world would be in a better state to-day if crude oil had never been discovered." Discuss.

5 Distinguish between, and explain the purpose of, (a) polymerisation, (b) isomerisation, (c) alkylation.

6 Write notes on (a) shale oil, (b) natural gas, (c) synthetic fuels.

7 Write an account of either (a) Energy from Chemicals, or (b) The Uses of Petroleum Products, or (c) British Refineries.

8 "The production of chemicals from crude oil has had and is having a profound effect on chemical industry." Discuss.

9 Give one example of each of the following processes: hydrogenation, dehydrogenation, dehydrocyclisation, alkylation, dealkylation, isomerisation.

10 'Petroleum has become the major industrial source of aliphatic and aromatic hydrocarbons for the production of organic chemicals'. Discuss this statement as fully as you can, illustrating your answer by reference to four examples of chemicals of your own choice. (L)

6

Alkanes

1 Introduction

The alkanes, with a general formula C_nH_{2n+2} or RH, provide the simplest homologous series of saturated hydrocarbons. The straight-chain isomers of the first ten members of the series are listed below. Methane, ethane and propane only exist in the straight-chain form but higher alkanes have branched-chain isomers the naming of which is described on p. 12.

The alkanes are generally unreactive, deriving their older name of paraffins from the Latin parum = a little and affinis = affinity. They do, however, burn well and play a large part in petroleum chemistry (p. 83); they undergo some important substitution reactions when hot or in the presence of light (p. 93); and their molecules can undergo some unusual elimination (p. 96) and rearrangement reactions particularly at high temperatures.

2 Natural sources of alkanes

Natural gas (p. 88) provides the main source of methane together with smaller quantities of C_2-C_6 alkanes. Hydrogen sulphide, if present, is first removed by absorption in a solution of a base. The alkanes are then separated by fractional distillation, and the sulphur is recovered for conversion into sulphuric(VI) acid.

Some methane is also obtained as a by-product in sewage treatment where it is formed by the action of anaerobic organisms on organic matter. Similar action gives methane in marsh gas and fire damp, and coal gas contains about 30 % methane.

The gas fraction from the distillation of crude oil also provides C_1-C_5 alkanes, and fractions of higher boiling points contain many higher alkanes as listed on p. 85.

3 Synthesis of alkanes

As alkanes can be obtained so readily from natural sources, synthetic methods are mainly used for replacing an $-X$ group in a compound by

–H. Typical methods are summarised below:

a. Alkanes from alkenes. Alkenes are readily reduced to alkanes by catalytic hydrogenation (p. 110),

$$C_nH_{2n} + H_2 \rightarrow C_nH_{2n+2}$$

$$R—CH{=}CH_2 + H_2 \rightarrow R—CH_2—CH_3$$

Alkane	b.p./K	m.p./K	Standary enthalpy of combustion/ kJ mol^{-1}
Methane, CH_4	111.7	90.7	−890.4
Ethane, C_2H_6	184.5	89.9	−1 560
Propane, C_3H_8	231.1	85.5	−2 220
Butane, C_4H_{10}	272.7	134.8	−2 877
Pentane, C_5H_{12}	309.2	143.4	−3 510
Hexane, C_6H_{14}	341.9	177.8	−4 195
Heptane, C_7H_{16}	371.6	182.5	−4 854
Octane, C_8H_{18}	398.8	216.4	−5 513
Nonane, C_9H_{20}	423.9	219.6	−6 125
Decane, $C_{10}H_{22}$	447.3	243.5	−6 778

b. Alkanes from halogenoalkanes. Iodoalkanes can be reduced (p. 161) by hydrogen, e.g.

$$R—I + [2H] \rightarrow R—H + HI$$

or coupled (p. 159) in the presence of sodium (*Wurtz's reaction*), e.g.

$$2R—I + 2Na \rightarrow R—R + 2NaI$$

The yields are not always good and they depend on the alkane and reducing agent used. Wurtz's reaction is only suitable for synthesising alkanes with an even number of carbon atoms. If it were used to make C_3H_8, for example, the starting materials would have to be CH_3I and C_2H_5I; they would yield C_2H_6 and C_4H_{10} as well as C_3H_8.

Halogenoalkanes can also be converted into alkanes via Grignard reagents (p. 341).

c. Alkanes from carboxylic acids. Carboxylic acids can be decarboxylated by heating their sodium salts with soda lime,* which functions as

---

* Soda lime is a mixture of sodium and calcium hydroxides made by adding concentrated sodium hydroxide solution to quicklime. It may be preferable to sodium hydroxide as a reagent because it is not deliquescent, it attacks glass less readily and it has a higher melting point.

sodium hydroxide, e.g.

$$CH_3—COO^-Na^+(s) + NaOH(s) \rightarrow CH_4(g) + Na_2CO_3(s)$$

$$R—COO^-Na^+(s) + NaOH(s) \rightarrow R—H + Na_2CO_3(s)$$

4 Physical properties of alkanes

a. Boiling point. The boiling points of straight-chain alkanes rise steadily (Fig. 46) as the number of carbon atoms increases due to the increasing strength of the van der Waals' forces. At room temperature and atmospheric pressure, alkanes up to and including butane are gases, pentane to pentadecane are liquids, and higher ones are solids.

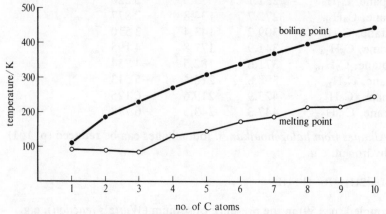

Fig. 46. Boiling and melting points of straight-chain alkanes.

For any one set of branched-chain isomers the boiling point decreases as the branching increases, e.g.

$$CH_3—CH_2—CH_2—CH_2—CH_3 \qquad CH_3—\overset{\displaystyle CH_3}{\underset{\displaystyle }{CH}}—CH_2—CH_3 \qquad H_3C—\overset{\displaystyle CH_3}{\underset{\displaystyle CH_3}{C}}—CH_3$$

Pentane　　　　　　　　　2-Methylbutane　　　　　2,2-Dimethylpropane
(b.p. = 309.2 K)　　　　　　(b.p. = 301 K)　　　　　　(b.p. = 282.7 K)

This arises because the more highly branched molecules are more nearly spherical so that they can pack together less closely with resultant decrease in strength of the van der Waals' forces.

b. Melting point. The melting points of straight-chain alkanes rise in a zig-zag pattern (Fig. 46) the ones with an even number of carbon atoms having slightly higher values than those of similar size with an odd number of carbon atoms. This is accounted for by closer packing in the alkanes with an even number of carbon atoms. The effect of branching on the packing and on the melting points is irregular.

c. Solubility. All alkanes are practically insoluble in water and, being less dense, the liquid and solid alkanes float on the surface of water. That is why water cannot be used for putting out petrol and oil fires. The lower members are soluble in organic solvents, but the solubility generally decreases in ascending the series.

d. Spectroscopic properties. The infra-red spectra of alkanes show strong absorptions at $2\,850$–$3\,000\,\mathrm{cm}^{-1}$ (C—H stretching) and $1\,400$–$1\,470\,\mathrm{cm}^{-1}$ (C—H bending in —CH_2 and —CH_3 groups).

The various molecular ions obtainable from alkanes (p. 79) also show up clearly in mass spectrometry which is widely used for analytical purposes.

5 Combustion of alkanes

Alkanes burn in an adequate supply of oxygen to form carbon dioxide and steam, in common with all other hydrocarbons, e.g.

$$CH_4(g) + 2O_2(g) \rightarrow CO_2(g) + 2H_2O(g)$$

$$C_xH_y + (x + \tfrac{y}{4})O_2 \rightarrow xCO_2 + \tfrac{y}{2}H_2O$$

The ease of burning accounts for the use of many alkanes as fuels (p. 85). Mixtures of alkanes with air or oxygen can, in certain proportions, be explosive; in a limited supply of air carbon monoxide is formed together with carbon dioxide.

The enthalpy changes on combustion of alkanes can be obtained from tables or are readily calculated, as below, from standard enthalpy of formation values:

$$\begin{array}{llll} CH_4(g) + 2O_2(g) \rightarrow CO_2(g) + 2H_2O(l) & \Delta H^{\ominus}(298\ \mathrm{K}) = -890.7\ \mathrm{kJ} \\ -74.8 \quad\ 0 \qquad\qquad -393.7\ \ 2(-285.9) \\ \qquad\qquad\qquad\qquad\quad -965.5 \end{array}$$

The enthalpy change on combustion becomes more and more negative as the number of carbon atoms increases (p. 91).

6 Substitution reactions of alkanes

a. Halogenation. One or more of the hydrogen atoms in an alkane can be replaced by Cl or Br with the elimination of HCl or HBr, e.g.

$$CH_4(g) + Cl_2(g) \rightarrow CH_3—Cl(g) + HCl(g)$$

$$CH_4 \xrightarrow{Cl_2} CH_3—Cl \xrightarrow{Cl_2} CH_2Cl_2 \xrightarrow{Cl_2} CHCl_3 \xrightarrow{Cl_2} CCl_4$$

Chloro-	Dichloro-	Trichloro-	Tetrachloro-
methane	methane	methane.	methane.(Carbon
		(Chloroform)	tetrachloride)

$$C_2H_6 \xrightarrow{Cl_2} C_2H_5—Cl \xrightarrow{Cl_2} C_2H_4Cl_2 \longrightarrow C_2Cl_6$$

| Chloro- | Dichloro- | Hexachloro- |
| ethane | ethane | ethane |

Ultra-violet light, or a temperature of about 400 °C, are necessary for the reaction to take place; the mechanism is discussed in section 7. A mixture of products tends to be formed but the monosubstituted alkane predominates if excess of the alkane is used and the fully substituted products predominates with an excess of chlorine.

Fluorine reacts explosively though direct fluorination can be carried out using a diluent of nitrogen; iodine does not react effectively.

b. Nitration. Hydrogen atoms in alkanes can be replaced by —NO$_2$ groups to form nitroalkanes. This can be done either by heating with concentrated nitric(V) acid at 140°C under pressure in a sealed tube (liquid phase nitration or by heating with the acid at about 450°C (vapour phase nitration), e.g.

$$C_3H_8(g) + HNO_3(g) \rightarrow C_3H_7—NO_2(l) + H_2O(g)$$

c. Sulphonation. Hydrogen atoms in higher alkanes can be replaced by —SO$_2$OH groups by treatment with chlorine and sulphur dioxide, or sulphur dichloride dioxide, SO$_2$Cl$_2$, in the presence of light, and subsequent hydrolysis, e.g.

$$R—H + SO_2Cl_2 \rightarrow R—SO_2Cl + HCl$$

$$R—SO_2Cl + H_2O \rightarrow R—SO_2OH + HCl$$

Some of the higher sulphonated alkanes are used in making detergents (p. 192).

7 The mechanism of alkane substitution

The substitution of alkanes takes place through free-radical chain reactions, and the chlorination of methane has been particularly well studied. The standard free energy change for the reaction (p. 37) at 298 K,

$$CH_4(g) + Cl_2(g) \rightarrow CH_3Cl(g) + HCl(g) \quad \Delta G^{\ominus}(298 \text{ K}) = -101.9 \text{ kJ}$$
$$-50.8 \quad 0 \qquad\qquad -57.4 \qquad -95.3$$

is highly negative so that the reaction is thermodynamically feasible; the calculated equilibrium constant, K, is 10^{18}.*

* The corresponding figures for the formation of CH$_3$Br are -28.3 kJ and 10^5; for CH$_3$I they are $+66.3$ kJ and 10^{-11} which explains why iodination is ineffective.

Chlorine and methane will not, however, react in the dark at room temperature as the activation energy (p. 41) is too high. In the presence of ultra-violet light or, in the dark at about 400 °C, the reaction may be explosive. These conditions provide a kinetically favourable chain-reaction path through the formation of free radical intermediates.

The initiating step is the homolytic fission (p. 46) of chlorine molecules by the light or thermal energy. The chlorine atoms formed possess an unpaired electron and they 'trigger off' a chain reaction made up of the following steps:

$$Cl—Cl \rightarrow Cl\cdot + Cl\cdot \qquad \text{Inititating}$$

$$\left.\begin{array}{l} Cl\cdot + CH_4 \rightarrow HCl + \cdot CH_3 \\ \cdot CH_3 + Cl_2 \rightarrow CH_3Cl + Cl\cdot \end{array}\right\} \text{Propagating}$$

$$\left.\begin{array}{l} Cl\cdot + Cl\cdot \rightarrow Cl_2 \\ Cl\cdot + \cdot CH_3 \rightarrow CH_3Cl \\ \cdot CH_3 + \cdot CH_3 \rightarrow C_2H_6 \end{array}\right\} \text{Terminating}$$

The concentration of free radicals in the reaction mixture is never very high, but the terminating steps lower it and hinder the reaction. As the concentration of molecules is so much higher than that of free radicals, however, the propagating steps, involving reaction between a molecule and free radical, are more favoured than the terminating steps which involve two free radicals. The homolytic fission of one molecule of chlorine can, therefore, lead to the formation of many thousands of molecules of chloromethane; the quantum yield or chain length of the reaction is said to be high.

Nitration and sulphonation of alkanes are also thought to take place via free radical chain reactions.

Evidence that free radicals are involved in a chain reaction is provided by the fact that addition of a substance that can increase the concentration of free radicals speeds up the reaction, whilst addition of substances that can remove free radicals, or convert them into less reactive radicals, slows down the reaction. Thus tetraethyllead(IV) speeds up the reaction by providing ethyl radicals and chlorine atoms,

$$(C_2H_5)_4Pb \rightarrow 4 \cdot C_2H_5 + Pb$$

$$\cdot C_2H_5 + Cl_2 \rightarrow C_2H_5Cl + Cl\cdot$$

whereas oxygen slows it down. The oxygen molecule contains two unpaired electrons so that it can react as a free radical, even though it is a stable one. It probably slows the reaction down by converting the reactive $\cdot CH_3$ radicals into less reactive $CH_3OO\cdot$ radicals,

$$\cdot CH_3 + \cdot O—O\cdot \rightarrow CH_3OO\cdot$$

Specialised physical techniques e.g. flash photolysis can also be used to detect free radicals in a reacting mixture despite their transient nature.

8 Dehydrogenation of alkanes

Alkanes can undergo industrially important elimination reactions in which they lose hydrogen. The reactions require a high temperature and, possibly, a catalyst. Typical examples are summarised:

a. Formation of carbon black. The incomplete combustion of methane (natural gas) or oil mixtures gives a sooty flame and the carbon obtained from it (carbon black) is an important product. Over 90 % of it is incorporated into rubber tyres, which may contain 30 % of carbon black. Smaller amounts are used in making printer's ink, black paints and black shoe polish.

b. Formation of alkenes. Many alkenes can be manufactured by high temperature dehydrogenation processes, e.g.

$$C_2H_6 \xrightarrow{850\,°C} H_2C{=}CH_2 + H_2$$
$$\text{Ethene}$$

$$C_3H_8 \xrightarrow[600\,°C]{Cr_2O_3 + Al_2O_3} H_3C{-}CH{=}CH_2 + H_2$$
$$\text{Propene}$$

$$C_4H_{10} \xrightarrow[600\,°C]{Cr_2O_3 + Al_2O_3} H_2C{=}CH{-}CH{=}CH_2 + 2H_2$$
$$\text{Buta-1,3-diene}$$

c. Formation of ethyne (p. 118). If methane is subjected to a temperature of about 1 500 °C for a very short period of time ethyne is formed,

$$2CH_4 \rightarrow HC{\equiv}CH + 3H_2$$
$$\text{Ethyne}$$

9 Use of alkanes

a. Methane. Methane in the form of natural gas is widely used as a fuel but it is also an important source of many other chemicals as summarised below:

Product	Reaction	Conditions
Chloromethanes	$CH_4 + Cl_2 \rightarrow CH_3Cl + HCl$	uv light or 400 °C
Synthesis gas	$CH_4 + H_2O \rightarrow CO + 3H_2$	900 °C. 30 atms. Ni catalyst
Methanol	$2CH_4 + O_2 \rightarrow 2CH_3OH$	200 °C. 100 atms. Cu tube
Ethyne	$2CH_4 \rightarrow C_2H_2 + 3H_2$	1 500 °C for short time

Product	Reaction	Conditions
Hydrogen cyanide	$2CH_4 + 2NH_3 + 3O_2 \rightarrow 2HCN + 3H_2O$	1 000 °C. Pt/Rh catalyst
Carbon disulphide	$CH_4 + 4S \rightarrow CS_2 + 2H_2S$	650 °C. Silica gel catalyst

b. Ethane. Ethane is used for making chloroethanes and, in U.S.A., as a source of ethene on dehydrogenation (p. 103).

c. Propane and butane. Propane and butane are used as fuels. For this purpose they are liquefied in cylinders under slight pressure, e.g. Calor gas (95 % butane and 2-methylpropane + 5 % propane).

Propane yields propene on dehydrogenation (p. 96) and it can be cracked (p. 87) into ethene and methane,

$$C_3H_8 \xrightarrow{850\,°C} C_2H_4 + CH_4$$

Butane is converted into buta-1,3-diene by dehydrogenation (p. 96) and into ethanoic acid by catalytic oxidation.

d. Higher alkanes. These are mainly important as components of fuel mixtures rather than as individual chemicals. The composition of such mixtures can be improved for some particular purpose by the important processes of cracking, isomerisation, alkylation and cyclisation described on p. 87.

10 Cycloalkanes
The cycloalkanes have a general formula C_nH_{2n}; they are alicyclic compounds with 3-, 4-, 5-, 6-, n-membered rings.

Cyclopropane Cyclobutane Cyclopentane Cyclohexane

The cyclopropane molecule is planar with a bond angle of 60 °; as sp^3 orbitals are involved, with a normal bond angle of 109 ° 28 ′ there is considerable strain in the ring. As a result cyclopropane is more reactive than other cycloalkanes and when it does react the ring tends to open, e.g.

$+ \; Br_2 \; \rightarrow \; Br—CH_2—CH_2—CH_2—Br$

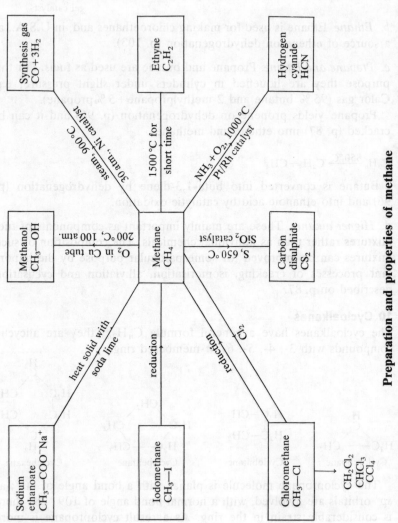

Preparation and properties of methane

Cyclobutane is not planar (p. 272) and there is considerably less strain in the ring than in cyclopropane. Cyclobutane is, therefore, less strained and less reactive; the ring opens less easily.

In higher cycloalkanes (p. 273) there is no significant strain and their reactions resemble those of the corresponding open-chain alkanes.

Questions on Chapter 6

1 What are the principal properties of the alkanes? $27.6 \, cm^3$ of oxygen were required to burn completely $8.7 \, cm^3$ of a mixture of methane (CH_4) and propane (C_3H_8). How much carbon dioxide would be formed? Calculate the composition of the mixture.

2 How would you distinguish between methane and ethane? How would you prepare propane from (*a*) methane, (*b*) ethane?

3 Discuss, critically, the methods of obtaining methane both in the laboratory and industrially.

4 Write a concise account of the methods of preparation and characteristic physical and chemical properties of the alkanes (containing not more than four carbon atoms). With the aid of examples taken from this class of compound, explain what is meant by (*a*) a homologous series, (*b*) structural isomerism. (JMB)

5 Outline the many methods by which methane may be obtained in chemical practice. Name precisely in each case the type of chemical reaction involved, giving equations where appropriate.

$16.1 \, cm^3$ of a mixture of methane, ethane and ethyne were added to $100 \, cm^3$ of oxygen and the mixture was exploded. On cooling to room temperature, a contraction in volume of $29.8 \, cm^3$ was measured. A further contraction of $26.1 \, cm^3$ ensued on treatment with potash. What were the partial pressures of the gases in the original mixture if the pressure throughout was $100 \, kPa$?

6 The enthalpies of combustion of hydrogen, carbon and methane are 292.9, 376.6 and $753.1 \, kJ \, mol^{-1}$ respectively. Calculate the enthalpy of formation of methane.

7 What pure alkane will give, on explosion with excess air, a decrease

in volume equal to the volume of carbon dioxide formed, all volumes being measured at s.t.p.?

8 Write down the structural formulae of (*a*) the five isomeric hexanes, (*b*) the nine isomeric heptanes.

9 Write structural formulae for (*a*) 2-methylpentane, (*b*) 3-methylpentane, (*c*) 2, 3-dimethylbutane, (*d*) 2, 2-dimethylbutane, (*e*) n-hexane. What is the relation between the five compounds?

10 Write structural formulae for (*a*) diethylmethane, (*b*) tetramethylethane, (*c*) methyltriethylmethane, (*d*) 2, 3-dimethylpentane, (*e*) 2, 2-dimethylbutane.

11 Suggest names for compounds with the following formulae: (*a*) $(CH_3)_2$-CH-CH-$(CH_3)_2$, (*b*) $(CH_3)_2$-CH-CH_2-CH_3, (*c*) $(C_2H_5)_2$-CH-CH_3, (*d*) $(CH_3)_2$-CH-C-$(CH_3)(C_2H_5)_2$.

12 What is a free radical? Give examples. How do free radicals differ from (*a*) ions, (*b*) carbanions?

13 Discuss the conversion of halogenated alkanes into alkanes by reduction from the point of view of (*a*) the halogenoalkane used and (*b*) the reducing agent used.

14 Illustrate the meaning of the terms isomerism, substitution and free radical using butane as the example.

15*Compare the enthalpies of combustion of (*a*) cyclopropane and propane and (*b*) cyclohexane and hexane. Comment on any points of significance.

16*Use bond energy values to find the enthalpy change in the reaction

$$CH_4(g) + Cl_2(g) \rightarrow CH_3Cl(g) + HCl(g)$$

17* Which is the easiest of the following processes thermodynamically? (*a*) The chlorination of ethane, (*b*) the iodination of methane, (*c*) the iodination of propane.

18*Calculate the equilibrium constants for the reaction

(*a*) $C(s) + 2H_2(g) \rightarrow CH_4(g)$

(*b*) $CH_4(g) + 2O_2(g) \rightarrow CO_2(g) + 2H_2O(g)$

from standard free energies of formation. Comment on the values obtained.

19*Explain why the enthalpy of combustion of alkanes becomes more and more negative as the number of carbon atoms in the molecule increases.

20*Take some specific examples and explain why high temperatures are necessary to bring about the dehydrogenation of alkanes.

7

Alkenes

1 Introduction

Alkenes are unsaturated hydrocarbons of general formula C_nH_{2n}. Their names are derived from the alkane with the same number of carbon atoms by replacing the *-ane* ending by *-ene*, e.g.

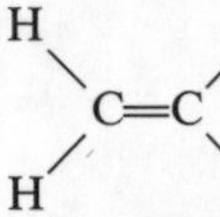

Ethene
(Ethylene)

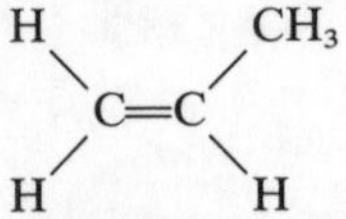

Propene
(Propylene)

Isomerism occurs in alkenes above C_3H_6, e.g.

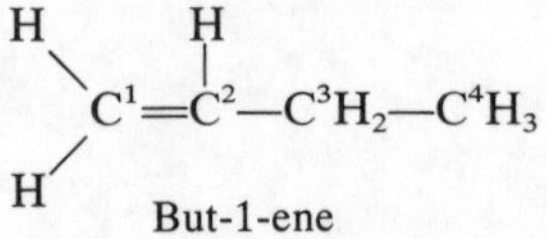

But-1-ene

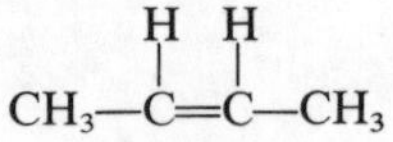

But-2-ene

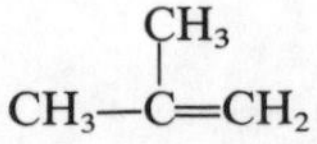

2-methylpropene

The position of the double bond is indicated numerically, the numbering of the longest carbon chain being such that the double bond position is indicated by the lowest possible number. The nature of the $C{=}C$ bond is discussed on p.21.

Three C_4H_8 isomers are shown above, but but-2-ene can itself have two isomeric forms,

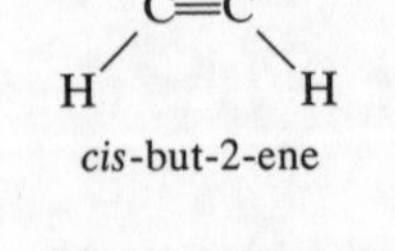

cis-but-2-ene

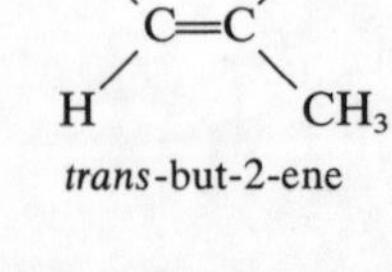

trans-but-2-ene

These geometric isomers, called *cis-* or *trans-* according to the symmetry of the molecule, arise because of the lack of free rotation about the C=C bond (p. 21).

Alkenes show the usual type of gradation of physical properties on ascending the series. Ethene, propene and the butenes are gases at room temperature, but higher members are liquids and, then, solids. The infra-red spectra of most alkenes shows a strong absorption between 1 680 and 1 620 cm^{-1} due to stretching of the C=C bond.

The main feature of the chemistry of the alkenes is their addition reactions in which the C=C bond is converted into a C—C bond. The products formed in these addition reactions with chlorine and bromine are oily liquids; hence the older name of *olefines* for the series.

2 Natural sources of alkenes
The commercial source of alkenes is natural gas in U.S.A. or the naphtha fraction from crude oil in Europe.

a. Cracking of alkanes. American, but not European, natural gas contains appreciable amounts of ethane, propane and butane and these alkanes can be cracked either thermally or catalytically (p. 87), e.g.

$$C_2H_6(g) \rightarrow C_2H_4(g) + H_2(g)$$
$$C_3H_8(g) \rightarrow C_2H_4(g) + CH_4(g)$$
$$C_3H_8(g) \rightarrow C_3H_6(g) + H_2(g)$$

Alkenes can be separated from the resulting gas mixture.

b. Cracking of naphtha. European natural gas does not contain many alkanes higher than methane, but naphtha provides an alternative source of alkenes, together with other useful by-products. A mixture of naphtha and steam is passed very rapidly through a coil at about 800 °C and the cracked mixture is quickly cooled. The mixture is then separated into a liquid fraction, which provides fuel-oil and petrol, and a gas fraction which contains hydrogen, C_1–C_4 alkanes and C_1–C_4 alkenes. These gases can be separated by distillation under pressure.

3 Synthesis of alkenes
The synthesis of an alkene generally involves an elimination reaction in which atoms attached to a C—C bond are removed to convert it into a C=C bond.

a. Dehydration of an alcohol. Alcohols with —OH and —H groups on adjacent carbon atoms can be dehydrated to alkenes,

$$-\overset{|}{\underset{H}{C}}-\overset{|}{\underset{OH}{C}}- \xrightarrow{-H_2O} ^{\diagdown}\!C\!=\!C^{\diagup}$$

Ethanol, for example, yields ethene using either excess concentrated sulphuric(VI) acid or aluminium oxide as the dehydrating agent,

$$C_2H_4 \xleftarrow[200\,^\circ C]{Al_2O_3} C_2H_5OH \xrightarrow[H_2SO_4 \text{ at } 170\,^\circ C]{Excess\ c.} C_2H_4$$

The reaction mechanism is discussed on p.180.

b. *Dehydrohalogenation of halogenoalkanes.* HX can be removed from a halogenoalkane molecule by refluxing with a concentrated solution of potassium hydroxide in ethanol,

$$-\overset{|}{\underset{H}{C}}-\overset{|}{\underset{X}{C}}- \xrightarrow{-HX} ^{\diagdown}\!C\!=\!C^{\diagup}$$

2-bromopropane, for example, yields propene,

$$CH_3-\overset{\overset{\displaystyle H}{|}}{\underset{\underset{\displaystyle Br}{|}}{C}}-\overset{\overset{\displaystyle H}{|}}{\underset{\underset{\displaystyle H}{|}}{C}}-H \xrightarrow[\text{alc. KOH}]{\text{Heat with}} CH_3-\overset{\overset{\displaystyle H}{|}}{C}=\overset{\overset{\displaystyle H}{|}}{C}-H$$

2-bromopropane   Propene

The yield of alkene depends on the halogenoalkane used and on the conditions of reaction and the method is not satisfactory for making ethene. The reaction mechanism is discussed on pp. 157–9.

4 Addition of bromine to ethene

Ethene and bromine react to form 1, 2-dibromoethane,

$$\overset{\displaystyle H}{\underset{\displaystyle H}{}}\!\!\!\!^{\diagdown}\!C\!=\!C\!^{\diagup}\!\!\!\!\overset{\displaystyle H}{\underset{\displaystyle H}{}} + Br_2 \longrightarrow H-\overset{\overset{\displaystyle H}{|}}{\underset{\underset{\displaystyle Br}{|}}{C}}-\overset{\overset{\displaystyle H}{|}}{\underset{\underset{\displaystyle Br}{|}}{C}}-H$$

1,2-dibromoethane

and the mechanism of the reaction is well understood.

In the presence of sunlight or initiating catalysts the reaction may take place via a free radical mechanism (p. 46) but under polar conditions in the absence of sunlight all the evidence points to a two-step, ionic mechanism.

a. Rate of reaction. There is little reaction between bromine and ethene in the dark under non-polar conditions. The dry gases will not react in a vessel with a non-polar, e.g. paraffin wax, surface and the rate of reaction in solution increases as the polar nature of the solvent increases or as Lewis acid catalysts are added.

b. Formation of mixed products. If the reaction is carried out in the presence of Cl^- or NO_3^- ions, $Br—CH_2—CH_2—Cl$ or $Br—CH_2—CH_2—NO_3$ are formed as well as $Br—CH_2—CH_2—Br$. The overall rate of the reaction is not changed which suggests that the X^- ion does not participate in the rate-determining step.

c. Trans-addition. Stereochemical studies show that the addition of bromine to a $C≡C$ bond in a more complex molecule than ethene, e.g. cyclohexene (p. 5) or cis-butenedioic (maleic) acid (p. 271) gives a trans-product in which the two bromine atoms have added on opposite sides of the molecular plane.

d. Suggested mechanism. The above facts are best accounted for by a mechanism involving the formation of a cyclic bromonium ion intermediate in the first-rate-determining step, i.e.

It is thought that the electron cloud of the π-bond in the C_2H_4 molecule first polarises an approaching Br_2 molecule and then forms a weak bond with the positive end of the polarised molecule. The product is sometimes referred to as a π-complex.

In the second step it is thought that the Br^- ion or, if present, a Cl^- or NO_3^- ions, attacks the bromonium ion. For steric reasons this attack will be on the side remote from the existing bromine atom, i.e.

Such a mechanism accounts for the formation of mixed products and the trans-nature of the addition.

There is no very strong evidence for the existence of the bromonium ion and an alternative suggestion involves the formation of a carbonium ion which contains the same atoms but a different arrangement of bonds and electrons. Thus

$$\begin{array}{c}
\text{H} \qquad\qquad \text{H} \\
\diagdown \quad\quad \diagup \\
\text{C}{=}\text{C} \\
\diagup \qquad\quad \diagdown \\
\text{H} \quad \text{Br}^{\delta+} \quad \text{H} \\
\big| \\
\text{Br}^{\delta-}
\end{array}
\longrightarrow
\begin{array}{c}
\text{H} \qquad\quad \text{H} \\
\diagdown \quad\quad \diagup \\
\text{C}{-}\text{C}^{+} \quad + \text{Br}^{-} \\
\diagup \quad \big| \quad \diagdown \\
\text{H} \quad \text{Br} \quad \text{H} \\
\text{Carbonium ion}
\end{array}$$

$$\begin{array}{c}
\text{H} \qquad\quad \text{H} \\
\diagdown \quad\quad \diagup \\
\text{C}{-}\text{C}^{+} \quad + \text{X}^{-} \\
\diagup \quad \big| \quad \diagdown \\
\text{H} \quad \text{Br} \qquad \text{H}
\end{array}
\longrightarrow
\begin{array}{c}
\quad\; \text{H} \quad \text{X} \\
\quad\;\; \big| \quad\; \big| \\
\text{H}{-}\text{C}{-}\text{C}{-}\text{H} \\
\quad\;\; \big| \quad\; \big| \\
\quad\; \text{Br} \quad \text{H}
\end{array}
\quad (\text{X}^{-} = \text{Br}^{-}, \text{Cl}^{-} \text{ or } \text{NO}_3^{-})$$

In the first step of both proposed mechanisms it is the positive end of the Br_2 molecule that attacks the ethene so that the bromine is acting as an electrophilic reagent; the reaction is, therefore, referred to as *electrophilic addition*. The ethene is acting as a nucleophile.

The nucleophilic nature of the C$=$C bond would be expected to increase as H atoms in ethene are substituted by $+I$ groups (p. 24), e.g. alkyl groups. As expected, addition of bromine to alkyl-substituted ethenes is more rapid than to ethene itself. Similarly, substitution of $-I$ groups, e.g. Br, leads to decrease in reaction rate.

5 Other simple addition reactions
A simple addition reaction of an alkene can be represented by the general equation

$$\begin{array}{c}
\diagdown \qquad\; \diagup \\
\text{C}{=}\text{C} \quad + \text{AB} \\
\diagup \qquad\; \diagdown
\end{array}
\rightarrow
\begin{array}{c}
\;\big| \quad\;\; \big| \\
{-}\text{C}{-}\text{C}{-} \\
\;\big| \quad\;\; \big| \\
\;\text{A} \quad\;\; \text{B}
\end{array}$$

In many cases, but not all, the mechanism follows that of the bromine-ethene reaction, the initial, rate-determining step being an attack on the C$=$C bond by the positive end of a polarised $A^{\delta+}$—$B^{\delta-}$ molecule or by an A^+ ion originating from AB. Subsequent reaction is with B^- ions.

a. Addition of halogens. Chlorine reacts with ethene, in the same way as bromine, to form 1, 2-dichloroethane. Fluorine tends to react explosively, and the reaction with iodine is very slow.

The decolorisation of a solution of bromine in tetrachloromethane serves as a good test for the presence of C=C bonds, and the amount of bromine used up measures the number of such bonds in a molecule.

b. Reaction with bromine or chlorine water. Ethene reacts on bubbling through bromine or chlorine water at room temperature to form, mainly, bromo- or chloroethanols. The overall effect is the addition of halogen(I) acids, HOX, e.g.

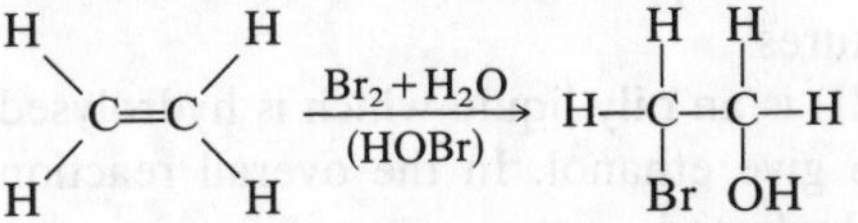

2-bromoethanol

As the product is colourless the reaction with bromine water can be used as a test for C=C bonds. The first step is the formation of a bromonium ion by attack of a polarised $Br^{\delta-}-Br^{\delta+}$ or a polarised $HO^{\delta-}-Br^{\delta+}$ molecule. This is then attacked by OH^-, Br^- or H_2O nucleophiles, but it is the water that is present in highest concentration so that that reaction predominates. Thus

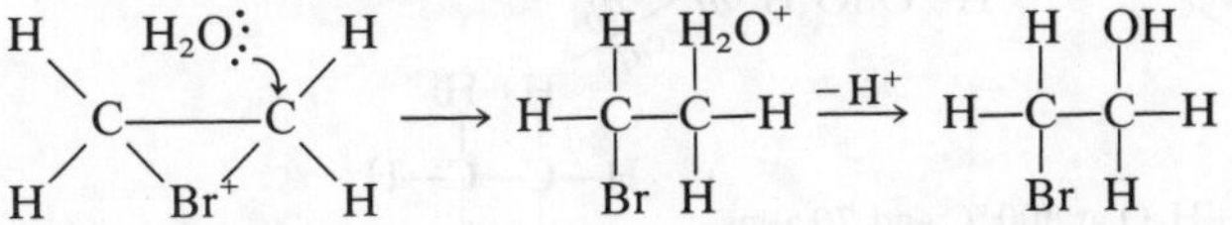

c. Addition of hydrogen halides. Hydrogen halides add on to ethene to form halogenoethanes, e.g.

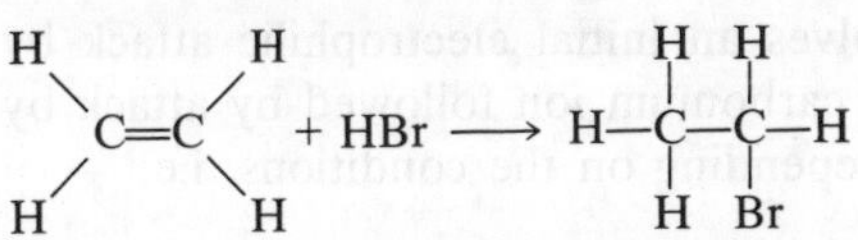

Bromoethane

The rate of reaction increases in passing from HF to HI, (in the order of increasing acid strength) and the rate is only satisfactory for HBr and HI.

The first step is the formation of a carbonium ion by attack on the C=C bond by H^+ or by the positive end of the polarised $H^{\delta+}-X^{\delta-}$ molecule

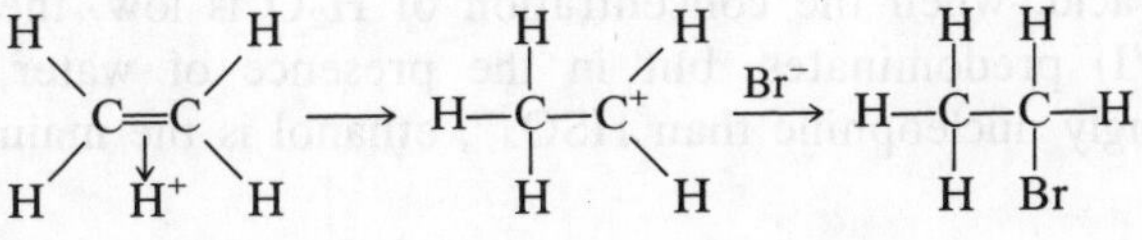

A cyclic protonium ion, analogous to the bromonium ion, is not likely to be formed because the hydrogen atom has insufficient electrons to form the necessary bonds.

As the HX molecule is unsymmetrical, two theoretical products are possible when it adds on to an unsymmetrical alkene (p. 111).

d. Addition of sulphuric(VI) acid. Hydration. Concentrated sulphuric(VI) acid reacts with ethene to form ethyl hydrogensulphate(VI). The reaction takes place at room temperature and can be used for removing alkenes from gas mixtures.

The ethyl hydrogensulphate(VI) is an oily liquid which is hydrolysed on warming with dilute acids to give ethanol. In the overall reaction the ethene has been effectively hydrated.

The hydration can also be brought about directly by passing ethene and steam over a phosphoric(V) acid catalyst at 300 °C and 70 atmospheres. The reaction provides an important method of making ethanol (p. 175).

The reaction mechanism involves an initial electrophilic attack by H^+ ions and the formation of a carbonium ion followed by attack by HSO_4^- ion or H_2O molecules depending on the conditions, i.e.

With concentrated acid, when the concentration of H_2O is low, the hydrogensulphate(VI) predominates, but in the presence of water, which is more strongly nucleophilic than HSO_4^-, ethanol is the main product.

e. Addition of hydroxyl groups. Hydroxylation. A 1 per cent aqueous solution of potassium manganate(VII) adds two hydroxyl groups onto ethene,

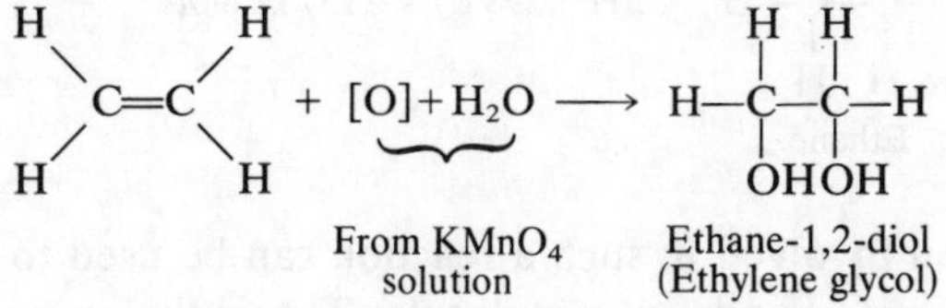

From KMnO$_4$
solution

Ethane-1,2-diol
(Ethylene glycol)

This reaction provides a further test for the C=C bond as the manganate(VII) solution is decolorised if it is acidified. A very dilute manganate(VII) solution at a low temperature is necessary to avoid oxidation of the diol.

Isotopic labelling shows that the oxygen atoms in the two added hydroxyl groups originate from the manganate(VII); a possible mechanism is outlined as follows:

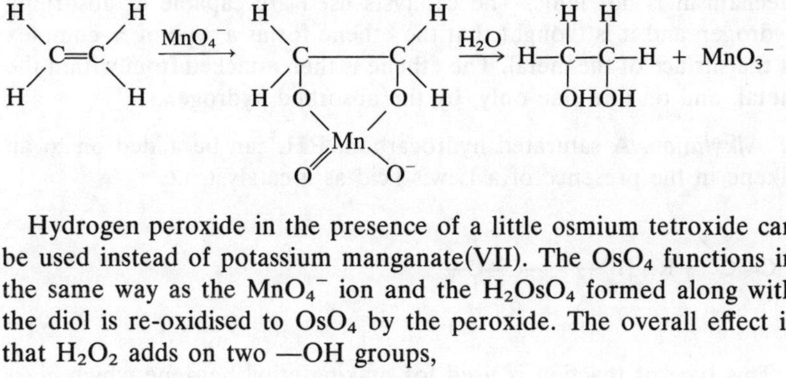

Hydrogen peroxide in the presence of a little osmium tetroxide can be used instead of potassium manganate(VII). The OsO$_4$ functions in the same way as the MnO$_4^-$ ion and the H$_2$OsO$_4$ formed along with the diol is re-oxidised to OsO$_4$ by the peroxide. The overall effect is that H$_2$O$_2$ adds on two —OH groups,

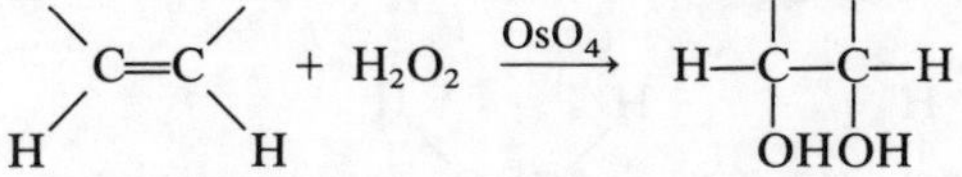

and the OsO$_4$ acts catalytically. The process is more controllable than that using potassium manganate(VII) but the osmium tetroxide is both very poisonous and expensive.

Neither of the two methods is used industrially as an alternative method of hydroxylation, via an epoxide (p. 113), is preferable.

f. Addition of hydrogen. Hydrogenation. A mixture of ethene and hydrogen passed over platinum black without heating, or over finely divided nickel at 140 °C, produces ethane,

$$
\begin{array}{c}
\text{H} \qquad\quad \text{H} \\
\diagdown \qquad\quad \diagup \\
\text{C}=\text{C} \qquad + \quad \text{H}_2 \longrightarrow \\
\diagup \qquad\quad \diagdown \\
\text{H} \qquad\quad \text{H}
\end{array}
\qquad
\begin{array}{c}
\text{H} \;\; \text{H} \\
| \;\;\;\; | \\
\text{H}-\text{C}-\text{C}-\text{H} \\
| \;\;\;\; | \\
\text{H} \;\; \text{H}
\end{array}
\qquad \Delta H^{\ominus}(298\ \text{K}) = -137\ \text{kJ mol}^{-1}
$$

  Ethene   Ethane

The amount of hydrogen involved in such a reaction can be used to measure the number of C=C bonds in a molecule. The enthalpy of hydrogenation ($-137\ \text{kJ mol}^{-1}$) is also significant for it is interesting to compare this value, found in ethene, with similar values for C=C bonds in other molecules (p. 139).

Catalytic hydrogenation is important industrially for it enables a mixture of inedible, unsaturated oils to be converted into saturated, edible fats. Margarine, for example, is made by mixing fats and oils and hydrogenated oils with milk and vitamins.

The addition of hydrogen to a C=C bond is *cis*-addition and the mechanism is not ionic. The catalysts used are capable of absorbing hydrogen and it is thought that the ethene forms a type of π-complex at the surface of the metal. The ethene is then attacked from within the metal, and on one side only, by the absorbed hydrogen.

g. Alkylation. A saturated hydrocarbon, RH, can be added on to an alkene in the presence of a Lewis acid as a catalyst, i.e.

$$
\begin{array}{c}
\diagdown \qquad\quad \diagup \\
\text{C}=\text{C} \qquad + \text{R}-\text{H} \longrightarrow \\
\diagup \qquad\quad \diagdown
\end{array}
\qquad
\begin{array}{c}
| \;\;\;\; | \\
-\text{C}-\text{C}- \\
| \;\;\;\; | \\
\text{R} \;\; \text{H}
\end{array}
$$

This type of reaction is used for making ethyl benzene which gives phenylethene (styrene) on dehydrogenation,

$$
\text{C}_2\text{H}_4 + \text{C}_6\text{H}_6 \xrightarrow[100\,°\text{C}]{\text{AlCl}_3}
\begin{array}{c}
\text{H} \quad\;\; \text{H} \\
| \qquad\; | \\
\text{H}-\text{C}-\text{C}-\text{H} \\
| \qquad\; | \\
\text{C}_6\text{H}_5 \;\; \text{H}
\end{array}
\xrightarrow{-\text{H}_2}
\begin{array}{c}
\text{H} \qquad\quad \text{H} \\
\diagdown \qquad\quad \diagup \\
\text{C}=\text{C} \\
\diagup \qquad\quad \diagdown \\
\text{C}_6\text{H}_5 \qquad \text{H}
\end{array}
$$

             Ethylbenzene   Phenylethene

and for converting benzene into (1-methylethyl) benzene (cumene) by reaction with prop-1-ene (p. 198). It is also widely used in the petroleum industry to convert small molecules into larger ones (p. 87).

6 Markownikov's rule

There are two possible products when a hydrogen halide, e.g. HCl, is added to an unsymmetrical alkene, e.g. propene, i.e.

$$
\begin{array}{ccccc}
\underset{\text{1-chloropropane}}{
\begin{array}{c}
\text{H} \quad \text{CH}_3 \\
\text{H—C—C—H} \\
\text{Cl} \quad \text{H}
\end{array}}
& \xleftarrow{\text{HCl}} &
\underset{\text{Propene}}{
\begin{array}{c}
\text{H} \qquad \text{CH}_3 \\
\text{C==C} \\
\text{H} \qquad \text{H}
\end{array}}
& \xrightarrow{\text{HCl}} &
\underset{\text{2-chloropropane}}{
\begin{array}{c}
\text{H} \quad \text{CH}_3 \\
\text{H—C—C—H} \\
\text{H} \quad \text{Cl}
\end{array}}
\end{array}
$$

Under normal conditions of ionic addition, 2-chloropropane is the major product in accordance with Markownikov's rule which states that *in the addition of HX to a C==C bond in an unsymmetrical alkene the H atom attaches itself to the carbon atom linked to the largest number of hydrogen atoms.* The rule was first put forward as an empirical statement but its theoretical basis is now understood.

In the first step of the reaction (the electrophilic attack by H^+) either a primary or a secondary carbonium ion might be formed, i.e.

$$
\begin{array}{ccccc}
\underset{\text{Primary}}{
\begin{array}{c}
\text{H} \quad \text{CH}_3 \\
\text{C}^+\text{—C—H} \\
\text{H} \quad \text{H}
\end{array}}
& \xleftarrow{H^+} &
\begin{array}{c}
\text{H} \qquad \text{CH}_3 \\
\text{C==C} \\
\text{H} \qquad \text{H}
\end{array}
& \xrightarrow{H^+} &
\underset{\text{Secondary}}{
\begin{array}{c}
\text{H} \quad \text{CH}_3 \\
\text{H—C—C}^+ \\
\text{H} \quad \text{H}
\end{array}}
\end{array}
$$

As it is the secondary ion that is most easily formed (p. 47) the final product is 2-chloropropane.

Markownikov's rule can be broadened by replacing HX by AB. It is, then, the more positive part of AB that becomes attached to the carbon atom linked to the largest number of hydrogen atoms. In a more modern version the rule states that *the electrophile adds so as to form the more stable carbonium ion.*

7 Deviation from Markownikov's rule

Markownikov's rule only applies to ionic addition via a carbonium ion intermediate. Hydrogen bromide, in the presence of peroxides, adds on via a free-radical chain reaction which replaces the ionic reaction as it is so much more rapid. So-called anti-Markownikov addition takes place, i.e.

$$
\begin{array}{c}
\text{H} \qquad \text{CH}_3 \\
\text{C==C} \\
\text{H} \qquad \text{H}
\end{array}
\; + \; \text{HBr} \;
\xrightarrow{\text{Peroxides}} \;
\underset{\text{1-bromopropane}}{
\begin{array}{c}
\text{H} \quad \text{CH}_3 \\
\text{H—C—C—H} \\
\text{Br} \quad \text{H}
\end{array}}
$$

Br· free radicals, formed as follows,

$$\text{R—O—O—R} \rightarrow 2\text{R—O·}$$

$$\text{R—O·} + \text{HBr} \rightarrow \text{ROH} + \text{Br·}$$

initiate the attack on the propene molecule. Two free radicals might be formed, i.e.

$$
\begin{array}{ccccc}
\underset{\text{Primary}}{
\begin{array}{c}
\text{H} \quad \text{CH}_3 \\
\text{·C—C—H} \\
\text{H} \quad \text{Br}
\end{array}}
& \xleftarrow{\text{Br·}} &
\begin{array}{c}
\text{H} \qquad \text{CH}_3 \\
\text{C==C} \\
\text{H} \qquad \text{H}
\end{array}
& \xrightarrow{\text{Br·}} &
\underset{\text{Secondary}}{
\begin{array}{c}
\text{H} \quad \text{CH}_3 \\
\text{H—C—C·} \\
\text{Br} \quad \text{H}
\end{array}}
\end{array}
$$

but it is the secondary free radical that is formed the more easily (p. 47). This radical produces 1-bromopropane in a chain-propagation reaction,

$$\underset{\substack{|\\Br}}{\overset{\substack{H\\|}}{H-C}}-\underset{\substack{|\\H}}{\overset{\substack{CH_3\\|}}{C}}\cdot \;+HBr \longrightarrow \underset{\substack{|\\Br}}{\overset{\substack{H\\|}}{H-C}}-\underset{\substack{|\\H}}{\overset{\substack{CH_3\\|}}{C}}-H + Br\cdot$$

The effect is sometimes referred to as the *peroxide effect*. Actual addition of peroxide is not always necessary as alkenes react slowly with air to form peroxides. The effect is also limited to addition of hydrogen bromide and can be prevented by the addition of radical acceptors to limit the chain reaction. With other halogen halides the rates of the necessary chain reactions are not high enough to compete favourably with the ionic addition.

8 Other reactions of alkenes

Other important reactions of alkenes may bear some relation to simple addition reactions but they do not involve a direct addition of an AB molecule.

a. Polymerisation. Ethene can, in effect, add on to itself to form an addition polymer, poly(ethene) or polythene, which is widely used (p. 326).

$$2n CH_2{=}CH_2 \longrightarrow \;\text{-}(\text{-}CH_2{-}CH_2{-}CH_2{-}CH_2\text{-})\text{-}_n$$
Ethene   Poly(ethene)
                                            (Polythene)

Higher or substituted alkenes can form similar polymers (p. 327).

b. Ozonolysis. Ozone reacts with alkenes dissolved in organic solvents to form ozonides, e.g.

$$\underset{\substack{H}}{\overset{\substack{H}}{}}C{=}C\underset{\substack{H}}{\overset{\substack{H}}{}} \;+O_3 \longrightarrow \; H_2C\underset{\substack{O{-}O}}{\overset{\substack{O}}{}}CH_2$$
Ethene ozonide

A low temperature is required as ozonides are unstable and may explode, like other compounds containing peroxo-, —O—O—, bonds; they are not generally isolated.

Their importance lies in the fact that they can be hydrolysed in solution by warming with water and zinc dust, e.g.

$$H_2C\underset{\substack{O{-}O}}{\overset{\substack{O}}{}}CH_2 + H_2O \longrightarrow 2HCHO + H_2O_2$$
Methanal

The zinc dust reduces the hydrogen peroxide formed and prevents it from oxidising the methanal to methanoic acid, some of which is formed in the absence of the zinc dust. The overall effect is a splitting of the C=C bond and the nature of the aldehydes formed can be used to discover the position of such a bond in a chain, e.g.

$$CH_3—CH=CH—CH_3 \rightarrow 2CH_3—CHO$$

$$CH_3—CH_2—CH=CH_2 \rightarrow CH_3—CH_2—CHO+HCHO$$

c. Reactions with oxygen. Alkenes, like other hydrocarbons, burn, possibly explosively, in air or oxygen,

$$2C_nH_{2n}+3nO_2 \rightarrow 2nCO_2+2nH_2O$$

Ethene can also be oxidised by oxygen in the presence of a silver catalyst. The product is epoxyethane (ethene oxide) which is readily converted into ethane-1, 2-diol by heating with water at 200 °C under pressure,

$$2CH_2=CH_2+O_2 \xrightarrow[300\,°C]{Ag\ cat.} 2\ \underset{\text{Epoxyethane}}{\overset{H_2C\text{———}CH_2}{\diagdown_{\ \ O\ \ }\diagup}} \xrightarrow[200\,°C]{2H_2O} 2\ \underset{\substack{\text{Ethane-1,2-}\\\text{diol}}}{\overset{H_2C—CH_2}{\underset{HO\ \ \ OH}{|\ \ \ \ \ |}}}$$

The two reactions together provide an industrial method of hydroxylation (p. 109).

d. The Wacker process. Oxidation of alkenes to aldehydes or ketones is brought about in the Wacker process. Alkenes and oxygen are passed into an aqueous solution containing palladium(II) and copper(II) chlorides. The palladium(II) chloride is converted into palladium in the reaction, e.g.

$$CH_2=CH_2+PdCl_2+H_2O \rightarrow CH_3—CHO+Pd+2HCl$$

$$CH_3CH=CH_2+PdCl_2+H_2O \rightarrow CH_3—COCH_3+Pd+2HCl$$

but is reconverted by reaction with hydrochloric acid and oxygen in the presence of copper(II) chloride,

$$2Pd+4HCl+O_2 \rightarrow 2PdCl_2+2H_2O$$

e. The Oxo process. Alkenes react with an equimolecular mixture of carbon monoxide and hydrogen to add on —H and —CHO groups; the process is known as *hydroformylation*,

$$\text{C=C} + CO + H_2 \xrightarrow[\text{130 °C, 150 atm}]{\text{Co(II) salt}} \underset{\underset{H}{|} \ \underset{CHO}{|}}{-C-C-}$$

The resulting aldehydes can be hydrogenated to alcohols (p. 191) and the process is useful for making higher ones (p. 221)

f. Carbonylation. —H and —COOH groups can also be added to an alkene by reaction with carbon monoxide and steam,

$$\text{C=C} + CO + H_2O \xrightarrow[\text{200 °C, 150 atm}]{\text{Ni(CO)}_4} \underset{\underset{H}{|} \ \underset{COOH}{|}}{-C-C-}$$

9 Dienes and cycloalkenes

Dienes are hydrocarbons containing two C=C bonds, the position of the bonds being shown, when necessary, numerically, e.g.

$CH_2\text{=}CH\text{—}CH_2\text{—}CH\text{=}CH_2$ $CH_2\text{=}CH\text{—}CH\text{=}CH_2$ $CH_2\text{=}C\text{=}CH_2$

Penta-1, 4-diene   Buta-1, 3-diene   Propadiene (Allene)

a. Bonding. The two C=C bonds in penta-1, 4-diene are sufficiently far apart to have little or no influence on each other; they function very much like the bonds in ethene.

In buta-1,3-diene, with what is known as a *conjugated* system of bonding, there is a considerable degree of delocalisation. This arises because the p orbitals which overlap to form the C=C bonds between the 1–2 and 3–4 carbon atoms can also overlap between the 2–3 atoms. The main chemical effect of this is that a mixture of addition products is formed, i.e.

$$CH_2\text{=}CH\text{—}\underset{\underset{X}{|}}{\overset{\overset{H}{|}}{C}}\text{—}\underset{\underset{X}{|}}{\overset{\overset{H}{|}}{C}}\text{—}H \quad \text{(1,2-addition)}$$

$$CH_2\text{=}CH\text{—}CH\text{=}CH_2 \xrightarrow{X_2}$$

$$H\text{—}\underset{\underset{X}{|}}{\overset{\overset{H}{|}}{C}}\text{—}\overset{\overset{H}{|}}{C}\text{=}\overset{\overset{H}{|}}{C}\text{—}\underset{\underset{X}{|}}{\overset{\overset{H}{|}}{C}}\text{—}H \quad \text{(1,4-addition)}$$

In propadiene (allene), with adjacent C=C bonds, there is no delocalisation as the two π-bonds are in planes at right-angles to each other. The two C=C bonds function, therefore, almost independently, but the formation of two adjacent bonds by the same carbon atom causes some strain and the compounds are not easy to make and not very stable. They tend to isomerise into ethynes, e.g.

$$CH_2\text{=}C\text{=}CH_2 \rightarrow CH_3\text{—}C\text{≡}CH$$

Substituted allenes provide a simple example of compounds which exhibit optical activity without the presence of an asymmetric carbon atom (p.280).

b. Buta-1,3-diene. This chemical, required for making synthetic rubbers (p. 336) and nylon (p. 332) is second only to sucrose in the world production figures for individual organic compounds.

It is obtained in Europe by the cracking of naphtha (p. 85) and in the U.S.A. by catalytic dehydrogenation of but-3-ene at 600 °C,

$$CH_3—CH=CH—CH_3 \xrightarrow{600°C} CH_2=CH—CH=CH_2 + H_2$$

c. *Cyclohexene.* This is a liquid alkene made by the dehydration of cyclohexanol,

or, in poor yield, by reaction between ethene and buta-1,3-diene,

The latter reaction provides a simple illustration of the *Diels-Alder* reaction which, using substituted alkenes, is widely applicable in making cyclic compounds.

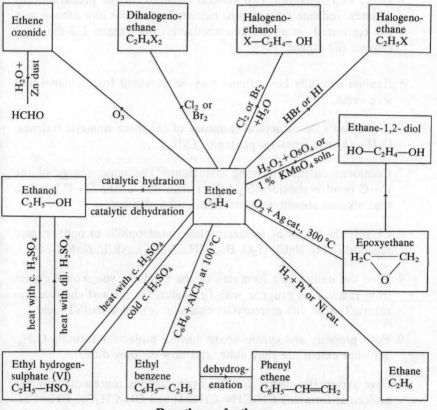

Reactions of ethene

Questions on Chapter 7

1 Ethane is a saturated, ethene an unsaturated compound. Explain the meaning of these terms and illustrate your answer by reference to the action of (*a*) chlorine, (*b*) hydrogen on ethene and ethane.

2 Explain, with varied illustrative examples, the meaning of the term addition reaction.

3 Outline two distinct methods for the preparation of ethene. State briefly how ethene can be converted to (*a*) ethane, (*b*) ethyne. Compare the properties of (*a*) and (*b*) and show how these properties may be accounted for by the structures assigned to the compounds. (L)

4 Give, with equations, two general methods for the preparation of alkenes. Indicate by means of balanced equations how ethene may be converted into (*a*) bromoethane, (*b*) ethane-1,2-diol, (*c*) ethyne, (*d*) ethane.

5 Explain carefully how ethene may be obtained from ethanol and vice versa.

6 Write down the structural formulae of (*a*) three isomeric butenes, C_4H_8, (*b*) five isomeric pentenes, C_5H_{10}.

7 Comment on the following statement: 'The bond energy of the $C{=}C$ bond is considerably higher than that of the $C{-}C$ bond so that alkenes should be more stable than alkanes'.

8 Classify the following as electrophilic, nucleophilic or neither (give reasons): NH_3, $NaCl$, H_2O, Br_2, CH_4, $NaBF_4$, $AlCl_3$, $NaNH_2$, O_3.

9 Give the names and formulae of the products you would expect from reaction of propene with (*a*) water, (*b*) nitrosyl chloride, (*c*) chloric(I) acid, (*d*) mercury(II) chloride, (*e*) sulphuric(VI) acid.

10 Both propene and cyclopropane have a molecular formula C_3H_6. To what extent are they alike, and how do they differ?

11 Give systematic names for the hydrocarbons represented by the molecular formulae $CH_3CH{=}CHCH_3$ and $CH_3CH_2CH_2CH{=}CH_2$ respectively. Outline a method of establishing the position of the

double bond in each molecule. State how each hydrocarbon reacts with hydrogen bromide and give the corresponding systematic name for the product. Give one general method of preparing alkenes and comment briefly on the importance of such compounds industrially. (W)

12 (*a*) State the Markownikov rule for predicting the direction of electrophilic addition of hydrogen bromide to alkenes. Explain in detail the rule in terms of the relative stabilities of primary, secondary, and tertiary carbonium ions. (*b*) Propene (propylene) reacts with hydrogen bromide to give a substance A, C_3H_7Br. Substance A, when heated with aqueous potassium hydroxide, gives an alcohol B. (i) Derive structures for A and B. (ii) Explain and illustrate the meaning of the terms base, nucleophile, and inductive effect by referring to the reactions of substance A with potassium hydroxide under various conditions. (JMB)

13 Some hydrocarbons are said to be 'saturated', others 'unsaturated'. Explain what these terms mean, both in terms of reactions and of structure. (Aromatic compounds should not be discussed). Give in account of the reactions of the two isomers of formula C_6H_{12} (cyclohexane and hex-1-ene) to illustrate the difference; indicate the possible mechanisms of the reactions you describe. (L)

14 9.8 g of a pure alkene having one C=C bond reacts with 16 g of bromine. What is the molecular formula of the product?

15 A mixture of propane with a gaseous hydrocarbon of the alkene series occupied $24 \, cm^3$. To burn the mixture completely, $114 \, cm^3$ of oxygen were required, and after the combustion $72 \, cm^3$ of carbon dioxide were left. Calculate (*a*) the formula of the alkene, (*b*) the composition of the mixture by volume. All volumes were measured at the same temperature and pressure.

16*Compare the thermodynamic feasibilities of the addition of (*a*) the halogens and (*b*) the hydrogen halides to ethene.

17*Compare the enthalpy changes in (*a*) the conversion of C=C into C—C, (*b*) the conversion of C≡C into C=C, (*c*) the conversion of C=O into C—O, all by reaction with hydrogen.

18*Water can, theoretically, be added onto ethene to form ethanol or to ethyne to form ethanal. Which is likely to be the more successful commercial process?

8

Alkynes

1 Introduction

Alkynes are unsaturated hydrocarbons with a general formula C_nH_{2n-2}. Their names are derived from the alkane with the same number of carbon atoms by replacing the *-ane* ending by *-yne*. If necessary, the position of the $C\equiv C$ bond is shown numerically, e.g.

H—C≡C—H   CH₃—CH₂—C≡CH   CH₃—C≡C—CH₃
Ethyne   But-1-yne   But-2-yne
(Acetylene)

Ethyne is much the most important member of the series. It is readily available in cylinders in which it is dissolved under pressure in propanone, the solution being absorbed in porous materials within the cylinder. The gas cannot be stored safely as a liquid for it is liable to decompose explosively into its elements. This arises because ethyne is a highly endothermic compound (p. 34).

The nature of the $C\equiv C$ bond is discussed on p. 22. When it is at the end of a molecule (a terminal bond) it causes an absorption in the infra red at about $2\,100\ cm^{-1}$; a C—H bond adjacent to the $C\equiv C$ bond causes a strong absorption at $3\,300\ cm^{-1}$. When the $C\equiv C$ bond is linked to two alkyl groups there is no C—H absorption at $3\,300\ cm^{-1}$ and the $C\equiv C$ absorption is very weak.

2 Manufacture of ethyne

Modern processes, still being developed, obtain ethyne from methane or naphtha by applying a temperature of about 1 500 °C for a fraction of a second, e.g.

$$2CH_4(g) \xrightarrow{\ 1\,500\,°C\ } C_2H_2(g) + 3H_2(g)$$

The temperature is achieved by partial combustion of methane, by using an electric arc, or by the use of heated refractory materials. The hot gases are immediately cooled to limit further decomposition and reaction and the ethyne present is extracted by a number of different selective solvents. The hydrogen is a very useful by-product.

These processes have replaced the older method of obtaining ethyne by reaction between calcium dicarbide and water,

$$CaC_2(s) + 2H_2O(1) \rightarrow Ca(OH)_2(s) + C_2H_2(g)$$

This reaction still serves as a satisfactory source of ethyne on a small scale but the cost of calcium dicarbide is too high for it to be commercially successful.

3 Addition reactions of ethyne

Ethyne takes part in addition reactions similar to those of the alkenes. The reactions take place in two stages, the $C\equiv C$ bond being converted into a $C=C$, and then into a $C—C$, bond.

a. Addition of hydrogen. Ethyne combines with hydrogen if a mixture of the two gases is passed over cold platinum black or over finely divided nickel at 100 °C,

$$C_2H_2 + 2H_2 \rightarrow C_2H_6$$

b. Addition of halogens. A mixture of chlorine and ethyne reacts explosively at room temperature producing carbon and hydrogen chloride. The reaction can be moderated by using solutions of the reagents and a metallic halide as a catalyst,

1,2-dichloroethene   1,1,2,2-tetrachloroethane

The tetrachloroethane is a good solvent but too toxic for practical use. It is converted into trichloroethene ($CHCl=CCl_2$) or perchloroethene ($CCl_2=CCl_2$) which are of practical commercial use as solvents as, for instance, in dry-cleaning.

Bromine reacts more slowly with ethyne, but the decolorisation of a solution of bromine in tetrachloromethane, or of bromine water, serves as a test for $C\equiv C$ bonds, as for $C=C$ bonds (p. 107).

c. Addition of hydrogen halides. Hydrogen halides react with ethyne, e.g.

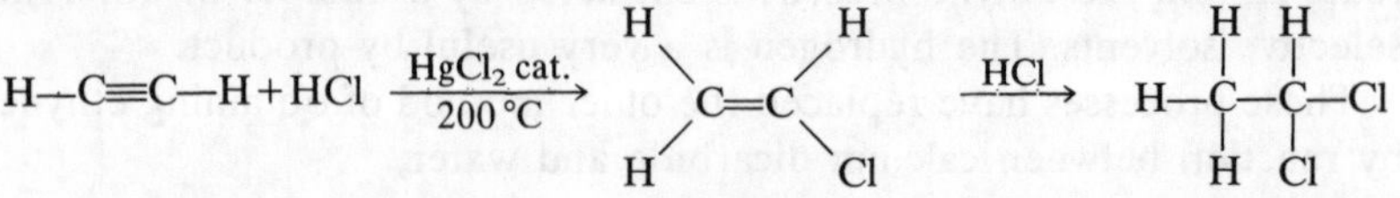

$$H\text{---}C{\equiv}C\text{---}H + HCl \xrightarrow[200\,°C]{HgCl_2\ cat.} \begin{array}{c} H \\ \diagdown \\ C{=}C \\ \diagup \quad \diagdown \\ H \qquad Cl \end{array} \xrightarrow{HCl} \begin{array}{c} H\ \ H \\ | \ \ | \\ H\text{---}C\text{---}C\text{---}Cl \\ | \ \ | \\ H\ \ Cl \end{array}$$

Chloroethene   1,1-dichloroethane
(Vinylchloride)

The chloroethene product is important in making PVC (p. 327). In the second stage of the addition process the addition is in accordance with Markownikov's rule (p. 111).

d. Hydration. Water can be effectively added on to ethyne by passing the gas into a warm solution of dilute sulphuric(VI) acid in the presence of a mercury(II) sulphate(VI) catalyst;

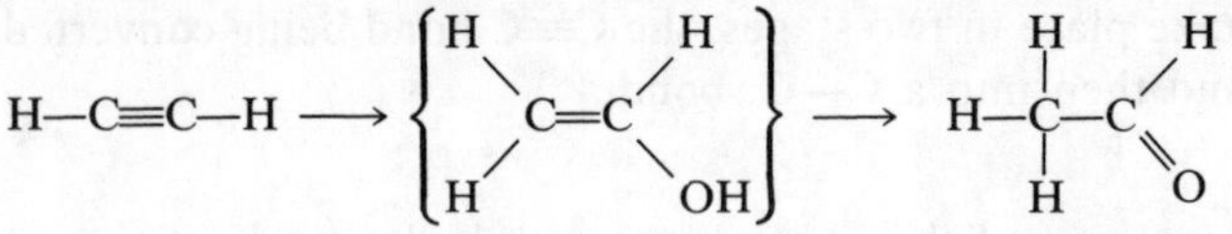

$$H\text{---}C{\equiv}C\text{---}H \longrightarrow \left\{\begin{array}{c} H \qquad H \\ \diagdown \quad \diagup \\ C{=}C \\ \diagup \quad \diagdown \\ H \qquad OH \end{array}\right\} \longrightarrow \begin{array}{c} H \quad H \\ | \quad \diagup \\ H\text{---}C\text{---}C \\ | \quad \diagdown \\ H \qquad O \end{array}$$

Ethanal

The method was, at one time, used as a commercial process for making ethanal.

4 Other reactions of ethyne

a. Combustion. Ethyne burns in air or oxygen, possibly explosively,

$$2C_2H_2(g) + 5O_2(g) \rightarrow 4CO_2(g) + 2H_2O(g)$$

A lot of air is needed for complete combustion so that an ethyne flame can be very sooty owing to incomplete combustion.

The oxy-acetylene flame, with a temperature of about 3 200 °C, is widely used in welding and cutting metals.

b. Formation of dicarbides (acetylides). If ethyne is passed into ammoniacal solutions of silver nitrate(V) or copper(I) chloride, precipitates of silver dicarbide, Ag_2C_2 (white), or copper(I) dicarbide, Cu_2C_2 (red-brown), are formed,

$$C_2H_2(g) + 2Ag(NH_3)_2^+(aq) \rightarrow Ag\text{---}C{\equiv}C\text{---}Ag(s) + 2NH_4^+(aq) + 2NH_3(g)$$

The precipitates are dangerously explosive when dry.

Their formation serves as a useful test for ethyne or for other ethynes with a terminal C≡C bond. The dicarbides can also be used for purifying impure ethyne as they re-liberate the gas on treatment with dilute hydrochloric acid or potassium cyanide solution.

In the formation of silver and copper(I) dicarbides, the hydrogen atoms in ethyne are being replaced by metals and to that extent ethyne is acting as an acid. Hydrogen gas can be displaced from ethyne by reaction with hot sodium, and a salt-like sodium dicarbide, which is ionic, is formed,

$$2C_2H_2(g) + 2Na(s) \rightarrow 2HC\equiv C^-Na^+(s) + H_2(g)$$

Ethyne is, however, only a very, very weak acid; much weaker than water, though stronger than ethene or ethane which exhibit no acidic properties.

The dicarbide ion is nucleophilic and it will undergo S_N2 reactions with halogenalkanes resulting in the formation of higher ethynes, e.g.

$$H\!-\!C\equiv C^-Na^+ + R\!-\!I \rightarrow H\!-\!C\equiv C\!-\!R + NaI$$

c. Polymerisation. Ethyne does not polymerise so readily as ethene. Two molecules of ethyne will, however, form a dimer in the presence of copper(I) chloride as a catalyst,

$$2CH\equiv CH \rightarrow CH_2\!=\!CH\!-\!C\equiv CH \xrightarrow{\text{HCl}} CH_2\!=\!CCl\!-\!CH\!=\!CH_2$$

2-chloro-1,3-butadiene
(2-chloroprene)

The dimer will add on hydrogen chloride and the product will itself polymerise to give neoprene rubbers (p. 336).

Ethyne will also polymerise to benzene,

$$3C_2H_2 \xrightarrow{\text{Heat}} C_6H_6$$

but the yield is poor.

For summarising charts see pp. 122–123

Questions on Chapter 8

1 How would you prepare a sample of pure ethyne? Describe how ethyne can be converted into (*a*) benzene, (*b*) ethane, (*c*) ethanoic acid. (O. and C.)

2 Taking ethane, ethene and ethyne, indicate how each compound can be obtained from the other two.

3 Compare and contrast the reactions of ethane, ethene and ethyne.

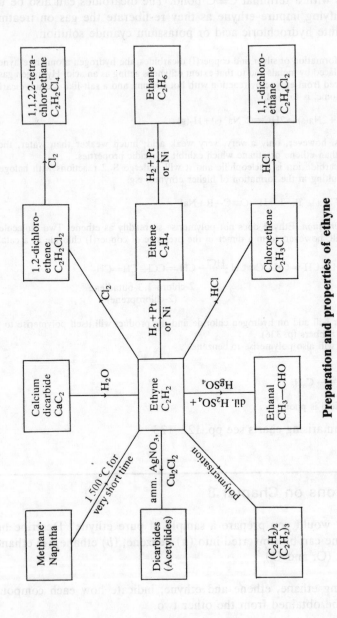

Preparation and properties of ethyne

Preparation and properties of ethane, ethene and ethyne

	C_2H_6	C_2H_4	C_2H_2
Commercial source	Crude oil. Natural gas in U.S.A.	Cracking of naphtha or alkanes	CH_4 at 1 500 °C
Methods of preparation	Reduction of C_2H_4 or C_2H_5I. $C_2H_5COO^-Na^+ + NaOH$	Dehydration of C_2H_5OH. Dehydrohalogenation of C_2H_5X	CaC_2 + water
Type of compound	Saturated hydrocarbon	Unsaturated hydrocarbon	Unsaturated hydrocarbon
Combustion		All burn, possibly explosively, to form CO_2 and H_2O if sufficient O_2. CO formed in limited supply of O_2.	
Substitution reactions	With Cl_2 and Br_2 + uv light	No substitution reactions	No substitution reactions
Addition reactions	No addition reactions	Many addition reactions with, for example, H_2, HX, X_2, HOX, H_2SO_4, O_3	
Dehydrogenation	Forms C_2H_4 or C at high temperature		
Polymerisation	No polymerisation	Forms poly(ethene)	Forms a dimer with Cu_2Cl_2 as catalyst
With ammoniacal solutions of $AgNO_3$ or Cu_2Cl_2	No reaction	No reaction	Forms dicarbides, Ag_2C_2 and Cu_2C_2

4 By means of selected typical reactions briefly compare the characteristic chemical properties of ethane, ethene and ethyne. When a
  given volume of ethane and ethene is completely oxidised, 0.440 g
  of carbon dioxide and 0.216 g of water are obtained. Calculate (*a*)
  the total volume of the original mixture at s.t.p., (*b*) its composition by volume. (JMB)

5 Give an account of the large-scale production of ethyne and
  describe how this substance is used in the synthesis of other organic
  compounds. How would you obtain and identify a sample of
  methane from a mixture containing hydrogen, carbon monoxide,
  ethene, ethyne and methane? (L)

6 Describe the uses of (*a*) methane, (*b*) ethene, (*c*) ethyne.

7 Ethyne has an acid dissociation constant of approximately 10^{-22}.
  What does this mean, how does the value compare with that for
  water, and what effect does it have on the properties of ethyne?

8 The gas ethyne, $HC\equiv CH$ may be made by a reaction between
  1,2-dibromoethane and a solution of sodium hydroxide in ethanol.
  The reaction is very rapid if the 1,2-dibromoethane is added a little
  at a time to the boiling sodium hydroxide in solution. Design a
  practical arrangement which might be used to make several boiling
  tubes of ethyne, illustrating the apparatus and giving practical
  details including suitable quantities of the reagents to use. (L.
  Nuffield)

9 What products would you expect to be formed when propyne
  reacted with hydrogen bromide?

10 State how you would distinguish chemically between hexane, hex-
  1-ene, and hex-1-yne ($C_4H_9C\equiv CH$). A hydrocarbon X, C_4H_6,
  (0.5 g), was shaken with hydrogen and palladium until uptake of
  hydrogen ceased; 415 cm^3 of hydrogen (measured at s.t.p.) were
  absorbed. Reaction of X with mercury(II) sulphate in dilute sulphuric acid yielded Y, C_4H_8O; Y gave a positive iodoform reaction. Polymerisation of X gave Z, $C_{12}H_{18}$. What are the structures
  of X, Y and Z? (O and C)

11 A mixture of methane, ethene and ethyne occupied 27 cm^3 and,
  after sparking it with 75 cm^3 of oxygen, the residual gas occupied

52.5 cm^3. Introduction of potassium hydroxide solution gave a further contraction and the volume was reduced to 7.5 cm^3. Calculate the percentage composition of the mixture.

12 The bond energy of the C—C link in ethane is 346 kJ mol^{-1}, that of the C=C link in ethene is 598 kJ mol^{-1}, and that of the C≡C link in ethyne is 835 kJ mol^{-1}. What light do these figures throw on the comparative reactivities of these compounds?

13 Write equations showing the complete combustion of (a) ethane, (b) ethene and (c) ethyne. The enthalpies of combustion of the three hydrocarbons are −1 559.8, −1 411 and −1 299.6 kJ mol^{-1} respectively. Relate these figures to the various bond energies involved. Why is it that ethyne, with the lowest enthalpy of combustion gives the highest flame temperature?

14*Compare the enthalpy change in the addition of 1 mol of bromine to 1 mol of ethyne with that of 1 mol of bromine to 1 mol of 1,2-dibromoethene. Comment on the result.

15*Calculate the equilibrium constants at 25 °C for (a) the hydrogenation of ethyne to ethene, (b) the hydrogenation of ethene to ethane, (c) the hydrogenation of ethyne to ethane.

16*Ethyne is much less stable with respect to its elements than ethene or ethane. Explain, quantitatively, what this means.

9

Arenes (*Aromatic Hydrocarbons*)

1 Introduction

There are a number of homologous series of arenes or aromatic hydrocarbons. The simplest, based on benzene, C_6H_6, and containing one ring of six carbon atoms have a general formula C_nH_{2n-6} where n is six or more. The nature of the bonding and the methods of representing the benzene ring are discussed on p. 26; further details regarding the structure of benzene are given on p. 139.

Successive replacement of hydrogen atoms by —CH_3 groups rise to the higher members of the series and isomers occur at C_8H_{10} by

Benzene   Methylbenzene   Ethylbenzene
             (Toluene)

1,2-dimethylbenzene   1,3-dimethylbenzene   1,4-dimethylbenzene
     (*o*-xylene)   (*m*-xylene)   (*p*-xylene)

replacing different hydrogen atoms in methylbenzene by a methyl group. Conversion of methyl- into ethyl-benzene is referred to as *side-chain substitution*; conversion into the xylenes is *ring-substitution*. There are three possibilities for ring-substitution, as shown; the older names used the prefixes *ortho-* (*o*-), *meta-* (*m*-) or *para-* (*p*-), but these are generally replaced, nowadays, by the 1,2-, 1,3- and 1,4-usage.

The homologous series based on benzene is paralleled by other arene series of polycyclic hydrocarbons based, for example, on naphthalene, anthracene and phenanthrene.

Naphthalene, $C_{10}H_8$ Anthracene, $C_{14}H_{10}$ Phenanthrene $C_{14}H_{10}$

2 Manufacture of benzene and methylbenzene

a. From coal. When coal is heated in a gas-works (mainly to make coal-gas) or in coke-ovens (mainly to make coke) it splits up into a mixture of gases, a black, viscous liquid (coal-tar), a watery fluid (ammoniacal liquor) and coke. Benzene and methylbenzene, together with other useful by-products, are separated from the gas mixture and the coal-tar.

The aromatic hydrocarbons present in the gas mixture are removed by absorption in oil and the mixture of hydrocarbons obtained, consisting mainly of benzene and methylbenzene, is known as *benzole*. A similar mixture can also be obtained by distillation of coal-tar. The benzole is used as a component of petrol and as a solvent, or it is separated into benzene and methylbenzene by distillation. Benzene may be further purified by crystallisation (m.p. = 5.5 °C).

b. From crude oil. Benzene and methylbenzene can be obtained by reforming (p. 88) the alkanes present in the gasoline and naphtha fractions from crude oil distillation. The process (platforming) involves cyclisation and dehydrogenation in the presence of platinum and aluminium oxide catalysts, e.g.

$$C_6H_{14} \xrightarrow{-4H_2} \qquad \qquad C_7H_{16} \xrightarrow{-4H_2}$$

Hexane   Benzene   Heptane

Methylbenzene

Some xylenes are also formed and all the aromatic hydrocarbons can be extracted by selective solvents.

The process tends to produce too much methylbenzene and xylene, and too little benzene. The balance is restored by a *hydrodealkylation* process. This involves treatment with hydrogen at 650 °C under pressure and in the presence of metal or metallic oxide catalysts, e.g.

Methylbenzene   Benzene

Some benzene can also be extracted from the products of naphtha cracking.

3 Physical properties of arenes

The arenes are liquids or low melting point solids with characteristic 'aromatic' odours. The vapours are, however, quite toxic and inhalation should be avoided. The arenes are very insoluble in water, and less dense.

The boiling points rise fairly regularly with increasing relative molecular mass but the melting points are irregular depending to a large extent or molecular symmetry. Benzene, for example, has a much higher melting point than methylbenzene, and 1,4-dimethyl benzene melts at a higher temperature than its two isomers.

	Benzene	Methyl-benzene	1,2-di-methyl benzene	1,3-di-methyl benzene	1,4-di-methyl benzene
m.p./°C	5.5	−95	−25	−47	13
b.p./°C	80	111	144	139	138

The nature of a substituted benzene can generally be ascertained from its infra-red spectrum (p. 72). This contains bands close to 1 500 and 1 600 cm^{-1} caused by bond stretching within the benzene ring, sharp bands near 3 030 cm^{-1} caused by C—H stretching, and other bands between 700 and 900 which correlate well with the number and the position of the substituted groups.

4 Substitution reactions of benzene

The nature of the π-bonds in benzene (p. 26) enables it to act as a nucleophilic reagent and its main reactions are with electrophilic reagents which replace one or more of the hydrogen atoms in the ring. A typical reaction, with an electrophile, X^+, can be summarised as follows:

$$\text{C}_6\text{H}_5 + X^+ \longrightarrow \underset{\text{I}}{[\text{C}_6\text{H}_5\text{HX}]^+} \longrightarrow \text{C}_6\text{H}_5\text{X} + H^+$$

The representation of the intermediate, I, indicates that the delocalisation of the π-electrons is limited to five carbon atoms, the ion carrying a positive charge which is distributed over these five atoms. The substitution of one group into the benzene ring affects both the ease and the position of substitution of a second group as described in Chapter 10 (p. 146).

a. Nitration. A mixture of concentrated nitric(V) and sulphuric(VI) acids (nitrating mixture) reacts with benzene, at 60 °C, to form nitrobenzene. At 100 °C, the substitution is carried further to 1,3-dinitrobenzene, and at higher temperatures and using fuming acids some 1,3,5-trinitrobenzene is formed (p. 290).

The reaction mechanism involves the nitryl cation (nitronium ion), NO_2^+, which is formed in the nitrating mixture. This electrophile then attacks the benzene ring in the rate-determining step, the ion intermediate subsequently losing a proton by reaction with HSO_4^- ion. In summary,

$$HNO_3 + 2H_2SO_4 \rightarrow NO_2^+ + H_3O^+ + 2HSO_4^-$$

$$\text{C}_6\text{H}_6 \xrightarrow{NO_2^+} [\text{C}_6\text{H}_6\text{NO}_2]^+ \xrightarrow{HSO_4^-} \text{C}_6\text{H}_5\text{NO}_2 + H_2SO_4$$

The existence of the NO_2^+ ion is shown by the fact that nitric(V) acid dissolved in sulphuric(VI) acid causes a depression of freezing point which corresponds to the presence of four ions, by the isolation of salts such as $NO_2^+ClO_4^-$, and by spectroscopic analysis. It is the strength of sulphuric(VI) acid as an acid that causes the nitric(V) acid to act as a base, by accepting protons from the acid,

$$H^+ + HNO_3 \rightarrow H_2O + NO_2^+$$

and other strong acids, e.g. $HClO_4$, serve the same purpose.

b. Halogenation. Benzene will react with chlorine or bromine to form chloro- or bromo-benzene. The reaction must be carried out in the absence of sun-light for that initiates a different, free-radical addition reaction (p. 132). Catalysts, referred to as halogen carriers are also necessary. These may be metals or their chlorides, e.g. Fe, Al, $FeCl_3$, $AlCl_3$; if the metals are used they are converted into the chloride by reaction with chlorine. The initial substitution product is the monoderivative, but further substitution can take place.

The catalyst functions as a Lewis acid causing polarisation in the halogen molecule. The positive end of this polarised molecule then attacks the benzene ring to form an ionic intermediary. In summary

$$2Fe + 3Cl_2 \rightarrow 2FeCl_3$$

$$Cl-Cl + FeCl_3 \rightarrow Cl^{\delta+}-Cl^{\delta-} \cdots FeCl_3$$

That the initial step is an electrophilic attack is supported by the fact that reaction with iodine monochloride, $I^{\delta+}-Cl^{\delta-}$, causes iodination and not chlorination. Direct reaction between iodine and benzene is a reversible reaction with an unfavourable equilibrium constant. It is only possible to get a reasonable yield in the presence of an oxidising agent or a base which will remove the hydrogen iodide formed. Direct fluorination of benzene is too rapid.

c. Sulphonation. Benzene will react with concentrated sulphuric(VI) acid to form benzenesulphonic acid. The reaction is slow, partly due to the immiscibility of the reagents, and reversible; prolonged heating for up to 24 hours is required to get a good yield. The mechanism is thought to involve electrophilic attack by SO_3,

$$2H_2SO_4 \rightarrow SO_3 + H_3O^+ + HSO_4^-$$

Benzene sulphonic
acid

That this mechanism is likely is supported by the fact that fuming sulphuric(VI) acid, containing dissolved SO_3, gives a better yield.

d. Alkylation. Benzene and halogenoalkanes react, in the presence of $AlCl_3$ or BF_3 as catalysts, to form alkylated benzenes. The reaction is known as the *Friedel-Crafts reaction* and a typical mechanism is summarised below:

$$CH_3—Cl + AlCl_3 \rightarrow H_3C^{\delta+}—Cl^{\delta-} \cdots AlCl_3$$

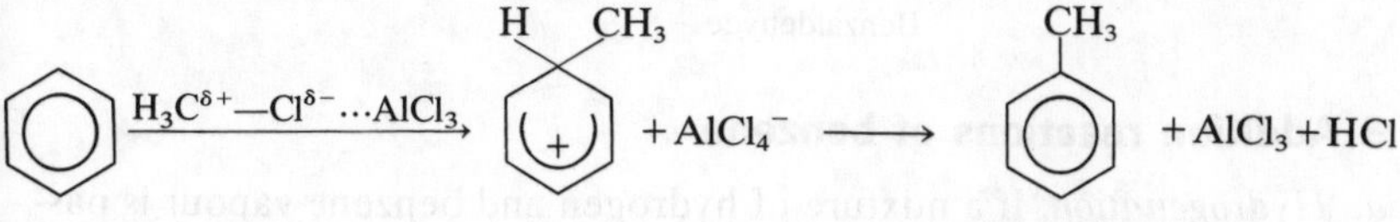

Methylbenzene

To ensure a good yield of the monoalkylbenzene it is necessary to start with an excess of benzene. This is because it is easier to alkylate an alkyl benzene than benzene itself (p. 144).

Alkylation can also be achieved by using alkenes in the presence of acid catalysts, e.g.

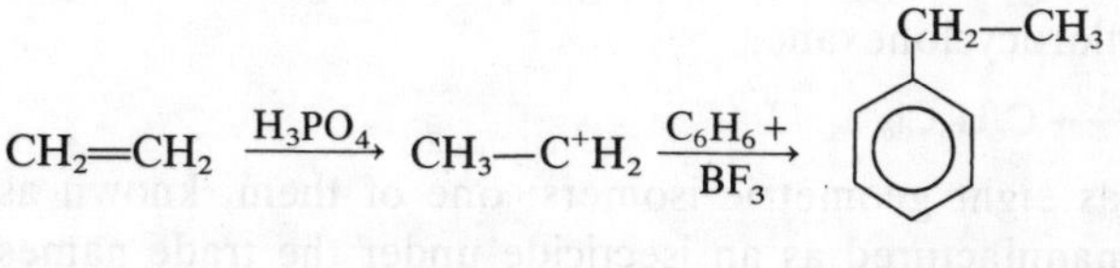

Ethylbenzene

e. Acylation. Acylation is the substitution of a hydrogen atom by an acyl, $R \cdot CO$, group. When R is CH_3 it is *ethanoylation (acetylation)*; when R is H it is *methanoylation (formylation)*, and so on. Acylation of benzene can be brought about by the Friedel-Crafts reaction using acid chlorides, $RCOCl$, or acid anhydrides, $R \cdot CO \cdot O \cdot OC \cdot R$ as the reagents and $AlCl_3$ or BF_3 as catalysts, e.g.

$$R—COCl + AlCl_3 \rightarrow R—C^+\!\!=\!\!O + AlCl_4^-$$

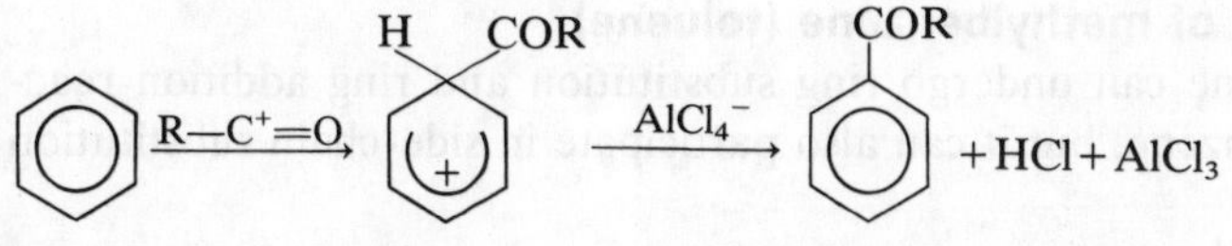

A ketone

Methanoyl (formyl) chloride, HCOCl, does not exist, but a mixture of carbon monoxide and hydrogen chloride serves the same purpose and converts benzene into benzaldehyde (the Gattermann–Koch reaction),

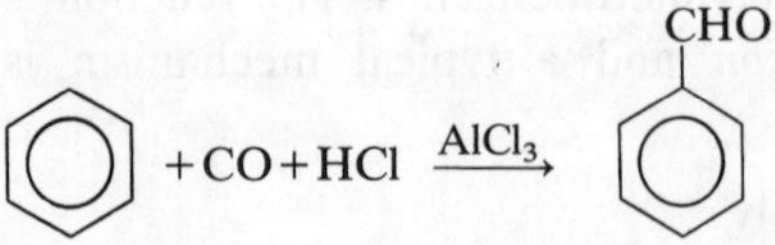

$$\bigodot + CO + HCl \xrightarrow{AlCl_3} \bigodot^{CHO}$$

Benzaldehyde

5 Addition reactions of benzene

a. Hydrogenation. If a mixture of hydrogen and benzene vapour is passed over finely-divided nickel at 150 °C the benzene is reduced to cyclohexane (p. 5),

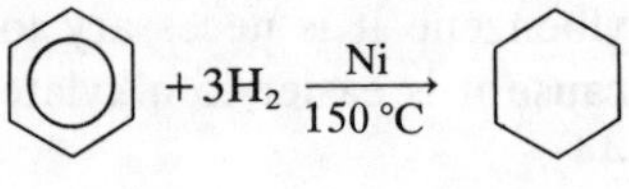

$$\bigodot + 3H_2 \xrightarrow[150\ °C]{Ni} \bighexagon$$

Cyclohexane

b. Addition of halogens. Chlorine and bromine add on to benzene in the presence of sunlight or ultra-violet light producing, for example, 1,2,3,4,5,6-hexachlorocyclohexane,

$$C_6H_6 + 3Cl_2 \xrightarrow{uv\ light} C_6H_6Cl_6$$

The compound has eight geometric isomers; one of them, known as the γ-isomer, is manufactured as an isecticide under the trade names of Gammexane or Lindane (p. 170).

The addition of chlorine to benzene involves a free-radical chain,

$$Cl_2 \xrightarrow{uv\ light} 2Cl\cdot$$

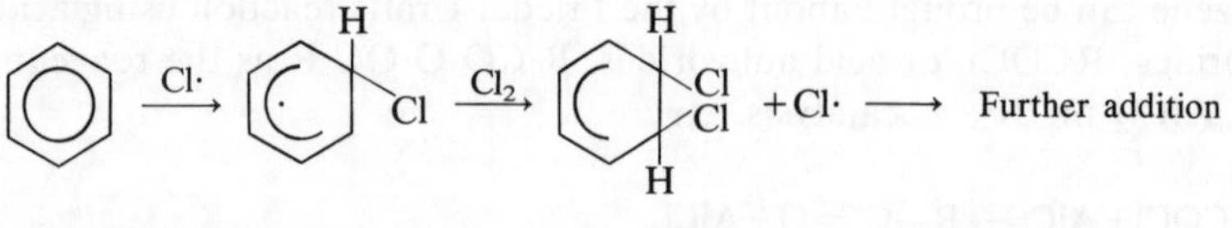

Further addition

6 Reactions of methylbenzene (toluene).

Methylbenzene can undergo ring substitution and ring addition reactions like benzene, but it can also participate in side-chain substitution reactions.

a. Ring substitution reactions. Methylbenzene can be substituted in the ring by all the electrophilic reagents which react with benzene. The mechanism of the reactions are like those of benzene but they take place more readily (p. 144). As the methylbenzene molecule is unsymmetrical there are also various possibilities for the entry position of a substituted group. It is found, experimentally, that a mixture of the 1,2- (*ortho*-) and 1,4- (*para*-) isomers is formed; the 1,3- (*meta*-) isomer is formed in only very small amounts. The methyl group is said to be *o-p*-directing and the reason for this is discussed on p. 146.

Typical monosubstitution reactions are as follows:

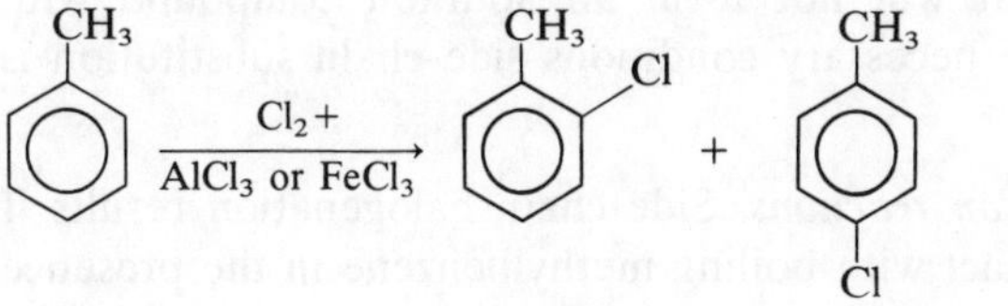

Chloro-2-   Chloro-4-
methylbenzene   methylbenzene

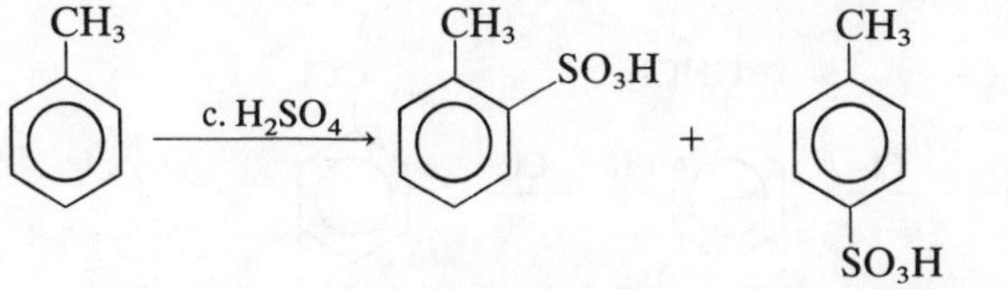

2-methylbenzene   4-methylbenzene
sulphonic acid   sulphonic acid

Substitution may proceed beyond the mono-stage and this is particularly well known in the formation of TNT (trinitrotoluene or methyl-2,4,6-trinitrobenzene), a yellow solid used as an explosive.

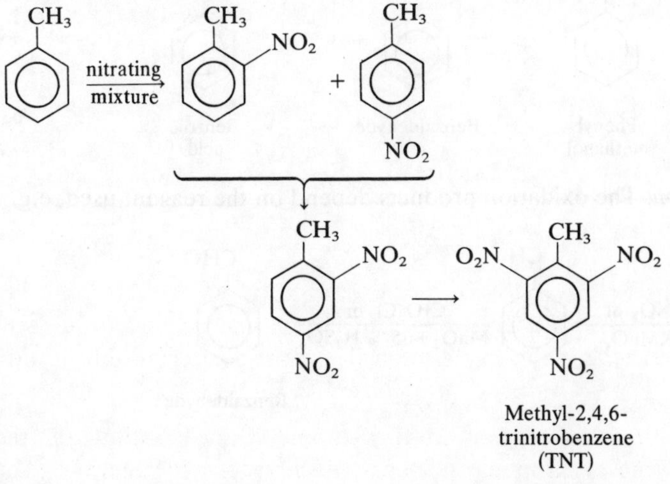

Methyl-2,4,6-
trinitrobenzene
(TNT)

b. Addition reactions. Methylbenzene can be hydrogenated at 150 °C in the presence of a nickel catalyst,

Methylcyclohexane

but, unlike benzene, it will not form an addition compound with chlorine, for under the necessary conditions side-chain substitution is easier.

c. Side-chain substitution reactions. Side-chain halogenation results if chlorine or bromine react with boiling methylbenzene in the presence of ultra-violet light. The reaction is a free-radical chain reaction like the halogenation of methane (p. 95). Mono-, di-, or tri-substituted products can be formed, and each product can be hydrolysed as shown.

(Chloromethyl)   (Dichloromethyl)   (Trichloromethyl)
benzene   benzene   benzene

|hydrolysis |hydrolysis |hydrolysis

CH₂OH   CHO   COOH

Phenyl   Benzaldehyde   Benzoic
methanol   acid

d. Oxidation. The oxidation products depend on the reagent used, e.g.

Benzoic   Benzaldehyde
acid

The stronger oxidising agents which oxidise the —CH_3 group to —COOH are capable of oxidising longer side-chains, e.g.

CH_2—CH_2—CH_3 COOH

$\xrightarrow{\text{HNO}_3 \text{ or } \text{KMnO}_4}$

This provides a useful method for establishing the position at which a side chain is attached to a ring.

For summarising charts see pp. 136–137.

Questions on Chapter 9

1 Compare and contrast the properties of benzene with those of (a) methane, (b) ethene, (c) hexane.

2 In what ways is the Kekulé formula a satisfactory representation of the benzene molecule and in what ways is it unsatisfactory? How does the isomerism of benzene derivatives support the idea of a closed-ring formula?

3 How many position isomers are there of (a) bromophenol, (b) dichlorophenol, (c) chloroiodophenol?

4 Write structural formulae for (a) 1,2-diphenylethene, (b) 1,3,5-trimethylbenzene, (c) (1-methylethyl) benzene, (d) methylcyclohexane, (e) benzene-1,2-dicarboxylic acid.

5 How many disubstitution products are there for (a) naphthalene, (b) anthracene if the two substituted groups are (i) alike and (ii) different?

6 How would you obtain a pure sample of methylbenzene from benzene? Suggest feasible routes for the conversion of benzene into 4-methylphenol, phenylacetic acid, 4-nitrobenzoic acid, and 3-nitrobenzoic acid.

7 Discuss the reaction of methylbenzene with chlorine.

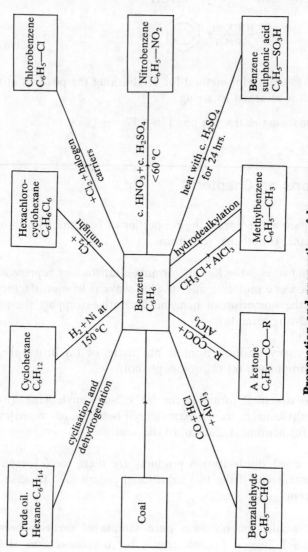

Preparation and properties of benzene

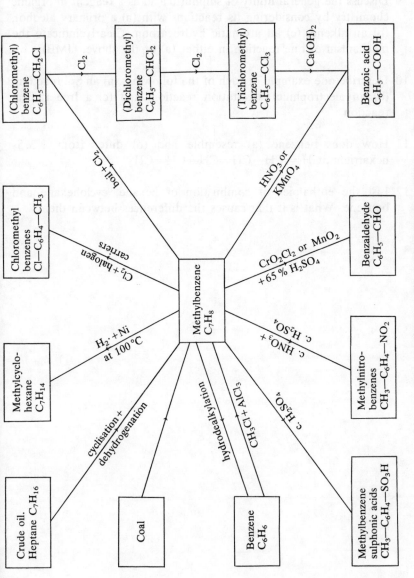

Preparation and properties of methylbenzene

8 How would you attempt to make monodeuterobenzene, C_6H_5D?

9 Discuss the general utility of sulphuric acid as a reagent in organic chemistry by considering its reactions with (a) a primary alcohol, (b) an alkene, (c) an aromatic hydrocarbon. Clearly indicate the mechanism of one reaction in either (a) or (b) above. (JMB)

10 Describe one example of each of the following: (a) an S_N1 reaction, (b) an electrophilic substitution reaction, and (c) a free radical reaction.

11 How does benzene (a) resemble and (b) differ from 1,3,5-hexatriene, $CH_2{=}CH{-}CH{=}CH{-}CH{=}CH_2$?

12*List the enthalpies of combustion of hexane, cyclohexane and benzene. What is it that causes the differences between them?

10

Aromaticity and Aromatic Substitution

1 Aromaticity

Aromatic compounds were originally so named because many of them were fragrant but the typical properties of benzene are also found in many derivatives with no smell and in other types of molecule. These properties are referred to, collectively, as aromatic character or aromaticity.

In benzene, they are best accounted for in terms of structures involving resonance hybrids (p. 29) or delocalised π-bonds (p. 26), each method of representation having some advantages over the other.

a. Delocalisation, stabilisation or resonance energy. Benzene is unexpectedly stable. It has no acidic or basic properties; it will not react with hot concentrated potassium hydroxide or hydrochloric acid; and it reacts only very slowly with hot solutions of chromic(VI) acid or potassium manganate(VII). Such properties cannot be accounted for on the basis of a single Kekulé formula (p. 29) containing C=C bonds and the benzene molecule is more stable than such a single Kekulé formula would indicate.

The enthalpy of hydrogenation of cyclohexene, containing one C=C bond, is $120.5 \text{ kJ mol}^{-1}$,

$$+ H_2 \longrightarrow \qquad \Delta H = -120.5 \text{ kJ mol}^{-1}$$

Cyclohexene, C_6H_{10} Cyclohexane, C_6H_{12}

If benzene had a structure containing three C=C bonds it would be expected that its enthalpy of hydrogenation would be 3×120.5, i.e. $361.5 \text{ kJ mol}^{-1}$. The measured value for the enthalpy of hydrogenation of benzene is, however, $208.4 \, \text{kJ} \, \text{mol}^{-1}$. The actual molecule of benzene is, therefore, more stable than a single Kekulé formula by 361.5–208.4, i.e. $153.1 \, \text{kJ} \, \text{mol}^{-1}$; this value is known as the delocalisation, stabilisation or resonance energy. The value can also be obtained by comparing the calculated enthalpy of formation for a single Kekulé structure with the actual enthalpy of formation of benzene (p. 37).

The greater the possibilities of delocalisation the higher the resulting delocalisation, stabilisation, or resonance energy. In naphthalene and anthracene (p. 127), for example, the values are 297 and $435 \, \text{kJ} \, \text{mol}^{-1}$ respectively. When delocalisation can occur between a benzene ring and a group substituted into the ring (p. 141) there is also a higher delocalisation, stabilisation or resonance energy than in benzene itself.

b. Substitution versus addition reactions. The stability of benzene arises from the delocalisation of the π-electrons over the ring of six carbon atoms and the more the electron-density in a molecule can be delocalised the more stable the molecule.

The delocalisation of the π-electrons also accounts for the fact that benzene undergoes electrophilic substitution reactions rather than addition reactions. It might be expected that the high electron-density associated with the π-orbitals in benzene would be attacked by electrophilic reagents in the same way as in ethene to form addition compounds. It is, however, only by forming substitution compounds that benzene can maintain its fully delocalised π-orbitals; formation of addition compounds would lead to less delocalisation. If bromine, for example, formed an addition compound with benzene, the delocalisation would be limited to four carbon atoms, whereas it is maintained at six carbon atoms when a substitution compound is formed, i.e.

Delocalisation over four
C atoms in addition
compound

Delocalisation over six
C atoms in substitution
compound

c. Possible delocalisation in substituted benzenes. The six p-orbitals of the six C atoms in a benzene ring overlap to give delocalised π-bonding in the ring. If the H atoms in the ring are substituted by atoms with available p-orbitals they too will overlap with the orbitals in the ring as, for example, in chlorobenzene (Fig. 47). This will cause an increase or decrease in the electron availability within the ring (p. 144) or on the substituted atom.

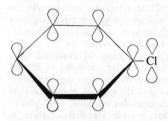

Fig. 47. Showing the *p*-atomic orbitals on six C atoms and one Cl atom that can overlap to form delocalised bonding. Compare Figs. 18 (p. 25) and 19 (p. 26).

Delocalisation between the ring and a substituted atom is only possible when that atom has *p*-orbitals available. It will not occur, for example, in methylbenzene, but it will occur in chlorobenzene, phenylamine and phenol where the atoms adjacent to the ring are Cl, N and O, all with lone pairs.

Delocalisation will not occur when such groups are substituted into alkanes or cycloalkanes because there are no available *p*-orbitals in the alkyl or cycloalkyl part of the molecule. It is this interaction of a substituted group with a benzene ring that makes the properties of some aryl-X compounds significantly different from those of the corresponding alkyl-X or cycloalkyl-X compounds.

When X is linked to a —C=C— bond in a straight-chain compound, e.g. chloroethene, CH_2=CH—Cl, the properties of the X group resemble those of the aryl-X compound as delocalisation can once again occur between X and the —C=C— grouping (Fig. 18 on p. 25)

In summary,

Delocalisation possible if X has a *p*-atomic orbital available

No delocalisation is possible

The term conjugated (p. 37) is sometimes used to describe a species in which delocalisation can occur.

d. Spectra of arenes and substituted arenes. The delocalisation within the benzene ring causes distinctive spectra both in the infra-red and ultra-violet regions and in NMR spectra.

In the infra-red region bands occur close to 1 500 and 1 600 cm^{-1} due to stretching of the carbon–carbon bonds in the ring and at 3 030 cm^{-1} due to the C—H bonds.

Arenes and substituted arenes also exhibit several bands in the ultra-violet region around 40 000 cm^{-1}. These arise from excitation of the relatively loosely held π-electrons. Excitation of the electrons in a σ-bond requires energy corresponding to a wave number of about 80 000 cm^{-1} which is too high for normal ultra-violet measurement. The more delocalised the π-electrons are in a conjugated system the more easily are they excited and the lower the corresponding wave number of absorbed light. Absorption occurs in ethene, buta-1,3-diene and benzene, for example, at around 56 000, 46 000 and 40 000 cm^{-1} respectively. In substituted arenes when the substituted group can participate in delocalisation with the electrons in the ring the absorption is at still lower wave numbers. Phenylethene and phenylamine, for example, both absorb around 36 000 cm^{-1}. When the substituted groups cause even greater delocalisation, as in methyl orange (p. 348), the absorbed wavelength becomes long enough to be in the visible region of the spectrum. Conversely, no significant ultra-violet spectrum occurs with molecules in which delocalisation is impossible.

In NMR spectra the chemical shift is approximately 2 ppm lower than that for protons attached to ordinary C=C bonds. This is because the delocalised electrons in an aromatic ring can sustain a current induced by an applied magnetic field. This current causes diamagnetism and deshielding of aromatic protons which are outside the ring. In annulenes there are both 'outside' and 'inside' protons; the former are deshielded but the latter are shielded so that they have widely different chemical shifts.

e. Hückel's rule. Hückel showed that the conditions necessary for aromaticity occur in any *planar,* cyclic compound in which the atoms in the ring have $(4n+2)$ π-electrons. Such compounds undergo the typical electrophilic substitution reactions of benzene.

In benzene, n is 1; in naphthalene, 2; in anthracene and phenanthrene, 3. Cyclopentadiene, C_5H_6, has only 4 π-electrons; it exhibits the properties of a diene with two C=C bonds and is not aromatic. Cyclopentadiene is, however, acidic losing a proton to form the cyclopentadienyl anion, $C_5H_5^-$. This ion has 6 π-electrons and is aromatic, the aromaticity being evident in the readiness with which the sandwich compound, ferrocene, undergoes electrophilic substitution reactions such as the Friedel–Crafts reaction

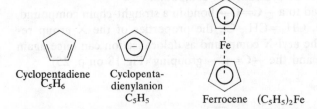

| Cyclopentadiene C_5H_6 | Cyclopenta-dienylanion C_5H_5 | Ferrocene $(C_5H_5)_2Fe$ |

f. Summary. The main, simple criteria of aromaticity are a high delocalisation, stabilisation or resonance energy and a strong tendency to undergo electrophilic substitution rather than addition reactions. The table on p. 143 summarises the differences between various types of hydrocarbon.

2 Activating influence of substituent groups

The electron-density distribution in the benzene ring is greatly affected when a hydrogen atom is substituted by some other atom or group. If the group donates electrons to the ring the electron-density will

Compound	Benzene, C_6H_6	Cyclohexane, C_6H_{12}	Cyclohexene, C_6H_{10}	Hexane, C_6H_{14}
Type of substance	Aromatic hydrocarbon	Saturated alicyclic hydrocarbon. Cycloalkane.	Unsaturated alicyclic hydrocarbon. Cycloalkene.	Saturated aliphatic hydrocarbon.
Delocalisation in molecule	Yes	No	No	No
H_2 + Ni catalyst	Addition reaction to form cyclohexane	No reaction	Addition reaction to form cyclohexane	No reaction
Cl_2 + sunlight	Forms addition compound, $C_6H_6Cl_6$	Forms one mono-substituted compound, $C_6H_{11}Cl$	Forms addition compound, $C_6H_{10}Cl_2$	Forms 3 isomeric mono-substituted compounds, $C_6H_{13}Cl$
Cl_2 + halogen carrier	Forms substitution compounds, e.g. C_6H_5Cl	No reaction	No reaction	No reaction
c. HNO_3 + c. H_2SO_4	Forms substitution compound, $C_6H_5.NO_2$	No reaction	No reaction	No reaction
c. H_2SO_4	Forms substitution compound, $C_6H_5.SO_3H$	No reaction	Forms addition compound, $C_6H_{11}.HSO_4$	No reaction
Ozone	Forms triozonide which gives OHC—CHO with Zn and H_2O	No reaction	Forms mono-ozonide which gives $OHC(CH_2)_4CHO$ with Zn and H_2O	No reaction
Dil. $KMnO_4$ or Br_2 water	No reaction	No reaction	Decolorised	No reaction

Properties of benzene, cyclohexane, cyclohexene and hexane

increase and electrophilic substitution in the ring will be easier than with benzene; the ring is said to be *activated*. Methylbenzene, phenol and phenylamine (aniline), for example, are all substituted more easily than benzene because the —CH_3, —OH and —NH_2 groups are electron-donating. An electron-withdrawing group will lower the electron-density in the ring causing deactivation and greater difficulty in electrophilic substitution. Nitrobenzene and benzoic acid, for example, are less easily substituted then benzene because the —NO_2 and —COOH groups are electron withdrawing.

Chemical properties can indicate, as above, the nature of the substituent group, but this can also be determined by measurement of dipole moments, (p. 24). If the negative end of the dipole of a molecule C_6H_5—X is towards the ring, X must be electron-donating; if away from the ring it must be the electron-withdrawing.

Common groups can be classified as follows:

Activating groups   —CH_3(—R)   —OH   —NH_2
(Electron-donating)

Deactivating groups   —NO_2 —SO_3H —CHO   —COOH   —CN
(Electron-withdrawing)

The donation or withdrawal of electrons comes about through an inductive and/or a mesomeric effect.

a. Inductive effect. A $+I$ group (p. 24) will donate electrons to the ring, whilst a $-I$ group will withdraw electrons from the ring. A —CH_3 (or —R) group will, therefore, activate the ring, whilst a —CCl_3 group will deactivate it. A negatively charged group will exert a strong $+I$ effect and activate the ring, whilst a positively charged group will be $-I$ and deactivate the ring. For example

Activation by   Deactivation by
inductive effect   inductive effect

The inductive effect may be enhanced or diminished by an associated mesomeric effect.

b. The mesomeric effect. The mesomeric effect is caused by electron-pair shifts when the substituent group has unshared electron pairs or multiple bonds. The effect may be electron-donating or electron-withdrawing; it may work in the same or opposite direction as an associated inductive effect; and it may be weaker or stronger than such an inductive effect.

The —OH group, for example, has a −I (electron-withdrawing) inductive effect but the overall effect is electron-donating because of a strong mesomeric effect in which the unshared pair of electrons on the oxygen atom interact with the delocalised π-orbitals in the benzene ring. The situation can be represented as follows:

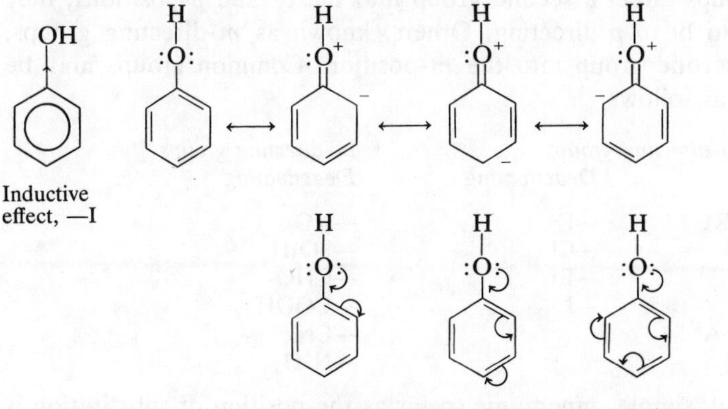

Inductive effect, −I

Mesomeric effects (electron-donating)

It will be seen that there is a build up of negative charge in the *ortho*- and *para*-positions. The same situation exists for the —NH$_2$ group but with —Cl or other halogen groups the mesomeric effect is weaker than the inductive effect so that the group, overall, is electron-withdrawing and deactivating though the electron-density in the ring is still concentrated in the *o*- and *p*-positions.

The —NO$_2$ group is electron-withdrawing both through a −I inductive effect and a mesomeric effect. The situation is summarised below:

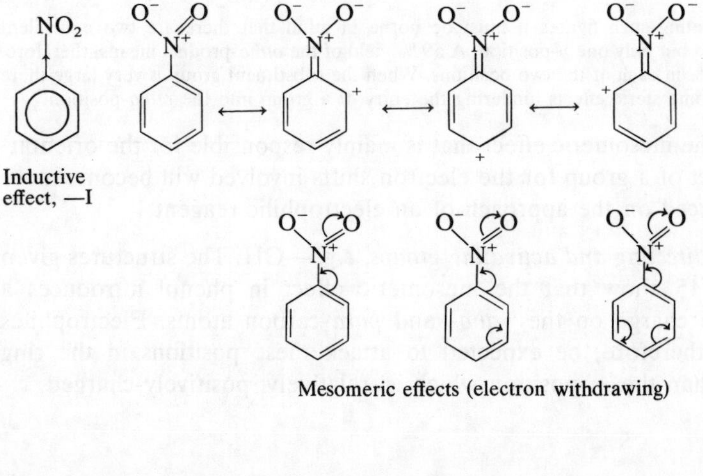

Inductive effect, −I

Mesomeric effects (electron withdrawing)

The overall electron-withdrawal builds up positive charge in the *ortho*- and *para*-positions, and a similar situation exists for —SO₃H, —CHO,—COOH and —CN groups.

3 Orientating influence of substituent groups

A substituent group in the benzene ring not only activates or deactivates the ring; it also controls the position of entry of a second group. Some groups direct a second group into the *o*- and *p*-positions; they are said to be *o-p*-directing. Others, known as *m*-directing groups, direct a second group into the *m*-position. Common groups may be classified as follows:

o-p-directing groups		*m*-directing groups
Activating	*Deactivating*	*Deactivating*
—CH₃ (—R)	—F	—NO₂
—OH	—Cl	—SO₃H
—O⁻	—Br	—CHO
—NH₂	—I	—COOH
—C₆H₅ (—Ar)		—CN
		—N⁺H₃

A useful, simple mnemonic so far as the position of substitution is concerned is provided by an empirical rule first suggested by Vörlander. It states that groups containing double or triple bonds between any atoms are *m*-directing, whilst others are *o-p*-directing. It applies reasonably accurately in simple cases.

The products formed may not be exclusively *meta*- or *ortho*- and *para*- but they are predominantly one or the other, e.g.

	% ortho	% meta	% para
Nitration of C₆H₅—CH₃	59	4	37
Nitration of C₆H₅—NO₂	6	93	1

In interpreting such figures it must be borne in mind that there are two equivalent *o*-positions but only one *p*-position. A 59 % yield of the *ortho*-product means, therefore, about 30 % in each of the two positions. When the substituent group is very large there are also some steric effects hindering the entry of a group into the *ortho*-position.

It is the mesomeric effect that is mainly responsible for the orientating effect of a group for the electron shifts involved will become more pronounced on the approach of an electrophilic reagent.

a. o-p-directing and activating groups, e.g. —OH. The structures given on p. 145 show that the mesomeric effect in phenol introduces a negative charge on the *ortho*- and *para*-carbon atoms. Electrophiles would, therefore, be expected to attack these positions in the ring rather than the *m*-position which is, relatively, positively-charged.

The matter can also be considered from the point of view of the stability of the intermediate carbonium ion. In the nitration of phenol, for example, the resonance hybrids of the various possible nitrophenols are summarised as follows:

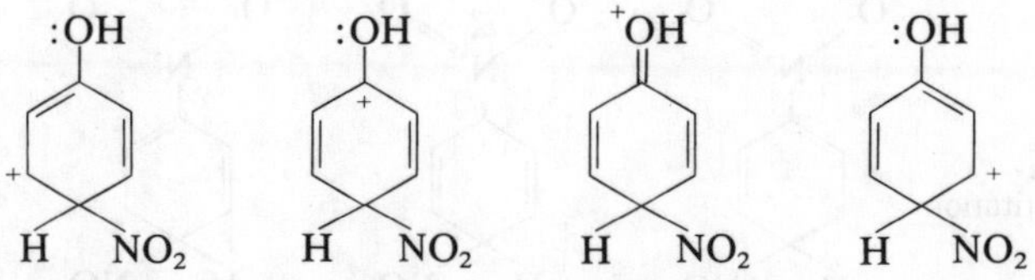

It will be seen that the *o*- and *p*-intermediates have four resonance hybrids as compared with three for the *m*-intermediate and this, in itself, will stabilise the *o*- and *p*-intermediates. They also have one resonance hybrid carrying positive charge on the ring carbon atom adjacent to the —OH group. As the —OH group is electron-donating it will be able to delocalise this charge more readily than the corresponding charge in the *m*-intermediate which is never adjacent to the —OH group.

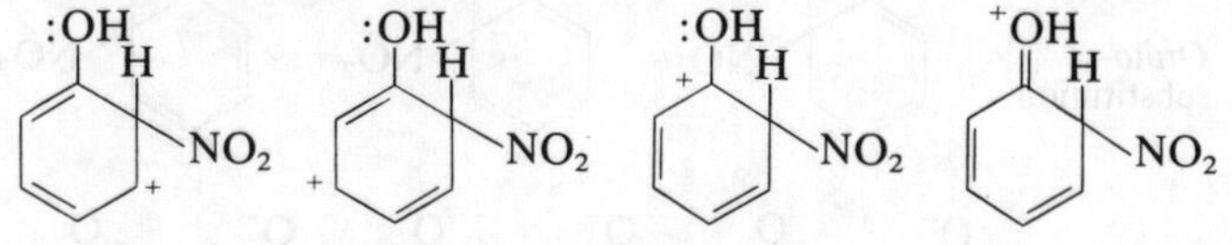

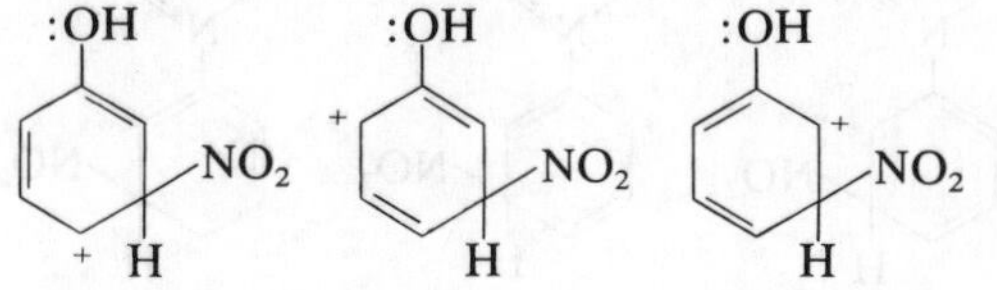

b. o-p-directing and de-activating groups, e.g. —Cl. Nitration of chlorobenzene gives carbonium ions like those formed from phenol above. The —Cl group is, therefore, *o-p*-directing, but, unlike the —OH group, it deactivates the ring because the —I effect of —Cl is greater than the mesomeric effect.

c. m-directing and de-activating groups, e.g. —NO₂. These groups are electron-withdrawing and the structures given on p. 145 show that the effect of this is to build up a positive charge on the *o*- and *p*-carbon atoms. The whole ring is deactivated but the deactivation is least in the *meta*-position where the relatively high electron density favours attack by an electrophile.

Alternatively, the situation can be considered from the point of view of the stability of the carbonium ions formed as intermediates. On further nitration, for example, the following carbonium ions might be formed:

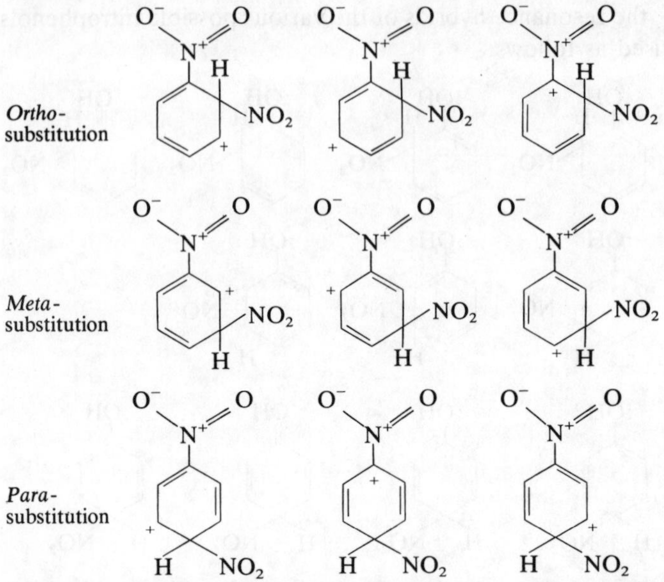

Ortho-substitution

Meta-substitution

Para-substitution

The *meta*-intermediate is the stablest because it is the only one in which no resonance hybrid has a positive charge on the C atom adjacent to the N atom. The existence of two adjacent positive charges, as in some of the *o*- and *p*-hybrids, would cause instability.

4 Nucleophilic substitution in benzene

The commonest substitution reactions of benzene are electrophilic but it can undergo nucleophilic substitution if a strongly electron-withdrawing group is present to lower the electron-density within the ring. Nitrobenzene, for example, will react with fused potassium hydroxide in the presence of air to form 2-nitrophenol and 4-nitrophenol. The nitro-group is *meta*-directing when electrophilic substitution is concerned but *o-p*-directing for nucleophiles.

Nitro groups in a benzene ring also greatly facilitate the readiness with which other substituents might be replaced. The Cl atom in chlorobenzene, for example, (p. 164) is not easily replaced but the introduction of nitro groups *ortho* or *para* to the chlorine makes the product undergo S_N2 reactions quite readily. That is why, for instance, 2–4 dinitrofluorobenzene (DNFB) is used in the N-terminal analysis of proteins (p. 308).

Questions on Chapter 10

1 Explain clearly what is meant by (a) delocalisation energy, (b) mesomerism, (c) hyperconjugation, (d) resonance.

2 Hückel's rule states that the conditions for aromaticity occur in any planar cyclic compound in which the atoms in the ring have $(4n + 2)$ π-electrons. Examine this statement.

3 What do you understand by the term 'aromatic compound'. Distinguish clearly, by means of examples, between aliphatic and aromatic compounds. (O Schol)

4 The introduction of a second substituent into a benzene nucleus is influenced, as regards position as well as the ease with which substitution occurs, by the nature of a substituent already present. Discuss this statement using as examples the influence of each of the following groups on subsequent substitution: $-CH_3$, $-NO_2$, $-OH$, $-Cl$. Suggest reaction schemes, naming the reagents used and reaction conditions (without experimental conditions), for the preparation of the following starting from benzene in each case: *p*-nitroaniline (4-nitrophenylamine), *m*-nitroaniline (3-nitrophenylamine), *o*-nitrophenol (2-nitrophenol). (W)

5 Dipropargyl has the formula

$$H-C\equiv C-CH_2-CH_2-C\equiv C-H$$

The heat of combustion of benzene vapour is 3 294 kJ and that of dipropargyl vapour is 3 570 kJ. The heat of formation of water is 286 kJ and that of carbon dioxide is 403 kJ. Give any explanation that you can to explain the differences in values of the heats of formation and the heats of combustion of the two hydrocarbons. (L)

6 Explain why the $-COOH$ group is deactivating and *m*-directing whereas the $-Cl$ group is deactivating and *o-p*-directing.

7 Explain carefully what it is that causes a group to be activating or deactivating and what causes it to be *m*-directing.

8 Why does a Cl atom linked to a benzene ring or to a $C\equiv C$ bond differ from one linked to a $C-C$ bond? Why is the situation different when a $-COOH$ group is concerned?

9 Borazole is isoelectronic with benzene. What does this mean and how does borazole resemble benzene?

10*Compare the values of the enthalpy of formation of benzene from
its atoms as calculated from the bond energies of a typical Kekule
structure and from the measured values of the enthalpies of combustion of benzene, carbon and hydrogen together with the enthalpies of atomisation of carbon and hydrogen. Why do the two values
not agree?

11 Aromaticity requires a molecule with a conjugated system and
atoms lying in the same plane. Comment on this statement.

11

Monohalogenoalkanes
Alkyl Halides

1 Introduction

Monohalogenoalkanes are formed from alkanes by the replacement of one hydrogen atom by one halogen atom. They have general formula C_nH_{2n+1}—X or R—X where X is used conventionally to represent any halogen atom. Typical examples are given below

Chloromethane   Bromoethane   2-iodopropane   2-chloro-2-methyl propane

They are classified as *primary*, *secondary* or *tertiary* depending on the nature of the C atom to which the halogen atom is linked.

The simplest monohalogenoalkanes are colourless gases but the higher members which are colourless, sweet smelling liquids are more convenient to use. The simplest chloro-compound which is a liquid is chloropropane; the simplest bromo-compound is bromoethane; and the simplest iodo-compound is iodomethane. All the compounds are practically insoluble in water but soluble in organic solvents.

Fluorocompounds are significantly different from the others and are discussed separately on p. 160. Chloro-, bromo- and iodo-alkanes are, chemically, very much alike, and they undergo many important substitution and elimination reactions. In general, the iodo-compound is the most reactive, and the chloro-compound the least.

The halogen in a monohalogenoalkane can be discovered by infrared spectroscopy. C—I and C—Br bonds do not cause any absorption but C—Cl and C—F bonds give strong absorption at 600–800 and $1\,000$–$1\,400\ \text{cm}^{-1}$ respectively.

2 Methods of preparation

a. Direct halogenation of an alkane. This method is only available for chloro- and bromo-alkanes (p. 94) and to obtain good yields of the mono-substituted compound it is necessary to use a large excess of the alkane so that substitution beyond the mono-stage is limited.

The method is used industrially for chloroethane,

$$C_2H_6 + Cl_2 \rightarrow C_2H_5\text{—}Cl + HCl$$
(excess)

the hydrogen chloride produced being used as in (b), below.

b. Addition of hydrogen halide to an ethene. Hydrogen halides add on to alkenes (p. 107) to form halogenoalkanes, e.g.

$$C_2H_4 + HCl \rightarrow C_2H_5\text{—}Cl$$

c. Replacement of —OH *group in alcohols by* —X. The hydroxyl group in an alcohol can be replaced by a halogen by treatment with hydrogen halides, i.e.

$$R\text{—}OH + HX \rightarrow R\text{—}X + H_2O$$

The reaction mechanism is discussed on p. 179. The reaction is slow and reversible for primary alcohols but takes place more readily and fully in passing to secondary- and tertiary-alcohols. For tertiary alcohols, concentrated solutions of hydrochloric, hydrobromic or hydriodic acids can be used. For primary alcohols a mixture of the alcohol and excess acid must be heated in the presence of anhydrous zinc chloride as a catalyst. For the chloro- and bromo-alkanes the acid can be prepared in situ by using a mixture of potassium chloride or bromide and concentrated sulphuric(VI) acid, but for the iodoalkane a mixture of potassium iodide and phosphoric(V) acid is needed, for sulphuric(VI) acid reacts with the iodide to give iodine and not hydrogen iodide.

Replacement of the hydroxyl group by chlorine can also be brought about using sulphur dichloride oxide, $SOCl_2$

$$R\text{—}OH + SOCl_2 \rightarrow R\text{—}Cl + SO_2 + HCl$$

Phosphorus tribromide or triiodide also react to give bromo- or iodoalkanes, e.g.

$$3R—OH + PBr_3 \rightarrow 3R—Br + H_3PO_3$$

$$3R—OH + PI_3 \rightarrow 3R—I + H_3PO_3$$

The phosphorus halides are prepared in situ by using mixtures of red phosphorus and bromine or iodine,

$$2P + 3Br_2 \rightarrow 2PBr_3 \qquad 2P + 3I_2 \rightarrow 2PI_3$$

In all these reactions pyridine, which is basic, may be added to lower the acidity and improve the yield.

3 Nucleophilic substitution, S_N, reactions

The halogen atom in halogenoalkanes can be replaced by many other atoms or groups as shown in the summary on p. 161. It is the ease of conversion of the R—X bond into R—O, R—C and R—N bonds that makes the halogenoalkanes so useful in organic synthesis.

Most of the reactions summarised are nucleophilic substitution reactions in which the nucleophile is either an anion, e.g.

$$R—X + Y^- \rightarrow R—Y + X^-$$

or a neutral molecule with an unshared pair of electrons, e.g.

$$R—X + H—Y: \rightarrow R—Y + H—X$$

This type of reaction has been extensively studied both for halogenoalkanes and for other compounds in which X represents other monovalent groups e.g. —OH, —OR, —N≡N.

a. $S_N 1$ or $S_N 2$ mechanisms. The general reaction

$$R—X + Y^- \rightarrow R—Y + X^-$$

can take place in two different ways.

If it takes place in two steps, via an R^+ carbonium ion, i.e.

$$R—X \xrightarrow{\text{slow}} R^+ + X^- \qquad R^+ + Y^- \xrightarrow{\text{fast}} R—Y$$

it is called an $S_N 1$ (*substitution, nucleophilic, unimolecular*) reaction.

The reaction is unimolecular (p. 43) as only one species is involved in the rate-determining step, and the overall rate of the reaction is independent of the Y^- concentration but proportional to the concentration of RX, i.e.

Rate $\propto$ [RX]

The reaction of 2-bromo-2-methylpropane with potassium hydroxide solution is of this type,

$$H_3C-\underset{\underset{CH_3}{|}}{\overset{\overset{CH_3}{|}}{C}}-Br + OH^- \rightarrow H_3C-\underset{\underset{CH_3}{|}}{\overset{\overset{CH_3}{|}}{C}}-OH + Br^-$$

Rate $\propto$ [(CH$_3$)$_3$CBr]

Alternatively, the reaction may involve an S_N2 (*substitution, nucleophilic, bimolecular*) mechanism in which an intermediate transition state (p. 43) is formed, i.e.

$$Y^- + R-X \rightarrow Y^- \cdots R \cdots X \rightarrow Y-R + X^-$$
$$\text{Transition state}$$

The formation of the transition state involves two species so that it is bimolecular, and the rate is given by

Rate $\propto$ [RX] $\times$ [Y$^-$]

The reaction of bromomethane with potassium hydroxide solution is of this type,

$$CH_3-Br + OH^- \rightarrow HO \cdots \underset{\underset{H_3}{}}{C} \cdots Br \rightarrow HO-CH_3 + Br^-$$

Rate $\propto$ [CH$_3$Br] $\times$ [OH$^-$]

The essential difference is that the S_N1 reaction involves the initial splitting of the R—X bond, whereas in the S_N2 mechanism Y^- begins to attach itself to R—X before the R—X bond actually splits. Which mechanism occurs depends on the nature of R, X and Y^- and on the solvent in which the reaction is carried out. There are also interesting stereochemical issues involved (p. 281).

b. The nature of R. In the reaction of a series of halides, e.g.

$$CH_3—Br \quad C_2H_5—Br \quad (CH_3)_2CH—Br \quad (CH_3)_3C—Br$$

——— S_N1 rate increases ———→

←——— S_N2 rate increases ———

with OH^- ions there is an increase in the rate of the S_N1 reaction and a decrease in the rate of the S_N2 reaction in passing along the series. The reaction of bromomethane is S_N2; that of 2-bromo-2-methylpropane is S_N1; the other compounds are intermediate.

The S_N1 mechanism is favoured on passing along the series as tertiary carbonium ions are more stable than primary ones (p. 47). The S_N2 mechanism is inhibited in passing along the series both by steric effects as the alkyl group increases in size and by the increased build up of negative charge, through the $+I$ inductive effect of methyl groups, on the carbon atom linked to bromine. This will make the carbon atom less open to attack by OH^- in forming the transition state.

c. The effect of the solvent. Solvolysis. The S_N1 mechanism, involving the formation of R^+ and Br^- ions is favoured by a solvent of high relative permittivity (dielectric constant) with high ionising power. Such a solvent both aids the initial formation of the ions and is then able to solvate them. Thus a reaction mechanism almost exclusively S_N1 in aqueous solution may become so much slower in a solvent of lower dielectric constant, e.g. ethanol, that the S_N2 mechanism begins to predominate.

If the solvent is itself nucleophilic it may participate in the reaction as a reagent; if so the reaction is known as *solvolysis*. It is to prevent such solvolysis that many substitution reactions of halogenoalkanes are carried out in ethanol or other non-aqueous solvents. These solvents, moreover, can dissolve both the halogenoalkane and the nucleophilic reagent.

d. The nature of X. X is referred to as the leaving group and its nature affects the rate of both S_N1 and S_N2 reactions. As the $C—X$ bond has to be split, the weaker it is the more readily will the substitution take place. Alternatively, X is a good leaving group if HX is a strong acid, or X^- a weak base.

For halogenaoalkanes this means that the reactivity increases in passing from RF to RI,

R—F R—Cl R—Br R—I

$\longrightarrow$ Increasing reactivity $\longrightarrow$

F^- Cl^- Br^- I^-

$\longrightarrow$ Decrease in basic strength $\longrightarrow$

C—F	C—Cl	C—Br	C—I
(485 kJ mol^{-1})	(327)	(285)	(213)

$\longrightarrow$ Decrease in bond energy $\longrightarrow$

Hydrogen atoms in hydrocarbons are not readily substituted by nucleophiles as the C—H bond is so strong (413 kJ mol^{-1}) and H$^-$ is such a strong base. Similarly it is not easy to replace —OH, —OR or —NH$_2$ groups as the leaving groups concerned (OH$^-$, OR$^-$ and NH$_2^-$ are such strong bases). In acid solution, however, (p. 178) the leaving groups become H$_2$O, ROH and NH$_3$ which are less basic so that substitution is easier.

e. The nature of Y$^-$. The nucleophile, Y$^-$, is sometimes referred to as the entering group. Its nature has no effect on S$_N$1 reactions which are independent of the nucleophile concentration but in S$_N$2 reactions the rate does depend on the nucleophilic strength (the nucleophilicity) of Y$^-$. Nucleophilic strength may coincide with basic strength but it is not always so for the former refers to donation of electrons to a carbon atom and the latter to a hydrogen atom.

For the same attacking atom the correlation between nucleophilic and basic strength is reasonably close, e.g.

H$_2$O RCO$_2^-$ C$_2$H$_5$O$^-$ OH$^-$ RO$^-$

$\longrightarrow$ Increase in nucleophilic power $\longrightarrow$
$\longrightarrow$ Increase in basic strength $\longrightarrow$

For different atoms within the same periodic table group it is the larger atom that has the highest nucleophilic strength, e.g.

F$^-$ Cl$^-$ Br$^-$ I$^-$

$\longrightarrow$ Increase in nucleophilic power $\longrightarrow$
$\longrightarrow$ Decrease in basic strength $\longrightarrow$

Here, the higher ease of polarisation and the lower degree of solvation of I^- makes it a stronger nucleophile than F^- even though the latter is the much stronger base.

The fact that I^- is both a good leaving and entering group means that it can function as a catalyst for other substitutions, e.g.

$$R—Br + H_2O \xrightarrow{\text{slow}} R—OH + H^+ + Br^- \quad \text{Uncatalysed}$$

$$\left.\begin{array}{l} R—Br + I^- \xrightarrow{\text{fast}} R—I + Br^- \\[2ex] R—I + H_2O \xrightarrow{\text{fast}} R—OH + H^+ + I^- \end{array}\right\} \text{Catalysed}$$

4 Elimination reactions

Halogenoalkanes can also take part in elimination reactions resulting in the formation of an alkene, e.g.

These reactions, which are the opposite of the addition of hydrogen halide to an alkene (p. 107), are brought about by heating with concentrated, alcoholic solutions of potassium hydroxide and good yields can be obtained from secondary and tertiary halogenoalkanes.

As the hydroxide is a base the eliminated acid reacts with it so that the overall reaction is

$$R—CH_2—CH_2—Br + OH^- \rightarrow R—CH=CH_2 + H_2O + Br^-$$

These elimination reactions compete with the nucleophilic substitution reactions which occur with OH^- ions, and, as with them, there are two possible mechanisms; E1 (*elimination, unimolecular*) and E2 (*elimination, bimolecular*).

In the E1 mechanism, the same carbonium ion is formed in the rate determining step as in the S_N1 reaction. This then loses a proton to the basic OH^- ion, e.g.

The reaction is unimolecular and the rate is proportional to the concentration of the halogenoalkane, i.e.

Rate $\propto [(CH_3)_3 CBr]$.

but independent of the OH^- ion concentration.

In the E2 mechanism, the loss of the Br^- ion and the removal of the proton take place simultaneously, e.g.

$$HO^- \; H$$
$$H-\overset{\displaystyle H}{\underset{\displaystyle H}{C}}-CH_2-Br \rightarrow H_2O + CH_2{=}CH_2 + Br^-$$

The reaction is bimolecular and the rate is given by

Rate $\propto [CH_3{-}CH_2{-}Br] \times [OH^-]$

a. Conditions favouring elimination. The essential difference between the elimination and substitution reactions is that a nucleophile attacks the carbon atom adjacent to the bromine (the α-carbon atom) in substitution whereas it attacks the neighbouring carbon atom (the β-atom) in elimination reactions. Moreover, the nucleophile is acting as a base in the elimination reactions, removing a proton from the β-carbon atom.

Whether the E1, E2, S_N1 or S_N2 mechanism will predominate under any one set of conditions depends on many factors. As the rate-determining step is the same for both E1 and S_N1 mechanisms they are affected by change of conditions in the same sort of way. If the rate of the elimination reaction is increased so is that of the substitution reaction so that, in many cases the ratio of elimination to substitution product remains fairly constant.

The ratio can be increased, however, under bimolecular conditions, i.e. the elimination mechanism and the yield of alkene can be maximised. This is done by using hot concentrated solutions of potassium hydroxide in ethanol. The use of ethanol as solvent favours the bimolecular mechanism for it is weakly ionising so that the carbonium ion formation required in the unimolecular mechanisms is limited. The potassium hydroxide provides OH^- ions which are both nucleophilic and strongly basic; they, therefore favour the E2 as against the S_N2 mechanism. A high concentration increases the rate of the reaction for this concentration term enters into the rate expression for the bimolecular mechanisms. Finally, the high temperature assists, for the activation energy for elimination reactions is higher than for substitution reactions so that the former are more favoured by increase in temperature.

Elimination reactions are also favoured, in general by increased alkyl substitution. Bromoethane, for example gives only 1–2 % of ethene even under the most favourable circumstances; 2-bromopropane gives about 80 % of propene; and 2-bromo-2-methylpropane gives almost 100 % of butene.

5 Formation of metallic alkyls

Halogenoalkanes react, in ethereal or hydrocarbon solvents, with many metals, particularly the more electropositive ones. The products are metallic alkyls, an important type of organometallic compound (p. 342), e.g.

$$R—X + 2Li \rightarrow LiX + LiR$$

(A lithium alkyl)

$$R—X + Mg \rightarrow RMgX$$

(A Grignard reagent)

$$4C_2H_5Cl + 4Na/Pb \rightarrow Pb(C_2H_5)_4 + 4NaCl + 3Pb$$

(Alloy)   Tetraethyl lead(IV)

Grignard reagents are useful synthetic reagents (p. 340). Tetraethyl lead(IV) is used as an anti-knock in petrol, to limit pre-ignition.

a. The Wurtz reaction. This is a very old reaction which enables halogenoalkanes to be coupled, e.g.

$$2R—X + 2Na \rightarrow R—R + 2NaX$$

The reaction can be used to prepare alkanes but the yield is low except for some alkanes of high relative molecular mass.

If carried out in the vapour phase the reaction probably involves free radicals, e.g.

$$R—X + Na· \rightarrow NaX + R·$$

$$2R· \rightarrow R—R$$

In ethereal or hydrocarbon solution, however, a sodium alkyl is first formed and then reacts with more halogenoalkane, e.g.

$$R—X + 2Na \rightarrow NaR + NaX$$
$$R—X + NaR \rightarrow NaX + R—R$$

6 Monofluoroalkanes

Since fluorine differs considerably from the other halogens it is not surprising that fluoroalkanes also differ from the other halogenoalkanes.

a. Preparation. Direct reactions of alkanes with fluorine tends to be explosive for the reactions are highly exothermic. Dilution with nitrogen helps but cobalt(III) fluoride is a more controllable fluorinating agent, e.g.

$$C_2H_6 + 2CoF_3 \rightarrow C_2H_5—F + 2CoF_2 + HF$$

Monofluoroalkanes can also be prepared from other halogenoalkanes by reaction with inorganic fluorides, e.g.

$$2CH_3Br + HgF_2 \rightarrow 2CH_3F + HgBr_2$$

or by addition of hydrogen fluoride to alkenes.

b. Properties. The physical properties of the monofluoroalkanes are similar to the alkane from which they are derived. They are very unreactive and the fluorine atom is not easily replaced in S_N reactions. This is due to the high bond energy of the C—F bond (p. 3).

Fully fluorinated alkanes, e.g. C_7F_{16}, are used as lubricating oils; they are known as *fluorocarbons*. Other polyfluorinated hydrocarbons and mixed fluoro-halogeno-compounds are also useful (p. 168).

Questions on Chapter 11

1 How is iodoethane prepared and how does it react with dilute aqueous potassium hydroxide, concentrated alcoholic potassium hydroxide and potassium cyanide solution?

2 Write structural formulae and names for all the compounds having a molecular formula of (a) $C_5H_{11}Br$ and (b) $C_4H_8Br_2$.

3 Give the structural formulae of all the products obtainable by chlorination of ethane.

4 Write equations showing how chloroethane may be used in preparing ethane, ethene, butane, ethanol and propanonitrile.

5 Compare and contrast the properties of sodium chloride and chloromethane.

Reagent	Nucleophile in S_N reaction	Main product	
Hydrogen (H_2 + Pt or Zn/Cu couple in aqueous alcohol)		Alkane	R—H
LiAlH$_4$ in ether solution	H$^-$	Alkane	R—H
Na (Wurtz reaction)		Alkane	R—R
Hot water		Alcohol	R—OH
Aqueous KOH solution	OH$^-$	Alcohol	R—OH
Sodium alkoxide in alcohol	OR$^-$	Ether	R—O—R
Alcoholic solution of NH$_3$ (heat in sealed tube)	NH$_3$	Amine	R—NH$_2$
Hot KCN in propanone or alcohol	CN$^-$	Nitrile	R—C≡N
AgCN in alcohol		Isocyanoalkane	R—N≡C
KNO$_2$ or AgNO$_2$ in alcohol	NO$_2^-$	Nitroalkane or Alkylnitrate (III)	R—N⟨O / =O R—O—N=O
Na or Ag ethanoate	CH$_3$COO$^-$	Ester	R—O—C(—CH$_3$)=O
Aromatic hydrocarbon + AlCl$_3$ (Friedel-Crafts reaction)	Aromatic hydrocarbon	Alkylated aromatic hydrocarbon	
Hot, concentrated, alcoholic solution of KOH		Alkene	R—CH=CH$_2$
Mg in dry ether		Grignard reagent	R—MgX

Typical reactions of halogenoalkanes, R—X

6 Give examples of S_N1, S_N2, E1 and E2 reactions.

7 What quantities would have to be known to calculate the enthalpy of reaction for the ionisation of chloromethane in aqueous solution:

$$CH_3Cl(aq) \rightarrow CH_3^+(aq) + Cl^-(aq)$$

What difficulties would be encountered in trying to find these quantities?

8 How would you attempt to compare the nucleophilic strength of the Cl^-, Br^- and I^- ions?

9 To what extent are nucleophiles, bases and reducing agents alike? Give examples.

10 Explain how measurement of the rate of the reaction under different conditions can throw light on the mechanism of the reaction between chloroethane and potassium iodide.

11 10 cm^3 of a gaseous hydrocarbon, X, required 45 cm^3 of oxygen for complete combustion and formed an addition compound which contained 79.2 per cent of bromine. Suggest a structural formula for X.

12 The reactions of alkyl halides are often classified as (i) elimination reactions, (ii) displacement reactions, (iii) reactions with metals. Give a brief account of the reactions of alkyl halides under these three headings. (W)

13*Record the bond energies of C—X and H—X bonds where X is a halogen. Show how the values affect the enthalpies of the reactions in which methane is halogenated by reaction with the halogen.

14*If the conversion of a methyl radical into a methyl cation in the gaseous state requires 979 kJ mol^{-1} calculate the enthalpy of the reaction

$$CH_3X(g) \rightarrow CH_3^+(g) + X^-(g)$$

when X is Cl, Br and I. What conclusion can you draw from your result?

12

Other Halogenated Hydrocarbons

Replacement of hydrogen atoms by halogens in aromatic hydrocarbons gives halogenoarenes (aryl halides) and similar replacement in unsaturated hydrocarbons gives unsaturated halides. If more than one halogen atom is involved the product is known as a polyhalide. It is interesting to compare the functioning of the —X atom in all these compounds with that in monohalogenoalkanes.

Halogenoarenes

1 Introduction
Benzene forms only one monohalogenated derivative, i.e. chlorobenzene, but methyl benzene, and higher homologues may have ring- or side-chain-hydrogen atoms substituted by halogens so that they form isomeric products, e.g.

Chlorobenzene   Chloro-2-methyl benzene   Chloro-3-methyl benzene   Chloro-4-methyl benzene   (Chloromethyl) benzene

The ring-substituted compounds differ from halogenoalkanes in that they are much less open to attack by nucleophilic reagents in S_N reactions (p. 153). The side-chain compounds, which may be regarded as substituted alkanes, are very similar to halogenoalkanes and undergo typical S_N reactions even more readily.

2 Ring-substituted compounds

a. Preparation. Ring substitution of aromatic hydrocarbons can be brought about by direct halogenation (p. 129). Alternatively, an —NH_2 group in the ring can be replaced by a halogen via a diazonium salt (p. 346).

Chlorobenzene is manufactured by the Raschig process in which a mixture of benzene vapour, air (or oxygen) and hydrogen chloride is passed over hot copper(II) chloride,

$$2C_6H_6 + 2HCl + O_2 \xrightarrow[\text{CuCl}_2]{\text{Hot}} 2C_6H_5\text{—Cl} + 2H_2O$$

b. Properties. Ring-halogenated hydrocarbons can undergo further ring-substitution reactions, e.g. nitration, sulphonation, halogenation, with electrophilic reagents. The original halogen atom in the ring is *o-p* directing and deactivating (p. 147).

Substitution of the original halogen atom by S_N reactions, which is so easy for halogenoalkanes, is, however, very difficult for halogeno-arenes. The reason for this is the overlap of a *p*-orbital of the halogen atom with the π-orbital of the benzene ring (p. 141) so that structures exemplified below contribute to a resonance hybrid,

Some double bond character is introduced into the C—Cl bond; this shortens its length and increases its strength making substitution more difficult. The introduction of some negative charge onto the carbon atom of the C—Cl bond also makes it less open to attack by nucleophiles (p. 155).

Substitution can be brought about under sufficiently vigorous conditions as, for example, in the manufacture of phenol from chlorobenzene in which —Cl is replaced by —OH (p. 197). The Cl atom can also be substituted easily if metadirecting groups, particularly the nitro group, are positioned *ortho-* or *para-* to it.

3 Side-chain-substituted compounds

Side-chain substitution is brought about by reaction with halogens on heating and in the presence of ultra-violet light (p. 95).

The products, typified by (chloromethyl)benzene, undergo S_N reactions more readily than similar halogenoalkanes. This is due to the high stability of the aromatic carbonium ion intermediates as compared with similar aliphatic ions, and to the $-I$ inductive effect of the $-C_6H_5$ group as compared with the $+I$ effect of alkyl groups.

(Chloromethyl)benzene also undergoes many electrophilic ring-substitution reactions, e.g. nitration, sulphonation, halogenation, and the side-chain can be oxidised to $-COOH$ by potassium manganate(VII).

	C_2H_5-Cl	$\langle O \rangle -Cl$	$\langle O \rangle -CH_2-Cl$
Method of preparation	Direct chlorination of C_2H_6 + uv light	Direct chlorination of C_6H_6 + halogen carrier	Direct chlorination of $C_6H_5-CH_3$ + u.v. light
	Addition of HCl to C_2H_4	C_6H_6 vapour + O_2 + HCl + $CuCl_2$ (Raschig process)	
	From C_2H_5-OH by reaction with HCl, $SOCl_2$ or PCl_5	From $C_6H_5-NH_2$ via diazonium salt	From $C_6H_5-CH_2OH$ by reaction with HCl, $SOCl_2$ or PCl_5
Delocalisation of electrons	No delocalisation	Delocalisation between the ring and the Cl atom	Delocalisation within the ring
Nucleophilic substitution	Occurs readily with many nucleophiles (see p. 161)	Very difficult	Easier than for C_2H_5Cl
Elimination reactions	Loses HCl to form C_2H_4 in very poor yield	No elimination reactions	
Electrophilic substitution	Nil	Can be nitrated, sulphonated and halogenated in the ring in the *o*- and *p*-positions	
Mg in dry ether	Form	Grignard	reagents

Preparation and properties of chloroethane, chlorobenzene and (chloromethyl)benzene

Polyhalides and Unsaturated Halides

4 Polyhalogenated methanes
Methane can be chlorinated to give four chloromethanes,

CH_3Cl	CH_2Cl_2	$CHCl_3$	CCl_4
Chloro-methane	Dichloro-methane	Trichloromethane (Chloroform)	Tetrachloromethane (Carbon tetrachloride)

and corresponding fluoro-, bromo- and iodo-compounds are known.

a. Trichloromethane. Chloroform. Chloroform is a colourless liquid with a characteristic sweetish smell. It is a good solvent for organic compounds like many other chlorinated hydrocarbons. It was one of the earliest anaesthetics (Sir James Simpson, 1847) but is now mainly used as a solvent and in making PTFE (p. 327).

It can be made by reaction between an aqueous suspension of bleaching powder and propanone or ethanol. The bleaching powder decomposes to liberate chlorine and calcium hydroxide,

$$CaOCl_2 + H_2O \rightarrow Cl_2 + Ca(OH)_2$$
Bleaching
powder

Using propanone, the overall reaction involves chlorination and hydrolysis, i.e.

$$CH_3—CO—CH_3 \xrightarrow{Cl_2} CH_3—CO—CCl_3 \xrightarrow{OH^-} CHCl_3 + CH_3—COO^-$$

With ethanol there is an initial oxidation to ethanal, followed by chlorination and hydrolysis, i.e.

$$C_2H_5—OH \xrightarrow{Cl_2} CH_3—CHO \xrightarrow{Cl_2} CCl_3—CHO \xrightarrow{OH^-} CHCl_3 + H—COO^-$$
                                    Ethanal

Trichloromethane reacts with hot potassium hydroxide solution to form potassium methanoate and some carbon monoxide (p. 167),

$$CHCl_3 + 4KOH \rightarrow 3KCl + 2H_2O + HCOO^-K^+$$

b. Other trihalogenomethanes. Tribromomethane is a dense, colourless liquid very much like trichloromethane; triiodomethane (iodoform) is a yellow crystalline solid with a characteristic 'antiseptic' smell; trifluoromethane is a remarkably stable, non-toxic gas.

Triiodomethane is made by treating propanone or ethanol with iodine and sodium hydroxide or sodium carbonate solution. The reactions that take place are like those involved in making trichloromethane.

The formation of triiodomethane, which can be seen as yellow crystals, in this kind of reaction can be used as a test (the iodoform test) for propanane or ethanol, or, more widely, for any compound containing the $CH_3—CO—$ or $CH_3—CH(OH)—$ groups.

c. Tetrachloromethane. This is mainly manufactured from chlorine and carbon disulphide,

$$CS_2 + 3Cl_2 \xrightarrow{FeCl_3} CCl_4 + S_2Cl_2$$

It is a colourless liquid, very much less reactive than the other halogenomethanes. It reacts, for example, only very slightly with hot potassium hydroxide solution.

It is used as a solvent, in fire extinguishers and in making tetrachloroethene (p. 170) and chlorofluoromethane (p. 168).

d. Reactions of halogenomethanes with potassium hydroxide solution.
Chloromethane reacts with OH^- ions in an S_N2 reaction (p. 154). Dichloromethane reacts similarly, but less readily, the first product then undergoing an elimination reaction,

$$H-\underset{\underset{Cl}{|}}{\overset{\overset{H}{|}}{C}}-Cl \xrightarrow{OH^-} H-\underset{\underset{Cl}{|}}{\overset{\overset{H}{|}}{C}}-OH+Cl^- \xrightarrow[(-HCl)]{OH^-} \underset{\text{Methanal}}{H-\overset{\overset{H}{|}}{C}=O}+Cl^-+H_2O$$

Tetrachloromethane reacts only very slightly with hot potassium hydroxide solution even though the reaction to form carbon dioxide and hydrogen chloride is thermodynamically feasible and the corresponding silicon compound reacts with water. The lack of reactivity in tetrachloromethane is presumably due to the steric hindrance of four chlorine atoms around one carbon atom.

Trichloromethane reacts readily with potassium hydroxide solution (p. 166), the reaction involving an intermediate carbanion, $:C^-Cl_3$, which loses a Cl^- ion to form dichlorocarbene, CCl_2,

$$CHCl_3+OH^- \rightarrow :C^-Cl_3+H_2O$$

$$:C^-Cl_3 \rightarrow Cl-\ddot{C}-Cl+Cl^-$$

This carbene has only six electrons around the central carbon atom so that it is highly electrophilic reacting readily with water,

$$:CCl_2+H_2O \rightarrow CO+2HCl$$

$$:CCl_2+2H_2O \rightarrow HCOOH+2HCl$$

5 Dihalogenated ethanes
There are two isomers of dichloroethane,

$$H-\underset{\underset{H}{|}}{\overset{\overset{H}{|}}{C}}-\underset{\underset{Cl}{|}}{\overset{\overset{H}{|}}{C}}-Cl \qquad H-\underset{\underset{Cl}{|}}{\overset{\overset{H}{|}}{C}}-\underset{\underset{Cl}{|}}{\overset{\overset{H}{|}}{C}}-H$$

1,1-dichloroethane   1,2-dichloroethane
(a gem-isomer)   (a vic-isomer)

The one with both chlorine atoms linked to the same carbon atom is called a gem-compound (gemini = twin); the other is a *vic*-compound (vicinal = adjacent).

The *gem*-isomer is made by addition of hydrogen chloride to ethyne (p. 120) or by reaction between ethanal and phosphorus pentachloride,

$$\underset{\underset{H}{|}}{\overset{\underset{H}{|}}{H-C}}-\overset{H}{C}=O+PCl_5 \rightarrow H-\underset{\underset{H}{|}}{\overset{\underset{H}{|}}{C}}-\underset{\underset{Cl}{|}}{\overset{\underset{H}{|}}{C}}-Cl+POCl_3$$

It reforms ethanal on reaction with potassium hydroxide solution,

$$CH_3-CHCl_2+2KOH \rightarrow CH_3-CHO+2KCl+H_2O$$

The vic-isomer is made by adding chlorine to ethene (p. 104). Each of the chlorine atoms functions like the single atom in a monohalogenoalkane. Thus 1,2-dichloroethane undergoes both substitution reactions, e.g.

$$\underset{\underset{CN}{|}}{\overset{\underset{H}{|}}{H-C}}-\underset{\underset{CN}{|}}{\overset{\underset{H}{|}}{C}}-H \xleftarrow{CN^-} \underset{\underset{Cl}{|}}{\overset{\underset{H}{|}}{H-C}}-\underset{\underset{Cl}{|}}{\overset{\underset{H}{|}}{C}}-H \xrightarrow{OH^-} \underset{\underset{OH}{|}}{\overset{\underset{H}{|}}{H-C}}-\underset{\underset{OH}{|}}{\overset{\underset{H}{|}}{C}}-H$$

Ethane-1,2-diol

and an elimination reaction with hot, concentrated, alcoholic potassium hydroxide solution,

$$\underset{\underset{Cl}{|}}{\overset{\underset{H}{|}}{H-C}}-\underset{\underset{Cl}{|}}{\overset{\underset{H}{|}}{C}}-H \xrightarrow{-HCl} \underset{}{\overset{\underset{H}{|}}{H-C}}=\overset{\underset{H}{|}}{C}-Cl \xrightarrow{-HCl} H-C\equiv C-H$$

Chloroethene   Ethyne

The first stage of the elimination reaction can also be brought about by heating at 500 °C and this is used as a method of manufacturing chloroethene (p. 169).

6 Mixed halogeno-alkanes

The hydrogen atoms in an alkane may be replaced partly by one halogen and partly by another.

The compounds involving fluorine and chlorine are known as freons, e.g. CF_2Cl_2, $CFCl_3$. They are used as refrigerants and as the propellant in aerosols, and are made by reaction of tetrachloromethane with antimony trifluoride and hydrogen fluoride, e.g.

$$3CCl_4 + 2SbF_3 \rightarrow 3CF_2Cl_2 + 2SbCl_3$$
$$3CCl_4 + SbF_3 \rightarrow 3CFCl_3 + SbCl_3$$

The hydrogen fluoride reforms antimony trifluoride from the antimony trichloride.

Fluothane, 1-bromo-1-chloro-2,2,2,-trifluoroethane, is an anaesthetic.

7 Halogenated ethenes

a. Chloroethene (Vinyl chloride). This is a colourless gas made by reaction between ethyne and hydrogen chloride (p. 120) or by heating 1,2-dichloroethane (p. 168). It polymerises under the influence of heat and catalysts to form polyvinylchloride, PVC,

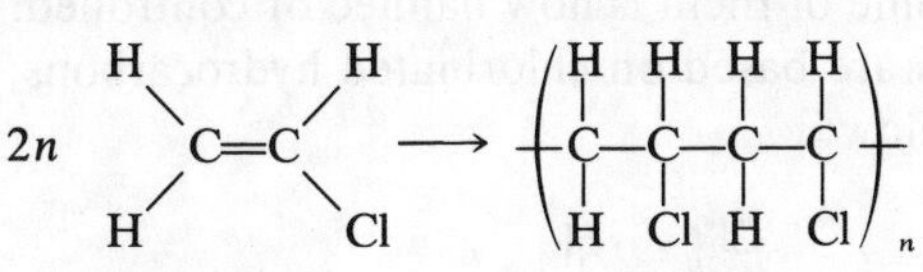

Chloroethene
(Vinyl chloride)

PVC

Chloroethene resembles chlorobenzene in its properties and will not undergo S_N reactions like chloromethane. A p-orbital of the chlorine atom can overlap with the π-orbital of the C=C bond. The C—Cl bond is, therefore, strengthened as in chlorobenzene (p. 164).

b. Tetrafluoroethene. This is a colourless gas made by heating chlorodifluoromethane,

$$CHCl_3 \xrightarrow{\;SbF_3 + HF\;} CHF_2Cl$$
$$2CHF_2Cl \longrightarrow CF_2{=}CF_2 + 2HCl$$

The product polymerises in the presence of a catalyst to give polytetrafluorethene, PTFE (p. 327).

c. Trichloroethene. This is made by heating tetrachloroethane at 400 °C in the presence of a barium chloride catalyst,

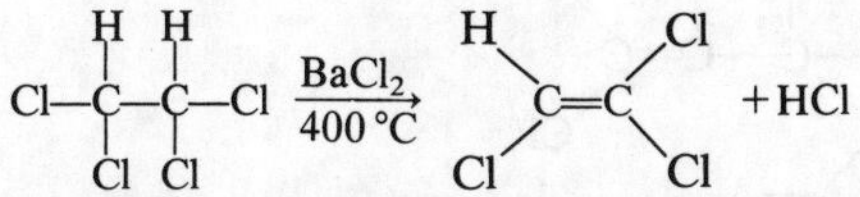

It is used as a commercial de-greasing solvent.

d. Tetrachloroethene. This is made by heating tetrachloromethane at 800–900 °C,

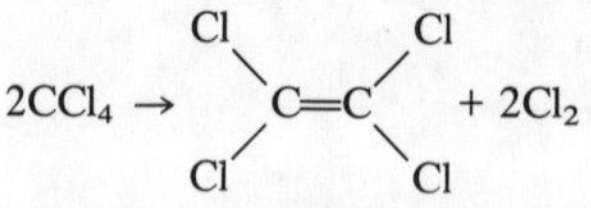

$$2CCl_4 \rightarrow \underset{Cl}{\overset{Cl}{}}C=C\underset{Cl}{\overset{Cl}{}} + 2Cl_2$$

It is used as a dry-cleaning solvent.

8 Organochlorine insecticides

Insecticides are very important both in limiting crop losses and the spread of diseases. Natural products such as pyrethrum (extracted from Chrysanthemum cinerariaefolium) and nicotine (obtained from tobacco) play a useful part but there has been extensive development of synthetic insecticides following the first use of DDT in the late 1930s. They have been used extensively until recent times but because of pollution hazards the use of some of them is now banned or controlled.

The commonest insecticides are based on chlorinated hydrocarbons; typical examples are given below:

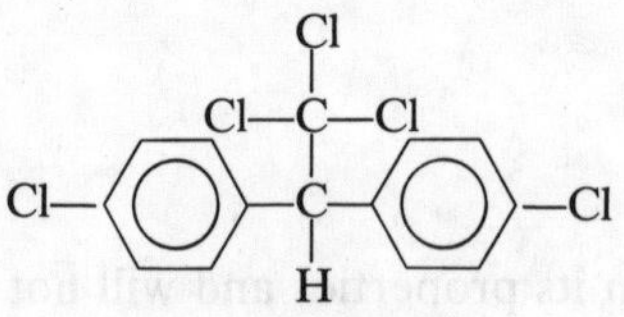

DDT
(Dichlordiphenyltri-
chloroethane)

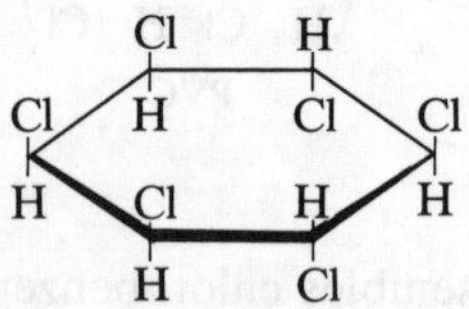

BHC
(Benzene hexachloride
Gammexane. Lindane)

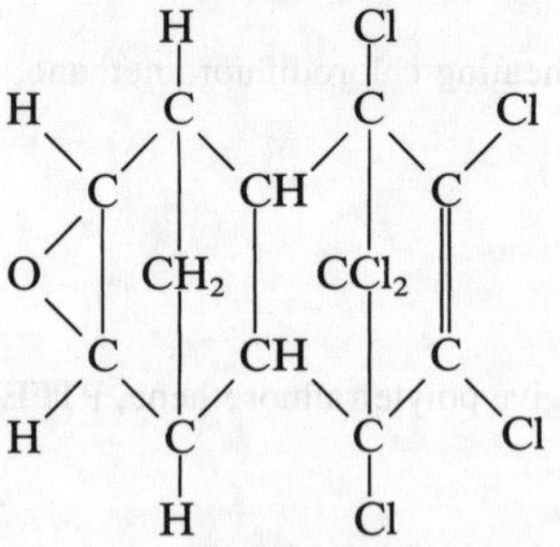

Dieldrin

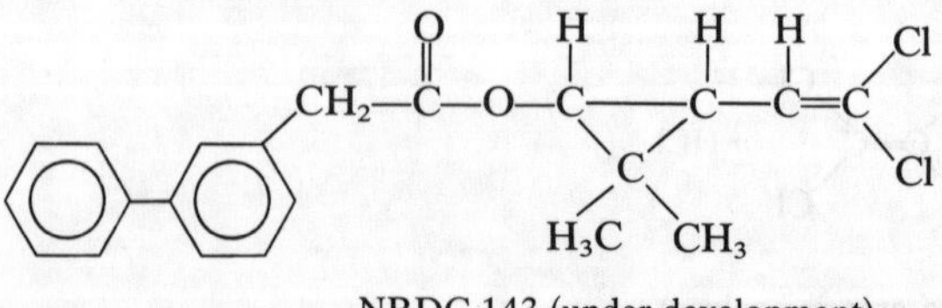

NRDC 143 (under development)

Questions on Chapter 12

1 Write structural formulae for all the isomeric trichlorobenzenes, chloronitrobenzenes and bromodinitrobenzenes.

2 Write formulae for the various possible mono-bromo derivatives of (a) ethylbenzene and (b) 1,2-dichlorobenzene.

3 Compare and contrast the preparation and properties of (a) chlorobenzene and chloromethane, (b) (chloromethyl)benzene and chloroethane.

4 'The properties of a chlorine atom attached to a carbon atom are markedly affected by the presence of other groups attached to the same or to a neighbouring carbon atom'. Discuss this statement.

5 Compare the reactivity of the chlorine atom in an S_N reaction in ethanoylchloride, monochloroethanoic acid, chloroethane, chlorobenzene, chloroethene.

6 The introduction of nitro groups *ortho-* or *para-* to the Cl atom in chlorobenzene increases the reactivity of the compound remarkably. Suggest why this should be so.

7 Describe the preparation of trichloromethane. A liquid is suspected of being chloroform. What steps would you take to establish its identity by chemical means?

8 Iodoform can be obtained by the electrolysis of a solution containing potassium iodide, sodium carbonate and propanone. Suggest a possible mechanism.

9 Discuss the replacement of hydrogen atoms in organic molecules by halogen atoms. Include in your answer an account of the reactions of chlorine with molecules such as methylbenzene (toluene), propanone (acetone) and ethanoic (acetic) acid. (O and C)

10 Under what conditions do (a) aromatic hydrocarbons (e.g. benzene) and (b) aliphatic hydrocarbons (e.g. methane) react with chlorine to give substitution products? Account for these requirements for different conditions in terms of the natures of the hydrocarbons

and the reaction mechanisms involved. Why does benzene undergo a substitution reaction with chlorine whereas, under the same conditions, ethene (ethylene) undergoes an addition reaction? (O and C)

11 Describe briefly how you would show the presence of chlorine in chlorobenzene. **A** is a liquid which, when chlorinated in ultraviolet light, gave **B** containing 44.10 % of chlorine. On hydrolysis **B** gave **C** which, on treatment with cold concentrated sodium hydroxide, gave two compounds **D** and **E**. In the presence of concentrated acid **D** and **E** reacted together to give **F**. Give the structures of **A**, **B**, **C**, **D**, **E** and **F**, and explain your reasoning. (AEB)

13

Aliphatic Monohydric Alcohols

1 Introduction

The simplest aliphatic alcohols are derived from alkanes by replacing one hydrogen atom by one hydroxyl group. They are known as monohydric alcohols, have a general formula $C_nH_{2n+1}OH$, and are named by replacing the last letter of the alkane name by -ol. When necessary, the position of the —OH group is shown numerically, e.g.

$$H-\overset{\displaystyle H}{\underset{\displaystyle H}{\overset{|}{\underset{|}{C}}}}-OH$$

Methanol

$$H-\overset{\displaystyle H}{\underset{\displaystyle H}{\overset{|}{\underset{|}{C}}}}-\overset{\displaystyle H}{\underset{\displaystyle H}{\overset{|}{\underset{|}{C}}}}-OH$$

Ethanol

$$H-\overset{\displaystyle H}{\underset{\displaystyle H}{\overset{|}{\underset{|}{C}}}}-\overset{\displaystyle H}{\underset{\displaystyle H}{\overset{|}{\underset{|}{C}}}}-\overset{\displaystyle H}{\underset{\displaystyle H}{\overset{|}{\underset{|}{C}}}}-OH$$

Propan-1-ol

$$H-\overset{\displaystyle H}{\underset{\displaystyle H}{\overset{|}{\underset{|}{C}}}}-\overset{\displaystyle H}{\underset{\displaystyle OH}{\overset{|}{\underset{|}{C}}}}-\overset{\displaystyle H}{\underset{\displaystyle H}{\overset{|}{\underset{|}{C}}}}-H$$

Propan-2-ol

Monohydric alcohols are classified as primary, secondary or tertiary depending on whether the hydroxyl group is attached to a primary, secondary or tertiary carbon atom, e.g.

$$R-\overset{\displaystyle H}{\underset{\displaystyle H}{\overset{|}{\underset{|}{C}}}}-OH \qquad R-\overset{\displaystyle R}{\underset{\displaystyle H}{\overset{|}{\underset{|}{C}}}}-OH \qquad R-\overset{\displaystyle R}{\underset{\displaystyle R}{\overset{|}{\underset{|}{C}}}}-OH$$

Primary   Secondary   Tertiary

It will be seen that primary alcohols contain the —CH₂OH group; secondary, the =CHOH group; and tertiary the ≡COH group.

Polyhydric aliphatic alcohols (p. 187) are derived from alkanes by replacing more than one hydrogen atom by hydroxyl groups. They are named as diols, triols etc. e.g.

$$
\begin{array}{ccc}
& H & H \\
& | & | \\
H- & C- & C- & H \\
& | & | \\
& OH & OH
\end{array}
\qquad
\begin{array}{cccc}
& H & H & H \\
& | & | & | \\
H- & C- & C- & C- & H \\
& | & | & | \\
& OH & OH & OH
\end{array}
$$

Ethane-1,2-diol Propane-1,2,3-triol
(Glycol)   (Glycerol)

2 General methods of preparation

a. By replacement reactions. A number of different groups can be replaced by —OH groups as summarised below:

Halogenoalkanes R—Cl $\xrightarrow{OH^-}$ R—OH (p.154)

Alkyl hydrogen
 sulphates(VI) R—HSO$_4$ $\xrightarrow[\text{dil. acid}]{\text{warm with}}$ R—OH (p. 108)

Alkylamines R—NH$_2$ $\xrightarrow[\text{HNO}_2]{\text{heat with}}$ R—OH (p. 296)

Esters R—OOCR $\xrightarrow[\text{NaOH}]{\text{warm with}}$ R—OH (p. 258)

b. Hydration of alkenes. Alkenes can be hydrated either directly by reaction with steam (p. 108) or indirectly via the hydrogen sulphate(VI), e.g.

$$C_2H_4 \xrightarrow{(H_2O)} C_2H_5OH$$

This process is used industrially for ethanol and for some higher alcohols (p. 191).

c. Reduction of carbonyl compounds. Aldehydes and ketones are readily reduced to form primary and secondary alcohols respectively, e.g.

$$
\begin{array}{ccc}
& H \\
& / \\
R-C & \xrightarrow{\text{reduction}} \\
\ \backslash \\
& O
\end{array}
\qquad
\begin{array}{c}
H \\
| \\
R-C-OH \\
| \\
H
\end{array}
\qquad
\begin{array}{c}
R \\
\backslash \\
\ \ \ C=O \xrightarrow{\text{reduction}} \\
/ \\
R
\end{array}
\qquad
\begin{array}{c}
R \\
| \\
R-C-OH \\
| \\
H
\end{array}
$$

Aldehyde   Primary   Ketone   Secondary
      alcohol   alcohol

Sodium amalgam and water, sodium and ethanol, zinc and ethanoic acid, lithiumtetrahydridoaluminate in ether, sodium tetrahydridoborate in water, or catalytic hydrogenation with platinum or nickel catalysts can all be used to effect the reduction.

Carboxylic acids and esters can also be reduced to alcohols, e.g.

$$R-C \overset{\displaystyle O}{\underset{\displaystyle OH}{\Big\langle}} \xrightarrow[\text{ether}]{\text{LiAlH}_4 \text{ in}} R-\overset{\displaystyle H}{\underset{\displaystyle H}{\overset{\displaystyle |}{\underset{\displaystyle |}{C}}}}-OH$$

$$R-C \overset{\displaystyle O}{\underset{\displaystyle O-R'}{\Big\langle}} \xrightarrow[\text{ether}]{\text{LiAlH}_4 \text{ in}} R-CH_2OH + R'-OH$$

but only lithiumtetrahydridoaluminate and diborane are effective as reducing agents.

d. Use of Grignard reagents. See page 340.

3 Manufacture of alcohols

a. Methanol. A little methanol is obtained as a by-product when wood charcoal is made by heating wood in the absence of air. Most methanol is manufactured, however, from mixtures of carbon monoxide and hydrogen,

$$CO + 2H_2 \xrightarrow[\text{3–400 °C.2–300 atm.}]{\text{ZnO + Cr}_2\text{O}_3 \text{ cat.}} CH_3-OH$$

The original gas mixture is obtained either from water gas or, more commonly nowadays, from synthesis gas,

$$\underset{\substack{\text{Coke} \quad \text{Steam} \quad \text{Water gas}}}{C + H_2O \rightarrow CO + H_2}$$

$$\underset{\substack{\text{Natural} \quad \text{Steam} \quad \text{Synthesis gas} \\ \text{gas}}}{CH_4 + H_2O \rightarrow CO + 3H_2}$$

b. Ethanol. Ethanol was at one time made from starchy materials by fermentation processes (p. 182) and some of it was then converted into ethene by dehydrogenation. Since ethene has been readily available as a petroleum product (p. 87), however, it has served as the major source of ethanol. The ethene is hydrated either directly or indirectly (p. 174).

c. Grades of alcohol. Industrial alcohol is a mixture of about 96% of ethanol with water. This is the constant boiling mixture (p. 58) that distils over from ethanol–water mixtures and it cannot be purified further by simple distillation. The 4 % of water can, however, be removed by distilling a mixture of the industrial alcohol with benzene.

Methanol is commonly mixed with ethanol to form *methylated spirits;* because methanol is poisonous, the mixture is undrinkable. Ordinary methylated spirits, available to the public, is known as mineralised methylated spirits; it contains about 90 % ethanol by volume, 9 % methanol and small amounts of naphtha and pyridine. A little methyl violet is also added so that the product is purple and, therefore, readily distinguishable. Other mixtures are used industrially.

4 Properties of alcohols

a. Physical properties. The physical properties of alcohols show the normal gradation associated with any homologous series. They are, however, considerably affected by hydrogen bonding between alcohol molecules. This causes some degree of association in alcohols, as it does in water, so that the boiling points of alcohols are higher than those of the corresponding thiols just as that of water is higher than hydrogen sulphide, e.g.

H_2O	100 °C	CH_3OH	64.5 °C	C_2H_5OH	78.0 °C
H_2S	−60.4 °C	CH_3SH	5.8 °C	C_2H_5SH	37.0 °C
		CH_4	−161.3 °C	C_2H_6	−88.5 °C

The boiling points of alcohols are also considerably higher than those of the corresponding alkanes in which there is no hydrogen bonding.

The lower alcohols are liquids completely miscible with water; the butanols are still liquids but not completely miscible with water; the higher alcohols are waxy solids almost insoluble in water. The solubility of any alcohol in water is much higher than the corresponding alkane because the alcohol is hydroxylic so that it can form hydrogen bonds with water as well as with itself. The solubility of alcohols decreases as the relative molecular mass increases for the hydroxylic nature decreases in relation to the alkane nature. The lower alcohols can be regarded as alkyl-substituted water; the higher ones as hydroxyl-substituted alkanes.

b. Spectra of alcohols. Free —OH groups give a strong infra-red absorption at about $3\,600\ \mathrm{cm}^{-1}$ but if the group is hydrogen bonded, as it commonly is, the O—H bond becomes weaker and the infra-red absorption occurs at a lower frequency. Study of the infra-red spectra can be used to estimate the extent of hydrogen bonding (see Fig. 40, p. 74). Alcohols also give a strong absorption at $1\,051\ \mathrm{cm}^{-1}$ caused by stretching of the C—O bond.

c. Acid-base properties. Alcohols can be regarded as alkyl-substituted water so that it is not surprising that, like water, they are amphoteric,

$$H_2O + H_2O \rightleftharpoons H_3O^+ + OH^-$$

Oxonium   Hydroxide
  ion   ion

$$R—OH + R—OH \rightleftharpoons R—O^+H_2 + OR^-$$

Alkyloxonium Alkoxide
    ion   ion

Alcohols are very weak acids (weaker, even, than water) and they will only react as acids with very strong bases, e.g.

$$C_2H_5OH + Na^+NH_2 \rightarrow C_2H_5O^-Na^+ + NH_3$$
$$C_2H_5OH + Na^+H^- \rightarrow C_2H_5O^-Na^+ + H_2$$

Sodium
ethoxide

or with reactive metals (p. 178). They are not strong enough to react with sodium or potassium hydroxide solutions.

The polarity, $O^{\delta-}—H^{\delta+}$, of the hydroxyl group in water is lowered in alcohols because of the inductive effect of the alkyl groups. The alcohols are, as a result, less ready to lose a proton and, therefore, weaker acids than water. As more alkyl groups are substituted the inductive effect builds up, as it does in carboxylic acids (p. 236), so that the acid strength gets less in passing from primary to secondary to tertiary alcohols. Thus

R (in ROH)	$(CH_3)_3C$	C_2H_5	CH_3	H
pK_a	19	18	16	15.7

As alcohols are weak acids, alkoxide ions are strong bases, and, therefore, highly nucleophilic.

Alcohols are also weak bases, the lone pairs of electrons on the O atom enabling them to accept protons from strong acids, e.g.

$$R—OH + H^+ \rightleftharpoons RO^+H_2$$

Alkyloxonium
ion

The basic strength increases in passing from primary to tertiary alcohols as tertiary carbonium ions are more stable than primary ones. The alklyoxonium ions are strong acids, and, therefore, highly electrophilic.

d. Types of reaction. Primary and secondary, but not tertiary alcohols can be easily oxidised, and most alcohols also undergo elimination (dehydration) and substitution reactions.

The substitution reactions can be of two kinds. The hydrogen atom of the hydroxyl group can be substituted with the O—H bond being split. In these reactions the alcohol acts as an acid and the R—O$^-$ ion is the intermediary, i.e.

$$R—O—H \xrightleftharpoons{-H^+} R—O^- \xrightarrow{X^+} R—O—X$$

As the O—H bond splits most readily in primary alcohols they react most quickly when substitution of the hydrogen atom is concerned.

Alternatively, the whole hydroxyl group may be substituted with the C—O bond splitting. Direct substitution of hydroxyl groups is not easy (p. 156) but it is facilitated in acid solution. The alcohol acts as a base and is protonated to form the R—O$^+$H$_2$ ion, which can undergo subsequent S$_N$1 or S$_N$2 reactions with nucleophiles, e.g.

$$R—O^+H_2 \xrightarrow{-H_2O} R^+ \xrightarrow{X^-} R—X \quad S_N1$$
$$R—O^+H_2 + X^- \longrightarrow R—X + H_2O \quad S_N2$$

The —O$^+$H$_2$ group is replaced more readily than the —OH group (p. 156).

Tertiary alcohols, which form the stablest R$^+$ ions, tend to undergo S$_N$1 reactions; primary alcohols tend to undergo S$_N$2 reactions (p. 155).

5 Reaction of alcohols with reactive metals

The acidic nature of alcohols shows up in their reaction with reactive metals to liberate hydrogen, e.g.

$$2R—OH + 2Na \rightarrow 2RO^-Na^+ + H_2$$

Sodium
alkoxide

The reaction can sometimes be conveniently used to produce hydrogen for reduction purposes (p. 174).

The alkoxides are crystalline salts easily reconverted to alcohols by reaction with water or dilute acids,

$$RO^-Na^+ + H_2O \rightarrow R—OH + NaOH$$

They also react in S$_N$ reactions with halogenoalkanes to form ethers (p. 161), e.g.

$$RO^- + R—I \rightarrow R—O—R + I^-$$

6 Replacement of —OH in alcohols by —X

This important reaction can be brought about by sulphur dichloride oxide (thionyl chloride) (p. 152), phosphorus halides (p. 153) or hydrogen halides (p. 152). With the strongly acidic hydrogen halides the alcohols first act as bases, e.g.

$$R—CH_2—OH + HX \rightleftharpoons R—CH_2—O^+H_2 + X^-$$

$$R—\overset{\displaystyle R}{\underset{\displaystyle R}{\overset{|}{\underset{|}{C}}}}—OH + HX \rightleftharpoons R—\overset{\displaystyle R}{\underset{\displaystyle R}{\overset{|}{\underset{|}{C}}}}—O^+H_2 + X^-$$

The primary alkyloxonium ion then undergoes an S_N2 reaction with X^-,

$$X^- \overset{\frown}{}CH_2 \overset{\frown}{}O^+H_2 \rightarrow X—CH_2 + H_2O$$

whereas the tertiary ion undergoes an S_N1 reaction via a carbonium ion,

$$R—\overset{\displaystyle R}{\underset{\displaystyle R}{\overset{|}{\underset{|}{C}}}}—O^+H_2 \xrightarrow{-H_2O} R—\overset{\displaystyle R}{\underset{\displaystyle R}{\overset{|}{\underset{|}{C}}}}{}^+ \xrightarrow{X^-} R—\overset{\displaystyle R}{\underset{\displaystyle R}{\overset{|}{\underset{|}{C}}}}—X$$

The difference arises because tertiary carbonium ions are more stable than primary ones (p. 47). Secondary alcohols are intermediate in behaviour.

7 Reaction of alcohols with concentrated sulphuric(VI) acid

Alcohols can react with concentrated sulphuric(VI) acid to form alkyl hydrogensulphates(VI), alkenes or ethers.

a. Formation of alkyl hydrogensulphates(VI). These are formed at low temperatures, the sulphuric(VI) acid acting as a strong acid in the same way as hydrogen halides, e.g.

$$R—OH + H_2SO_4 \rightarrow R—O^+H_2 + HSO_4^-$$

$$R—O^+H_2 + HSO_4^- \rightarrow R—OSO_3H + H_2O$$

Alkyl hydrogen
sulphate(VI)

b. Formation of alkenes. Most alkyl hydrogensulphates(VI) undergo elimination of sulphuric(VI) acid on heating, the overall result being the dehydration of the original alcohol to give an alkene, e.g.

$$R-\underset{\underset{H}{|}}{\overset{\overset{H}{|}}{C}}-\underset{\underset{H}{|}}{\overset{\overset{H}{|}}{C}}-OH \xrightarrow[\text{cold}]{c.\ H_2SO_4} R-\underset{\underset{H}{|}}{\overset{\overset{H}{|}}{C}}-\underset{\underset{H}{|}}{\overset{\overset{H}{|}}{C}}-OSO_3H + H_2O$$

$$\downarrow -H_2SO_4\,(170\,^\circ C)$$

$$\underset{H}{\overset{R}{}}C=C\underset{H}{\overset{H}{}}$$

The best yields of alkenes are obtained at 170 °C and by using excess concentrated sulphuric(VI) acid. Similar dehydration of alcohols can also be brought about by passing the alcohol vapour over aluminium oxide at 200 °C (p. 104).

The dehydration is only possible in molecules containing at least one H atom on the carbon atom next but one to the —OH group.

c. Formation of ethers. The elimination reaction which alkyl hydrogen-sulphates(VI) undergo at 170 °C in the presence of excess acid is largely replaced, at 140 °C and in the presence of excess alcohol, by a substitution reaction to yield an ether (p. 208) e.g.

$$R-\underset{\underset{H}{|}}{\overset{\overset{H}{|}}{C}}-OSO_3H + C_2H_5OH \xrightarrow{140\,^\circ C} R-\underset{\underset{H}{|}}{\overset{\overset{H}{|}}{C}}-O-C_2H_5 + H_2SO_4$$

8 Formation of esters

Alcohols react with acids to form esters. The acid concerned may be inorganic or organic but it is the esters formed from organic acids, which are usually sweet-smelling liquids, that are commonest. These esters can be regarded as being formed by replacing the H atom in the —OH group of an organic acid by an alkyl group, and this can be achieved by direct reaction between an alcohol and an acid. The names of the esters are related to the alcohol and the acid, e.g.

$$H-C{\overset{\displaystyle O}{\underset{\displaystyle O-H}{}}} + C_2H_5OH \rightarrow H-C{\overset{\displaystyle O}{\underset{\displaystyle O-C_2H_5}{}}} + H_2O$$

Ethyl methanoate

$$CH_3-C{\overset{\displaystyle O}{\underset{\displaystyle O-H}{}}} + CH_3OH \rightarrow CH_3-C{\overset{\displaystyle O}{\underset{\displaystyle O-CH_3}{}}} + H_2O$$

Methyl ethanoate

The reactions are both slow and reversible. The rates can be speeded up by using small amounts of strong acids, e.g. sulphuric(VI) acid or hydrogen halides, as catalysts. The yields can also be improved by using an excess of one reagent and by distilling off the ester as it is formed. Further details are given on p. 257.

9 Oxidation of alcohols

All alcohols can be oxidised, by combustion, to carbon dioxide and water, e.g.

$$C_2H_5OH(l) + 3O_2(g) \rightarrow 2CO_2(g) + 3H_2O(g)$$

Primary and secondary alcohols can also be oxidised, by dehydrogenation, to aldehydes and ketones respectively. The reactions involve the loss of the hydroxyl hydrogen together with a hydrogen atom from the adjacent alkyl group. With primary alcohols, the initial aldehyde product may be oxidised further to a carboxylic acid, particularly if excess oxidising agent is used, but ketones cannot readily be oxidised:

Primary alcohol   Aldehyde   Carboxylic acid

Secondary alcohol   Ketone

Tertiary alcohols have no hydrogen atom on the alkyl group adjacent to the hydroxyl group and they cannot readily be oxidised.

This difference between the three types of alcohol can be used to differentiate between them.

10 Uses of alcohols

a. Methanol. Methanol is used as a solvent, as a source of many organic chemicals, and in making methylated spirits (p. 176). It can be used as a substitute for petrol in internal combustion engines but it is more expensive. It can be distinguished from ethanol by the iodoform test (p. 166).

b. Ethanol. Ethanol is used as a solvent and in making organic compounds. It is also the major component of beers, wines and spirits.

c. Beers. In brewing beer, barley, or some similar cereal, is mixed with malt (p. 318), and the mixture is ground up and steeped in water at about 50 °C in a mashing-tun. Diastase in the malt converts starch in the cereal into maltose, and the resulting sweet solution is known as 'wort'.

After filtering off the 'wort' from the husky material, it is sterilised by boiling and hops are added to impart a slight bitter taste and to act as preservative. The mixture is then cooled to about 30 °C and yeast is added. This causes fermentation, maltase in the yeast converting the maltose into glucose, and zymase in the yeast converting this glucose into ethanol and carbon dioxide (p. 314).

The fermentation is continued until a solution containing 2–6 per cent of ethanol is obtained. The resulting beer is filtered off from the yeast, some carbon dioxide may be added, and finally the beer is put into barrels or bottles.

The process may be varied in many ways. Bitter beers are made by using more hops than in making mild beers; dark beers by using slightly burnt malt; lager beers by using a yeast which sinks. Moreover the alcoholic content and time of storage of a beer can vary considerably, colouring materials and other additives may be incorporated, and the kind of water used is a very important factor.

d. Wines. Wines are made by the fermentation of grapejuice. The juice contains glucose which is converted into ethanol by the living organisms which exist naturally on the grapes.

There are many varieties of wine. Portugal is famous for port and madeira; Spain for sherry; Germany for hock and moselle; Italy for chianti; France for claret, burgundy, chablis, sauterne and champagne; whilst Australia, South Africa and many other countries produce many wines.

Dry wines contain little or no grape sugar; sweet wines contain grape sugar and may have extra sugar added; sparkling wines, e.g. champagne, are bottled before fermentation is complete; red wines are made by including the skins of 'black' grapes in the manufacturing process.

The natural fermentation process will only produce wines with an alcohol content of about 15 per cent for the fermentation is inhibited by a greater alcohol content. Stronger wines such as port, sherry and madeira have ethanol added; they are said to be fortified.

Cider is made by the fermentation of apple-juice; perry from pear-juice; mead from honey; and saki from rice.

e. Spirits. Spirits are made by the distillation of fermented solutions and contain, on an average, 35–40 per cent of ethanol.

Fermented wort yields whisky and, when flavoured with extracts from juniper berries, gin; fermented grape-juice yields brandy; and fermented molasses, rum and arrack.

f. Proof-spirit. The concentration of ethanol-water mixtures is usually determined by measuring the density, using a hydrometer. Conversion tables give the corresponding percentage alcohol content.

The customs and excise standard is known as proof-spirit. This was originally the weakest spirit capable of firing gunpowder when mixed with it and ignited. Nowadays, proof spirit contains, by definition, 49.28 per cent by weight of alcohol or 57.1 per cent by volume at 60 °F. A spirit 20 under proof contains 80 volumes of proof spirit and 20 volumes of water at 60 °F; a spirit 20 over-proof gives 120 volumes of proof spirit when diluted with water.

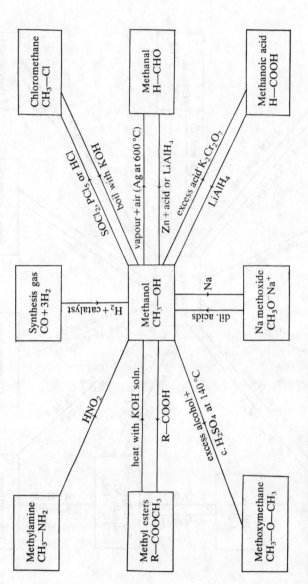

Preparation and properties of methanol

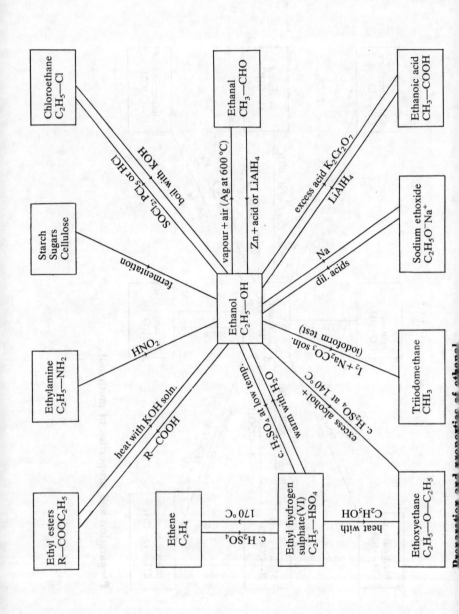

Preparation and properties of ethanol

Questions on Chapter 13

1 Give the names and formulae of a varied selection of chemicals which contain the hydroxyl group.

2 What isomers are there of C_4H_9OH?

3 On what evidence is the formula for ethanol represented by C_2H_5OH? How can this compound be distinguished from methanol?

4 In what ways does methanol (a) resemble, (b) differ from water?

5 What experimental evidence is there to support the view that hydrogen bonding exists in alcohols?

6 How would you attempt to obtain a specimen of absolute alcohol from whisky?

7 How would you expect the melting point of a mixture of methanol and hydrogen bromide to change as the composition changed?

8 How may the following conversions be brought about: (a) ethyne to methane, (b) ethene to ethanol, (c) methanol to ethanoic acid, (d) propan-1-ol to propanone, (e) ethyne to ethanol?

9 The enthalpies of combustion of graphite, hydrogen and ethanol are -393.5, -285.9 and $-1\,366.7\ \mathrm{kJ\,mol^{-1}}$ respectively at $25\,^\circ C$ and 1 atmosphere pressure. Calculate the enthalpy of formation of ethanol.

10 Compare the reactions of ethanol and diethyl ether with (a) halogen acids, (b) acidified sodium dichromate, (c) alkali and alkaline earth metals. From your knowledge of the differences between H_2O and H_2S predict how C_2H_5SH might differ from C_2H_5OH in (a) volatility, (b) in acidity, (c) on oxidation. Give your reasons. (O. Schol.)

11 Name the four alcohols represented by the molecular formula C_4H_9OH, and write their structural formulae. What is the effect of oxidation upon each of these compounds? Outline an experiment

by which, using an acidified dichromate solution as a relatively mild oxidising agent, you could differentiate as far as possible between these four alcohols by recognition of the character of their oxidation products. Outline the procedure by which pure ethanol can be obtained industrially from starch. (JMB)

12 What are the conditions for the interconversion of butan-1-ol and methyl butanoate? Discuss the mechanism of the reaction between butan-1-ol and hydrobromic acid. Under what conditions may butan-1-ol be dehydrated and what is the product? Write structural formulae for the other alcoholic isomers of butan-1-ol. (W)

13 A compound A, containing 59.95 % carbon, 13.42 % hydrogen and 26.63 % oxygen, gave a methyl ether of vapour density 37. On passage over heated alumina, A gave a gaseous compound B. When this was passed into concentrated sulphuric acid, an addition product C was formed which, after hydrolysis, gave rise to a compound D which was found to be isomeric with A and to give a positive iodoform reaction. (a) Explain the above reactions and identity the compounds A, B, C and D. (b) How would you synthesise compound A from ethanol as the starting material? (L)

14*Tabulate (a) the boiling points and (b) the enthalpies of combustion of selected alcohols and comment on any points of significance.

15*Use bond energy values to calculate the enthalpy of formation of ethanol from its free atoms.

16*Compare the theoretical ease of oxidation of methanol and ethanol to methanoic and ethanoic acids respectively.

14

Polyhydric Alcohols

Polyhydric alcohols are formed from alkanes by replacing more than one hydrogen atom by hydroxyl groups. They are known as diols, triols, etc. e.g.

$$\begin{array}{cc} & CH_2\text{---}OH \\ CH_2\text{---}OH & CH\text{---}OH \\ CH_2\text{---}OH & CH_2\text{---}OH \\ \text{Ethane-1,2-diol} & \text{Propane-1,2,3-triol} \\ \text{(Glycol)} & \text{(Glycerol)} \end{array}$$

1 Ethane-1,2-diol (glycol)

a. Preparation and manufacture. Ethane-1,2-diol is made from ethene. In the laboratory a 1 per cent solution of potassium manganate(VII), or hydrogen peroxide with a little osmium tetroxide can be used (p. 109),

$$\begin{array}{l} H \qquad\quad H \\ \diagdown\diagup \\ C\!=\!C \qquad +\quad \underset{\substack{\text{From KMnO}_4 \\ \text{solution}}}{[O]+H_2O} \quad\longrightarrow\quad H\text{---}\underset{\substack{| \\ OH}}{C}\text{---}\underset{\substack{| \\ OH}}{C}\text{---}H \\ \diagup\diagdown \\ H \qquad\quad H \end{array}$$

Industrially the diol is made via expoxyethane (p. 113),

$$CH_2\!=\!CH_2 + \tfrac{1}{2}O_2 \xrightarrow[300\,°C]{\text{Ag cat.}} \underset{\displaystyle O}{H_2C\text{------}CH_2} \xrightarrow[200\,°C]{H_2O} \begin{array}{l} CH_2\text{---}OH \\ CH_2\text{---}OH \end{array}$$

Epoxyethane

b. Properties and uses. Ethane-1,2-diol is a colourless liquid miscible with water in all proportions. It is viscous and has a relatively high boiling point due to hydrogen bonding between the hydroxyl groups.

The chemical properties resemble those of monohydric alcohols though reaction with the second hydroxyl group is more difficult than with the first. Mono- or di-alkoxides are formed with sodium; the —OH groups are replaced by —Cl with phosphorus pentachloride or sulphur dichlorideoxide (thionyl chloride); mono- or di-esters are formed with acids; and there are a number of oxidation products as each of the two —CH$_2$OH groups can be oxidised to —CHO or —COOH.

Ethane-1,2-diol is used as a solvent, as an anti-freeze in the cooling systems of motor-car and aeroplane engines, and in making Terylene (p. 332).

2 Propane-1,2,3-triol (glycerol)

This is obtained as a by-product in the manufacture of soap (p. 190) or by synthesis from propene.

a. Properties. Propane-1,2,3-triol is a hygroscopic, colourless liquid, miscible with water in all proportions. It is more viscous than ethane-1,2-diol as it has more —OH groups for hydrogen bonding.

Chemically the triol exhibits all the expected 'alcohol-like' properties reacting with sodium, phosphorus pentachloride and sulphur dichloride oxide, forming esters with acids (p. 189), and giving many different oxidation products. It can also be dehydrated by heating with phosphorus pentoxide or potassium hydrogensulphate(VI) to form propenal (acrolein),

$$
\begin{array}{l}
\text{CH}_2\text{—OH} \\
\text{CH—OH} \quad - 2\text{H}_2\text{O} \rightarrow \\
\text{CH}_2\text{—OH}
\end{array}
\qquad
\begin{array}{l}
\text{H} \qquad\qquad \text{H} \\
\quad \diagdown \qquad \diagup \\
\qquad \text{C}=\text{C} \\
\quad \diagup \qquad \diagdown \\
\text{H} \qquad\qquad \text{CHO}
\end{array}
$$

On treatment with dry hydrogen chloride at about 100 °C, two of the —OH groups are replaced by —Cl atoms and the resulting product loses hydrogen chloride on treatment with solid potassium hydroxide to give a chlorinated epoxypropane (epichlorohydrin) which is used in making epoxy resins (p. 330),

$$
\begin{array}{l}
\text{CH}_2\text{—OH} \\
\text{CH—OH} \\
\text{CH}_2\text{—OH}
\end{array}
\xrightarrow[100\,°\text{C}]{\text{dry HCl at}}
\begin{array}{l}
\text{CH}_2\text{—Cl} \\
\text{CH—OH} \\
\text{CH}_2\text{—Cl}
\end{array}
\xrightarrow[\text{KOH}]{\text{solid}}
\begin{array}{l}
\text{CH}_2\text{—CH—CH}_2\text{—Cl} \\
\qquad \diagdown\diagup \\
\qquad \text{O}
\end{array}
$$

b. Uses. Propane-1,2,3-triol is used in making medical and pharmaceutical products, cosmetics, leather polishes, glyptals (p. 329) and nitroglycerine explosives.

Nitroglycerine is the common name of the ester formed between propane-1,2,3-triol and nitric(V) acid; it is really a nitrate(V), not a nitro-compound, with a systematic name of propane-1,2,3-triyl trinitrate(V).

CH_2—O—NO_2
|
CH—O—NO_2
|
CH_2—O—NO_2

Propane-1,2,3-triyl
trinitrate(V)
(nitroglycerine)

It is made by adding propane-1,2,3-triol, to a mixture of concentrated sulphuric(VI) and nitric(V) acids at a temperature kept below 25 °C. The nitrate(V) ester is a heavy, colourless oil which is too sensitive to be used by itself as an explosive. Nobel (1867) found, however, that kieselguhr, a naturally occurring siliceous material, would absorb the liquid to form a useful explosive known as dynamite. The nitrate(V) ester can also be mixed with other materials to make widely used explosives. Typical examples are gelignite (a mixture of nitroglycerine, nitrocellulose and potassium nitrate(V)), cordite (a stabilised mixture of nitroglycerine and gun cotton), and blasting gelatine (nitroglycerine made up into a stiff jelly).

3 Soap

Soap is made by the hydrolysis of naturally occurring fats and oils (p. 192) by sodium or potassium hydroxide solutions (*saponification*). The fats and oils used are tri-esters of propane-1,2,3-triol with long-chain organic acids, e.g.

$C_{15}H_{31}$—COOH $C_{17}H_{35}$—COOH $C_{17}H_{33}$—COOH

Hexadecanoic acid   Octadecanoic acid   Octadec-9-enoic acid
(Palmitic acid)   (Stearic acid)   (Oleic acid)

CH_2—O—CO—$C_{15}H_{31}$ CH_2—O—CO—$C_{17}H_{35}$ CH_2—O—CO—$C_{17}H_{33}$
| | |
CH—O—CO—$C_{15}H_{31}$ CH—O—CO—$C_{17}H_{35}$ CH—O—CO—$C_{17}H_{33}$
| | |
CH_2—O—CO—$C_{15}H_{31}$ CH_2—O—CO—$C_{17}H_{35}$ CH_2—O—CO—$C_{17}H_{33}$

Tripalmitin   Tristearin   Triolein
(Occurs in palm   (Occurs in beef or   (Occurs in
oil)   mutton fat)   olive oil)

These, and other similar fats and oils can be easily hydrolysed. With steam under pressure or with dilute acids the products are propane-1,2,3-triol and free acids, and the free acids obtained in this way are used in making candles (p. 193).

With sodium or potassium hydroxide solutions propane-1,2,3-triol is produced together with the sodium or potassium salts of the acids, e.g.

$$
\begin{array}{l}
CH_2-O-CO-C_{17}H_{35} \\
| \\
CH-O-CO-C_{17}H_{35} \ + 3NaOH \rightarrow \\
| \\
CH_2-O-CO-C_{17}H_{35}
\end{array}
\qquad
\begin{array}{l}
CH_2-OH \\
| \\
CH-OH \ + 3C_{17}H_{35}-COO^-Na^+ \\
| \\
CH_2-OH
\end{array}
$$

The salts are used as soaps, the sodium salts being soft soaps and the potassium salts hard soaps.

In making a hard soap, a mixture of fats and oils is steam-heated with an aqueous solution of sodium hydroxide in a big vat. The mixture of propane-1,2,3-triol and sodium salts are separated (a process known as salting out) by adding sodium chloride. This causes the soap to separate as an insoluble mass on the surface whilst the propane-1,2,3-triol remains as a lower layer mixed with water and sodium chloride. The aqueous solution of propane-1,2,3-triol is drawn off from the vat, filtered and evaporated in vacuo. The sodium chloride present crystallises out and is filtered off and used again; the hot soap is run into boxes in which it sets solid. It can then be cut up into slabs and bars and used as household soap. In making toilet soap, special blends of fats and oils are chosen and the soap is cut up into thin shreds. These are dried, mixed with perfumes and dyes and forced into moulds.

In an alternative, continuous process a mixture of fats and oils, mixed with a catalyst, is passed up a tower down which hot water is passing. This hydrolyses the fats and oils, the resulting carboxylic acids passing to the top of the tower whilst the glycerol falls to the bottom. The acids from the top of the tower are then treated with sodium hydroxide.

4 Synthetic detergents

Soap is the traditional example of a detergent (cleansing agent). It depends for its action on the fact that its molecules have one ionic (polar) end and one covalent (non-polar) end. In brief, the polar (hydrophilic) end 'attracts' water whilst the non-polar (hydrophobic) end 'attracts' oils and greases so that soap molecules can make water and oils or greases come into an emulsion which can be washed away.

Soap is not, however, entirely satisfactory, particularly when used in hard or acidic waters and in recent years many synthetic detergents have been developed. They are usually more soluble in water than soap is; they do not form a scum in hard water as their calcium and magnesium salts are soluble; they enable water to spread and penetrate more fully over or through an article being cleaned; and they can generally be used equally well under alkaline or acidic conditions.

Like soap, these modern detergents have molecules with one polar and one non-polar end. The polar end generally consists of an $-SO_3H$

(sulphonic acid) or —OSO_3H (hydrogen sulphate(VI)) group, present in the form of its sodium salt to increase solubility. The non-polar end of the molecule is a long alkyl or alkyl-substituted benzene chain. It is important that this long chain should not be extensively branched. Experience has shown that detergents with branched chains are not biochemically degraded so that they cause pollution in rivers and sewage works.

Detergents are used either as liquids or solids and they may be mixed with inorganic phosphates (to soften hard water), sodium sulphate(VI) and sodium silicate (to increase the bulk and the handling properties), oxidising agents, e.g. sodium perborate (to act as a bleach), fluorescent materials (to act as whiteners) and other additives. Enzyme detergents contain added enzymes which help to remove biological stains by chemical action on the proteins concerned.

Typical examples of materials used as detergents are given below.

a. Alkyl sulphates(VI). These have a general formula

$$\begin{matrix} & H & & & CH_3 \\ & | & & & | \\ R& -C-O-SO_3^-Na^+ & & R& -C-O-SO_3^-Na^+ \\ & | & & & | \\ & H & & & H \end{matrix}$$

the alkyl group being a primary or secondary group with a carbon chain length between 10 and 14. They can be made from long chain alkenes which are first converted into alcohols by the Oxo process (p. 113). The alcohol is then converted into a hydrogensulphate(VI) ester by reaction with sulphuric(VI) acid and this is then neutralised with sodium hydroxide, e.g.

$$\begin{matrix} R' \\ \diagdown \\ C=CH_2 \xrightarrow[\text{(p.113)}]{\text{Oxo process}} R-CH_2OH \xrightarrow{H_2SO_4} R-CH_2-O-SO_3H \\ \diagup \\ H \end{matrix}$$

$$\downarrow NaOH$$

$$R-CH_2-O-SO_3^-Na^+$$

Typical examples are made from dodecan-1-ol (lauryl alcohol) and secondary dodecyl alcohols,

$$\begin{matrix} & & & CH_3 \\ & & & | \\ C_{12}H_{25}-O-SO_3^-Na^+ & & C_{10}H_{21}& -C-O-SO_3^-Na^+ \\ & & & | \\ & & & H \end{matrix}$$

They are usually liquids which are used in solution, e.g. Teepol.

b. Alkylbenzene sulphonates. These have the general formula

$$R-\bigcirc-SO_3^-Na^+$$

with an alkyl chain length of 9–15 carbon atoms. They are commonly made by a Friedel–Crafts type reaction (p. 131) between alkenes and benzene followed by sulphonation and neutralisation, e.g.

$$R-CH=CH_2 + \bigcirc \longrightarrow R-\underset{\displaystyle CH_3}{\overset{}{C}}H-\bigcirc$$

$$\big\downarrow H_2SO_4$$

$$R-\underset{CH_3}{CH}-\bigcirc-SO_3^-Na^+ \xleftarrow{NaOH} R-\underset{CH_3}{CH}-\bigcirc-SO_3H$$

They are usually solids, used, for example, in Omo, Surf and Tide.

c. Non-ionic detergents. The detergents described in (a) and (b) are known as *anionic detergents* for it is, essentially, the anion that is functioning as the detergent. *Cationic detergents*, e.g. quaternary ammonium salts $R-N^+(CH_3)_3X^-$ are known (p. 295) but not widely used.

Non-ionic detergents have general formulae

$$R-O-(CH_2-CH_2-O)_n-H \qquad R-\bigcirc-O-(CH_2-CH_2-O)_n-H$$

and are made from alkylated phenols and epoxyethane. The oxygen atoms in the $(CH_2-CH_2-O)_n$ groups, which can form hydrogen bonds with water, make them hydrophilic but the molecules are not ionised. Non-ionic detergents remain active in the presence of electrolytes e.g. acids and alkalis. They do not foam so that they are particularly useful in washing machines.

Fats and Oils

5 Introduction

The term oil is used in three different senses in organic chemistry. There are *mineral oils*, i.e. petroleum products such as paraffin oil and diesel oil; *essential oils*, i.e. volatile oils derived from plants such as oil of wintergreen (p. 245) and oil of cloves; and *fixed oils*, i.e. plant or animal products made up of esters of propane-1,2,3-triol.

The fats and oils considered here are fixed oils, a fat differing from an oil in being solid at room temperature. Animal fats and oils are usually obtained by treating the animal tissues with hot water. This breaks down the cell walls and allows the molten fat or oil to rise to the surface. Tallow, the well-known 'dripping', is obtained in this way from beef or mutton; pure lard from pigs. Vegetable oils are usually obtained by pressing the plant to squeeze out the oil.

Naturally occurring fats and oils do not generally consist simply of one tri-ester. Olive oil, for example, contains about 25 per cent tristearin, 70 per cent triolein and 5 per cent trilinolein (a tri-ester of octadeca-dienoic or linoleic acid, $C_{17}H_{31}COOH$). Mixed esters with three different alkyl groups in the one tri-ester also occur.

6 Uses of fats and oils

a. As foodstuffs. Fats are primarily important as foodstuffs for, together with carbohydrates, they provide a source of energy for animals. The fats are hydrolysed in the animal's body yielding carbon dioxide and water with the liberation of energy. Unlike carbohydrates, fats can be stored in the body acting as a reserve supply of energy. As they are poor conductors of heat they also serve a useful purpose as insulators.

b. In making soap. The fats and oils most commonly used for this purpose are tallow, coconut oil, bleached palm oil, soya bean oil and olive oil. Fats made by hydrogenating whale oil and rosin (p. 331) are also used.

c. In making candles. When tallow is hydrolysed by steam under pressure it gives a mixture of stearic and palmitic acids known as stearine (with an 'e'); this is used in making candles.

d. In making glycerol (p. 190).

e. In making paints, oilcloth and linoleum. Linseed oil from the flax plant, or tung oil (China wood-oil), are mainly used in making these materials. The oils, particularly after boiling, slowly absorb oxygen from the air and form a hard transparent solid.

Oil paints are merely a suspension of finely divided pigments in boiled linseed oil; oilcloth is made by covering cloth with coloured linseed oil and letting it dry; linoleum is made by mixing linseed oil and pulverised cork and pressing into a fabric background.

7 Hydrogenation of oils

Naturally occurring oils are usually esters of unsaturated acids, whereas the fats are esters of saturated acids. The oils can be converted into the more useful fats by reduction with hydrogen. Liquid triolein, for example, gives solid tristearin.

The reduction is carried out at 200 °C in the presence of finely divided nickel. The process is known as *catalytic hydrogenation* and oil treated in this way is said to be *hardened*. By using different oils and by controlling the hardening process it is possible to make different kinds of fat.

Margarine is made by mixing fats and oils and hydrogenated oils with milk. Vitamins and artificial colouring are added to make the margarine as good a substitute for butter as possible.

Questions on Chapter 14

1 What compounds would you expect to be able to obtain by the oxidation of (a) ethane-1,2-diol, (b) propane-1,2,3-triol?

2 What is soap, how is it made, and how does it act as a cleansing agent? Why has soap been replaced, to some extent, by other detergents?

3 The iodine number of a fat is measured in terms of the number of grammes of iodine which will react with the unsaturated acid present in 100 g of the fat. Calculate the iodine number of olein.

4 The saponification value of a fat is the number of milligrammes of potassium hydroxide required to hydrolise completely 1 g of the fat. Calculate the saponification value of tristearin.

5 Two oxidation products of propane-1,2,3-triol contain 40 % C and 6.7 % H, and have a relative molecular mass of 90. Write their structures and indicate how they might be distinguished experimentally.

6 What is the difference between hard and soft soaps? Why does soap cause a lather and why is this lathering diminished in hard water?

7 A neutral oil, A, $C_{15}H_{26}O_6$, is insoluble in water, but on boiling with sodium hydroxide it forms B ($C_3H_8O_3$) and C ($C_4H_7O_2Na$). B, with phosphorus pentachloride, forms a compound D, containing 24.4 % carbon, 3.4 % hydrogen, and 72.2 % chlorine by weight. Assign a structure to A and explain the reactions. Give alternative structures for A and say how you would distinguish between them. Outline a synthesis of B from propene (propylene). (SUJB)

15

Aromatic Alcohols and Phenols

1 Introduction

Aromatic alcohols contain an —OH group in a side-chain attached to a benzene ring. They may be regarded as aryl-substituted aliphatic alcohols and they are named on this basis, e.g.

Phenylmethanol   2-Phenylethanol

The hydroxyl group functions in very much the same way as in aliphatic alcohols, but the aromatic alcohols contain a benzene ring so that they exhibit many 'aromatic' properties, e.g. they can be nitrated, sulphonated or halogenated.

When the —OH group is attached directly to a benzene ring the compounds are known as phenols. They may be monohydric, Ar—OH, e.g.

Phenol   2-methylphenol   3-methylphenol   4-methylphenol
          (o-cresol)   (m-cresol)   (p-cresol)

or di- or tri-hydric, e.g.

Benzene-1,2-diol Benzene-1,3-diol Benzene-1,4-diol Benzene-1,2,3-triol
(Catechol)   (Resorcinol)   (Quinol)   (Pyrogallol)

The phenols have significantly different properties from the aliphatic or aromatic alcohols (p. 205).

2 Preparation of phenol

Phenol and methylphenols can be obtained from coal-tar but synthetic processes are used more commonly.

a. From benzenesulphonic acid. Benzene sulphonic acid forms sodium benzenesulphonate if neutralised with sodium hydroxide solution and phenol is obtained if the salt is heated with solid sodium hydroxide in nickel or iron pots at 300 °C,

$$C_6H_5—SO_3H + NaOH \rightarrow C_6H_5—SO_3^-Na^+ + H_2O$$
$$C_6H_5—SO_3^-Na^+(s) + 2NaOH(s) \rightarrow C_6H_5—O^-Na^+ + Na_2SO_3 + H_2O$$

The sodium phenoxide may be extracted from the mixture with water and on acidifying with sulphuric(VI) acid free phenol is formed which can be purified by distillation.

A modified version of this process was one of the original industrial methods of obtaining phenol.

b. From chlorobenzene. Chlorobenzene can be converted into phenol either by treating with steam at 425 °C in the presence of a catalyst (Raschig process) or by heating with sodium hydroxide solution at 360 °C under pressure and subsequently treating the sodium phenoxide with a dilute acid (Dow process),

$$C_6H_5—Cl + H_2O(g) \rightarrow C_6H_5—OH + HCl$$
$$C_6H_5—Cl + NaOH(aq) \rightarrow C_6H_5O^-Na^+ + HCl$$

Both these processes have been used industrially but are now being superseded.

c. From (1-methylethyl) benzene (cumene). This is the most important, modern process for making phenol. Benzene and prop-1-ene react at

250 °C and 25 atm pressure in the presence of phosphoric(V) acid as a catalyst to form (1-methylethyl) benzene,

(1-methylethyl)benzene

The product reacts with air at 100 °C to form a hydroperoxide which reacts with dilute acid to give a mixture of phenol and propanone (a valuable by-product),

The breakdown of the hydroperoxide involves a rearrangement (p. 49) with the phenyl group migrating from a C to an O atom.

3 General properties of phenol

a. Physical properties. Phenol is a colourless, hygroscopic crystalline solid (m.p. = 42 °C) which turns pink on exposure to air and light. It is corrosive, poisonous, and has a characteristic 'antiseptic' smell. It is completely miscible with water above 65.8 °C but only partially miscible at lower temperatures.

Phenol has a higher melting point, boiling point and solubility than the corresponding alcohol (cyclohexanol, p. 203). This is due to an increased polarity, $O^{\delta-}$—$H^{\delta+}$, in the hydroxyl bond in phenol which leads to stronger hydrogen bonding. The increased polarity arises because the unshared pair of *p*-electrons on the oxygen atom can overlap with the π-orbitals of the benzene ring to give some measure of delocalisation which does not exist in cyclohexanol, water or aliphatic alcohols. The delocalisation causes some electron shift into the ring and a consequent build up of positive charge on the hydrogen atom of the hydroxyl group. It also causes a shortening of the C—O bond distance; 0.136 nm in phenol as compared with 0.143 nm in alcohols.

b. Spectra of phenols. Phenols give a strong infrared absorption due to stretching of the O—H bond: the frequency of absorption is about 3 600 cm^{-1}. Absorption also occurs at lower frequencies when the OH group is hydrogen bonded either inter- or intra-molecularly (p. 201).

c. Acid–base properties. Phenol is a stronger acid than water, whereas aliphatic alcohols are weaker (p. 177), and the acidity of phenol is recognised in the old name of carbolic acid.

The polarity in the hydroxyl bond facilitates the loss of a proton and the formation of the phenoxide ion, and the delocalisation in the

Phenoxide ion

phenoxide ion also stabilises it as compared with RO$^-$ or OH$^-$ ions.

If electron-withdrawing groups ($-$I) are substituted into the benzene ring the polarity of the O—H bond is increased still further giving still stronger acids, e.g.

Ar (in Ar—OH)	CH$_3$	H	C$_6$H$_5$	NO$_2$C$_6$H$_4$	(NO$_2$)$_3$C$_6$H$_2$
pK$_a$	16	15.7	10	7.2	1

d. Types of reaction. The reactions of phenol are significantly different from those of aliphatic alcohols mainly because of the increased acidity and the existence of the benzene ring.

Where splitting of the O—H bond, and substitution of the hydrogen atom, is concerned phenols resemble aliphatic alcohols and react more readily. The splitting of the C—O bond, and substitution of the hydroxyl group, is, however, very difficult so that phenol does not react with hydrogen halides, phosphorus halides or sulphur dichloride oxide (thionyl chloride). The difficulty in substituting the —OH group is due to stabilisation caused by overlap of the *p*-orbital of the oxygen atom with the π-bonding in the ring. The situation is similar to that in chlorobenzene (pp. 141 and 164).

Phenol also differs from most aliphatic alcohols in that it will not undergo elimination (dehydration) reactions, gives complex oxidation products and can be reduced, and exhibits many 'aromatic' properties. It also participates in a number of important, unique reactions.

4 Formation of phenoxides

Phenol reacts as an acid with bases and reactive metals to form phenoxides, e.g.

$$C_6H_5OH + NaOH \rightarrow C_6H_5 O^-Na^+ + H_2O$$
$$2C_6H_5OH + 2Na \rightarrow 2C_6H_5 O^-Na^+ + H_2$$

Sodium phenoxide

It is not, however, a strong enough acid to react with sodium carbonate, i.e. it is a weaker acid than carbonic acid. This enables phenol to be distinguished from the stronger, carboxylic acids, and it also means that carbonic acid, as well as stronger acids, will convert a phenoxide into phenol,

$$2C_6H_5O^-Na^+ + CO_2 \rightarrow 2C_6H_5OH + Na_2CO_3$$

Phenoxides are crystalline salts. The phenoxide ion is highly nucleophilic, forming, for example, ethers in an S_N2 reaction with halogenoalkanes (p. 208), e.g.

$$C_6H_5O^- + CH_3I \rightarrow C_6H_5{-}O{-}CH_3 + I^-$$

Methoxybenzene
Anisole

5 Formation of esters

Phenol, unlike alcohols, will not react directly with acids to form esters but it will react with acid chlorides or acid anhydrides, e.g.

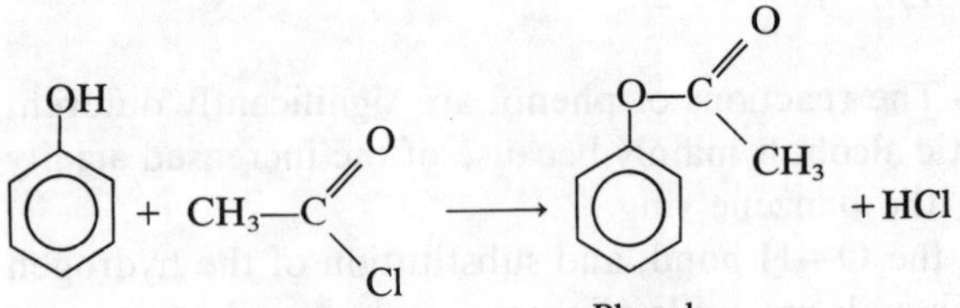

Phenyl
ethanoate

The reaction is particularly effective in sodium hydroxide solution (the *Schotten–Baumann* reaction) for the alkali first forms the phenoxide ion which then reacts with the acid chloride in an S_N2 reaction, e.g.

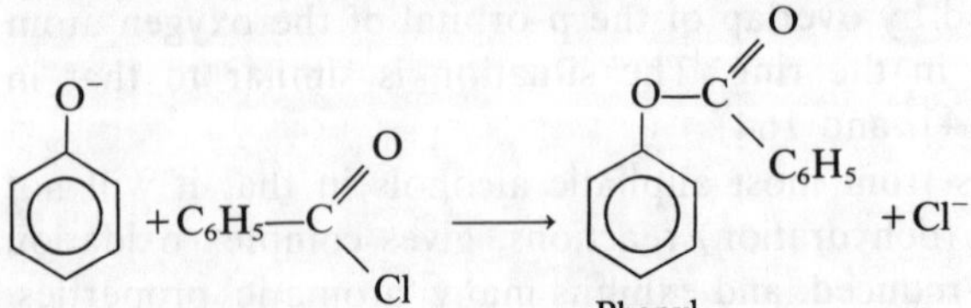

Phenyl
benozoate

6 Ring substitution of phenol

The —OH group in phenol is strongly activating, and the —O⁻ group in the phenoxide ion even more so (p. 144). Both are *o-p* directing so that ring substitution by electrophilic reagents is easy.

a. Halogenation. Both chlorine and bromine in aqueous solution halogenate phenol so readily that the 2,4,6-trihalogenated phenols are formed almost quantitatively at room temperature. They are white, insoluble solids, e.g.

2,4,6-tribromophenol

Chlorination of molten phenol can be controlled to give either the mono-, di- or tri-substituted product.

b. Nitration. Dilute nitric(V) acid reacts with phenol at room temperature to give a mixture of 2-nitro- and 4-nitrophenol,

2-nitrophenol

4-nitrophenol

The concentration of nitronium ions, NO_2^+, is very low in dilute nitric(V) acid and it is probable that the nitration involves the formation of nitrosophenols, by reaction of the nitrosyl cation, NO^+, which are then oxidised by the nitric(V) acid.

2-nitro- and 4-nitro-phenol differ in physical properties because the —OH and —NO₂ groups are suitably placed in the 2-nitrophenol to form an intramolecular hydrogen bond but can only form intermolecular bonds in 4-nitrophenol. The hydrogen bonding in 2-nitrophenol lowers the hydroxylic character and the water solubility; the intermolecular hydrogen bonding in 4-nitrophenol causes stronger intermolecular attraction and gives a higher boiling point. 2-nitrophenol is, therefore, the less soluble and more volatile of the two isomers. It can be separated from the 4-nitrophenol by steam distillation.

Concentrated nitric(V) acid nitrates phenol fully, giving 2,4,6-trinitrophenol (picric acid). This is a yellow solid, explosive when dry. The alternative name shows that it has very significant acidic properties (p. 199).

c. Other substitution reactions. —OSO_2H, —COOH and —CHO groups can also be substituted into the ring in phenol. Mixtures of isomers result, typical examples being summarised as follows:

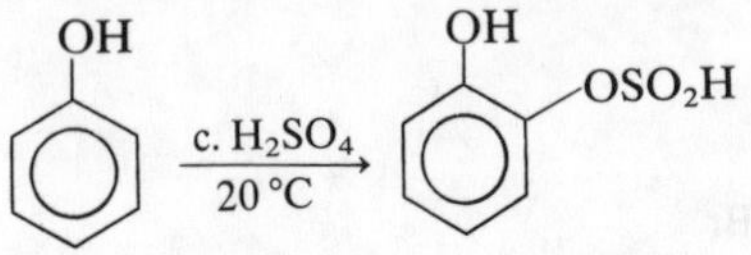

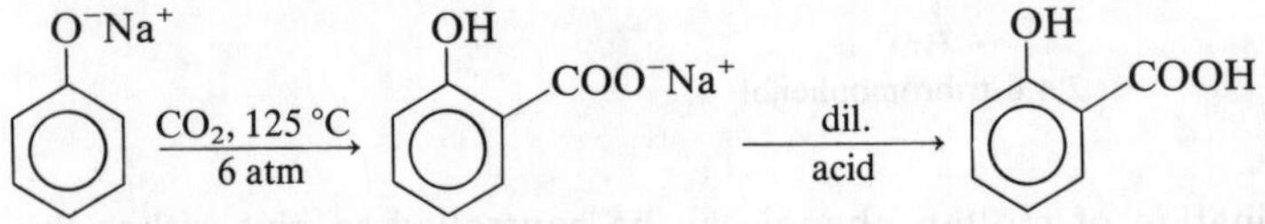

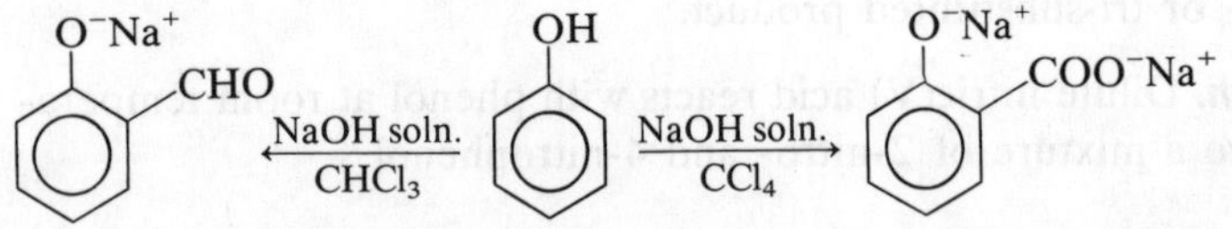

The reaction with carbon dioxide is known as the *Kolbé reaction;* that with sodium hydroxide solution and trichloro- or tetrachloromethane as the *Reimer–Tiemann reaction.*

Phenol can also be alkylated. The yield in the Friedel–Crafts reaction is not good and better results are obtained using alcohols or alkenes with sulphuric(VI) acid, e.g.

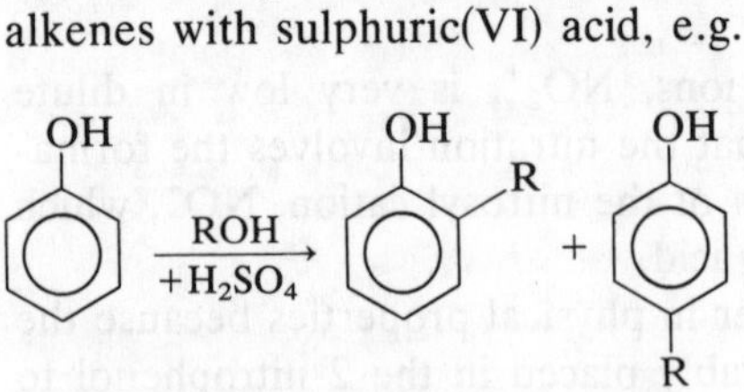

7 Other reactions of phenol

a. Oxidation. Phenol can be oxidised to a number of complex products.

b. Reduction. Passing phenol vapour over hot zinc dust gives some benzene, but a more important reduction occurs when the vapour is mixed with hydrogen and passed over a nickel catalyst at 160 °C; the product is cyclohexanol, an alicyclic alcohol,

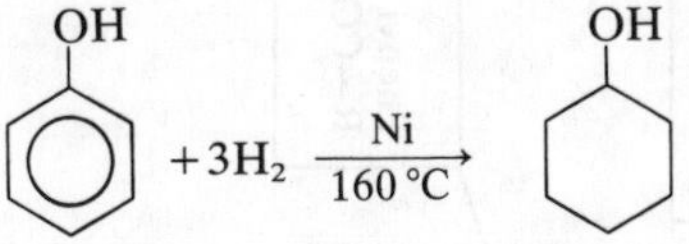

Cyclohexanol

c. Reaction with neutral iron(III) chloride solution. One or two drops of neutral iron(III) chloride solution added to a solution of phenol give a characteristic violet coloration due to the formation of a complex. Similar colorations are given by many other compounds containing the $=C=C-OH$ group and the reaction can be used for identification purposes.

d. Reaction with benzene-1,2-dicarboxylic anhydride (phthalic acid). Two molecules of phenol react with one of phthalic anhydride, with the elimination of water, to give phenolphthalein.

8 Uses of phenol

Phenol is an important industrial chemical. It is used in making Bakelite (p. 328), Nylon (p. 331), epoxy resins (p. 330), aspirin (p. 245) and dyes (p. 347).

Phenol is also well known as a powerful antiseptic. It was first used as such by Lister around 1865 and is a component of carbolic soaps and coal-tar disinfectants such as Jeyes fluid. Substituted phenols are also components of well known antiseptics, e.g.

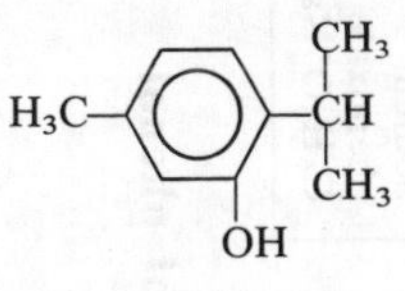

| Thymol | TCP (Trichlorophenol) | Dettol |

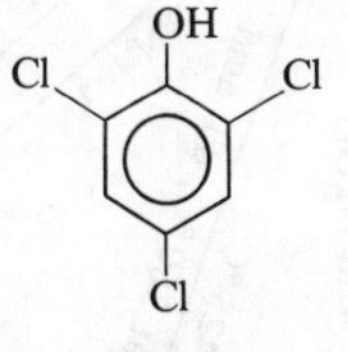

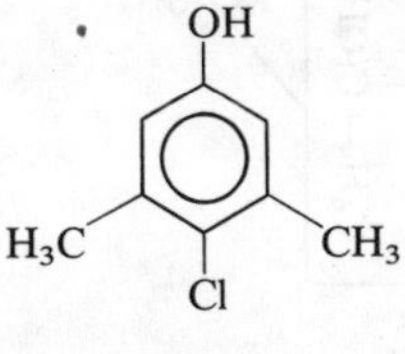

2,4-dichlorophenol reacts with chloroethanoic acid (p. 240) to form 2,4-D which is widely used in selective weed killers.

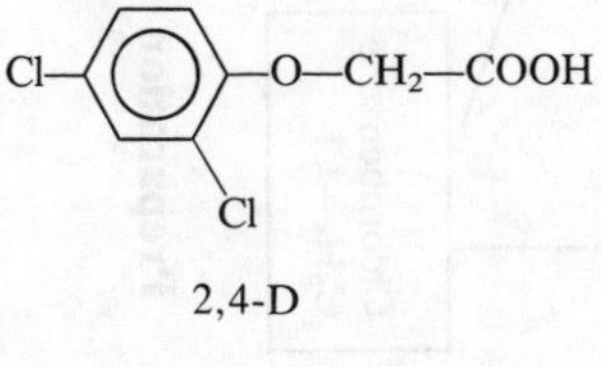

2,4-D

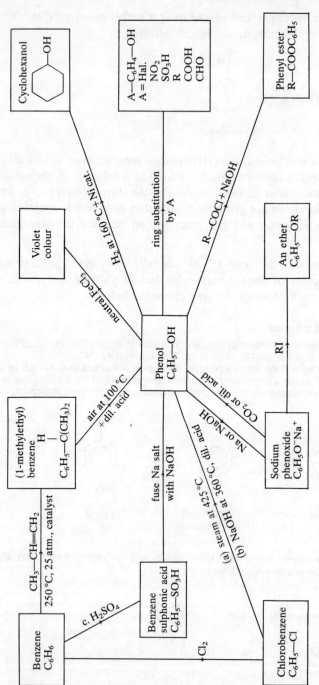

Preparation and properties of phenol

	C_2H_5-OH	⬡-OH	⬡-CH₂-OH
Methods of preparation	By replacement reactions from C_2H_5-Cl, $C_2H_5-HSO_4$, $C_2H_5-NH_2$, or esters	By replacement reactions from C_6H_5-Cl and $C_6H_5-SO_3H$	By replacement reactions as for ethanol
	Hydration of C_2H_4		$C_6H_5-CHO+NaOH$ (Cannizzaro's reaction)
	Reduction of ethanal or ethanoic acid		Reduction of benzaldehyde or benzoic acid
	Fermentation	From cumene	
Delocalisation of electrons	No delocalisation	Delocalisation between ring and O	Delocalisation within the ring
Acid strength	Very, very weak	A weak acid	Very, very weak
Replacement of —OH by —Cl	With HCl, SOCl₂ or PCl₅	Very difficult	With HCl, SOCl₂ or PCl₅
Formation of esters	With acids	With acid chlorides +NaOH	With acids
Oxidation	To aldehyde or acid	Complex	To aldehyde or acid
Reduction	Not reduced	To cyclohexanol	To substituted cyclohexanol
With FeCl₃ solution	No colour	Violet coloration	No colour
Ring substitution	Nil	Very easy	Very easy

Preparation and properties of ethanol, phenol and phenylmethanol

Questions on Chapter 15

1 How can phenol be made from benzene? Compare and contrast the properties of phenol and ethanol.

2 Summarise the methods used for introducing the —OH group into organic molecules.

3 How would you distinguish by chemical tests between phenol, 2-methylphenol and phenylmethanol?

4 Discuss the classification of alcohols and summarise the main differences between the various classes.

5 How and under what conditions will 2-methylphenol react (if at all) with (a) bromine water, (b) a mixture of nitric(V) and sulphuric(VI) acids, (c) a strong solution of sodium hydroxide in water, (d) an aqueous solution of sodium carbonate?

6 How may phenol be converted into (a) phenylethanoate, (b) benzene, (c) aspirin, (d) picric acid, (e) 2-hydroxybenzaldehyde (salicylaldehyde)?

7 Compare the actions of sodium, sodium hydroxide and chlorine on (a) water, (b) methanol, and (c) phenol.

8 Describe briefly one method by which benzene can be converted into phenol. How would you separate a mixture of benzoic acid and phenol? Bromine is added slowly to phenol (0.282 g) dissolved in water until a slight excess of bromine is present. Calculate the weight of bromine used. (O and C)

9 A by-product of coal gas manufacture contains phenol mixed with hydrocarbons and organic bases. What property of phenol could be used to isolate it from this mixture? Illustrate differences between phenol and ethanol by reference to their reactions with (a) chlorine, (b) ethanoic acid, (c) sulphuric(VI) acid. How can phenol be reduced and what products are obtained from its reduction? (O and C)

10 Describe briefly the various methods by which benzene may be converted into phenol. When phenyl benzoate is nitrated a compound A, $C_{13}H_9NO_4$, can be obtained. Explain how you could determine the structure of A. (O.Schol.)

11 Discuss (a) the boiling points, (b) the acidities, in each of the following groups of compounds: (i) water and hydrogen sulphide, (ii) water, ethanol and phenol. Give reasoned predictions about boiling points and acidities in the series hydrogen sulphide, ethanethiol (C_2H_5SH), benzenethiol (C_6H_5SH). (W)

16

Ethers

1 Introduction

Straight-chain ethers have a general formula R—O—R′ where R and R′ may be alkyl or aryl groups and may be alike or different. If R and R′ are alike the ether is known as a *simple* ether; if different, as a *mixed* ether.

The systematic names are based on the substitution of an —OR group into a hydrocarbon. Older names indicate the nature of the two groups linked to the oxygen atom or are completely unsystematic, e.g.

C_2H_5—O—C_2H_5

Ethoxyethane
(Diethylether)

CH_3—O—C_2H_5

Methoxyethane
(Ethylmethylether)

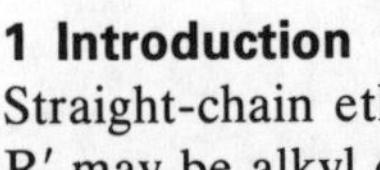

⟨benzene ring⟩—O—⟨benzene ring⟩

Phenoxybenzene
(Diphenylether)

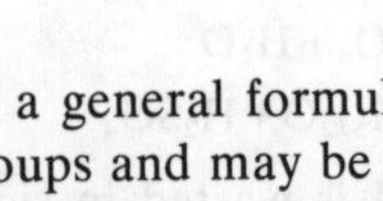

⟨benzene ring⟩—O—CH_3

Methoxybenzene
(Methylphenylether)
(Anisole)

These straight-chain ethers are isomeric with the corresponding alcohol or phenol.

Cyclic ethers (p. 210) contain the —O— bonding within a ring, e.g.

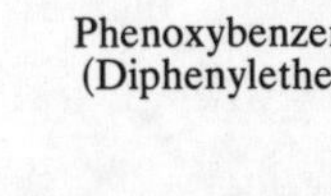

H_2C———CH_2
  \\ /
   O

Epoxyethane
(ethylene oxide)

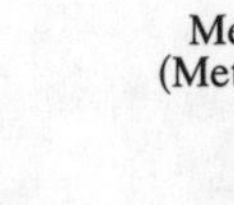

H_2C—CH_2
 | |
 O—CH_2

Trimethylene
oxide

H_2C———CH_2
 | |
H_2C CH_2
  \\   /
     O

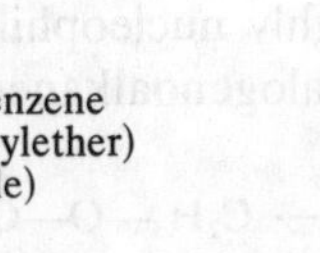

Tetrahydrofuran

2 Preparation of ethers

a. From alcohols. Alcohols can be dehydrated by reaction with sulphuric(VI) acid in two ways, e.g.

$$C_2H_5\text{—OH} \xrightarrow{-H_2O} C_2H_4$$
$$2C_2H_5\text{—OH} \xrightarrow{-H_2O} (C_2H_5)_2O$$

The first reaction, yielding an alkene, predominates at 170 °C using excess acid (p. 180); the second reaction is favoured at 140 °C using excess alcohol (p. 180).

Theoretically, no sulphuric(VI) acid is used up in the dehydration process, e.g.

$$C_2H_5OH + H_2SO_4 \rightarrow C_2H_5HSO_4 + H_2O$$
$$C_2H_5HSO_4 + C_2H_5OH \rightarrow (C_2H_5)_2O + H_2SO_4$$

but, in practice, some acid is wasted in side-reactions and has to be replaced.

In the laboratory, the alcohol is added to concentrated sulphuric(VI) acid at 140 °C and the ether is distilled off at the same rate. Industrially, alcohol vapour is passed into a hot mixture of alcohol and acid. The process is known as *continuous etherification;* it is best suited to the preparation of simple ethers.

Some ethoxyethane is obtained industrially as a by-product in the manufacture of ethanol from ethene, some of the ethanol being dehydrated to ethoxyethane in the normal course of the reaction.

b. From alkoxides or phenoxides. Williamson's synthesis. This method can be used for making both simple and mixed ethers. Alkoxides and phenoxides are highly nucleophilic and they will displace the halogen atom in primary halogenoalkanes, e.g.

$$C_2H_5\text{—O}^- \quad CH_3\text{—I} \rightarrow C_2H_5\text{—O—}CH_3 + I^-$$

Alkyl sulphates(VI) may be used instead of the halogenoalkanes.

3 Physical properties of ethers

Methoxymethane is a colourless gas and other lower aliphatic ethers are volatile liquids; ethers of higher relative molecular mass are solids.

The boiling points are close to those of the corresponding hydrocarbons and lower than those of the isomeric alcohols because there is no hydrogen bonding in the ethers as there is in the alcohols (p. 176). This is due to the lack of hydroxyl groups in ethers.

This lack of hydroxyl groups also means that many ethers are only sparingly soluble in water; others are miscible with water as their oxygen atom can act as an acceptor in hydrogen bond formation with water. They are very good solvents and their chemical inertness makes them particularly useful as such. They are used for extracting and purifying organic chemicals and they are also good solvents in which to carry out some reactions. In particular, they dissolve many organo-metallic compounds, e.g. $LiAlH_4$, $NaBH_4$ and Grignard reagents, so that reaction involving these reagents are commonly carried out in ethereal solution.

If a low boiling point solvent is needed ethoxyethane is generally used; for a higher boiling point solvent tetrahydrofuran may be chosen.

The ethers have a characteristic smell and some of them can act as anaesthetics. Ethoxyethane was first used for this purpose by Simpson in the years following 1847.

4 Reactions of ethers

Ethers are, in general, unreactive for the C—O bonds are not easily split, i.e. —OR groups are not easily substituted (p. 156).

a. Reaction with strong acids. Ethers like alcohols are weakly basic and will react with strong acids to form dialkyloxonium ions which then undergo substitution reactions with the anion from the acid, e.g.

$$C_2H_5-O-C_2H_5 \xrightleftharpoons{H^+} C_2H_5-\underset{\underset{H}{|}}{O^+}-C_2H_5 \xrightarrow{I^-} C_2H_5OH + C_2H_5I$$

The reaction is most effective with hot hydriodic acid for I^- is the strongest X^- nucleophile (p. 156). With excess hydriodic acid the alcohol produced reacts to give more iodoalkane, e.g.

$$C_2H_5-O-C_2H_5 + 2HI \rightarrow 2C_2H_5I + H_2O$$

A mixed ether will give a mixture of two iodoalkanes.

b. Reaction with very strong bases. An ether molecule can be fragmented by reaction with very strong bases such as alkali metal alkyls, e.g.

$$Na^+CH_3^- + H-CH_2-CH_2-O-C_2H_5 \rightarrow CH_4 + CH_2{=}CH_2 + C_2H_5O^-Na^+$$

c. Oxidation. Ethers can be oxidised, by combustion, to carbon dioxide and water, e.g.

$$(C_2H_5)_2O + 6O_2 \rightarrow 4CO_2 + 5H_2O$$

The ones with low boiling points are dangerously inflammable and must be handled with care.

Ethers may also undergo atmospheric oxidation in the presence of sunlight. The ultra-violet light produces a free radical which then reacts to give a peroxide, e.g.,

$$C_2H_5-O-\overset{\overset{\displaystyle H}{|}}{\underset{\underset{\displaystyle H}{|}}{C}}-CH_3 \xrightarrow[+O_2]{uv} C_2H_5-O-\overset{\overset{\displaystyle H}{|}}{\underset{\underset{\displaystyle O-OH}{|}}{C}}-CH_3$$

These peroxides explode if heated and ethers are stored in dark glass bottles to limit their formation.

5 Cyclic ethers. Epoxyethane

The nature of a cyclic ether depends very much on the ring size. Once the number of atoms in the ring becomes greater than four there is little or no strain in the ring and the properties of the cyclic ether resemble those of the corresponding straight-chain ether. But with 3- or 4-membered rings the considerable strain in the ring means that the C—O bonds are readily broken so that the cyclic ether is much more reactive than the corresponding straight-chain compound.

Epoxyethane, manufactured by the oxidation of ethene is the most important example (p. 113). It is a volatile liquid, soluble in water and in organic solvents.

It undergoes ring cleavage reactions with a number of reagents, producing useful products. In summary,

Reagent	*Product*
H_2O at 200 °C	HO—CH_2—CH_2—OH Ethane-1,2-diol (Glycol, p. 187)
R—OH	HO—CH_2—CH_2—O—R A mono-alkyl ether of glycol
HCl	HO—CH_2—CH_2—Cl 2-chloroethanol (Ethylene chlorohydrin)
NH_2	HO—CH_2—CH_2—NH_2 2-aminoethanol (Ethanolamine)

The main use of epoxyethane is in making ethane-1,2-diol but the ethers of glycol and 2-chloroethanol are useful solvents.

Questions on Chapter 16

1 How can ethoxyethane be made? Outline the arguments in support of its structural formula.

2 What are the physical and chemical properties of ethers?

3 Write an account on anaesthetics.

4 Describe the part played by (a) Simpson, (b) Pasteur, (c) Lister in the development of modern medicine.

5 Explain (a) how you would distinguish between ethanol and methoxymethane by chemical tests, (b) why a sample of ethoxyethane which has been exposed to the light for some time may be dangerous, and (c) the meaning of the term 'mixed ether'.

6 Describe the scene, and your personal reactions to it, when you have undergone any operation involving a local or general anaesthetic.

7 Compare the reactions of water, methanol and methoxymethane.

8 In what ways are epoxyethane and cyclopropane alike?

9 What would be the effect on the rate of Williamson's synthesis of ethers from alkoxides of using (a) different halogenoalkanes and (b) primary, secondary and tertiary halogenoalkanes?

10 Suggest how compounds having the following formulae might be prepared: (a) $CH(OEt)_3$, (b) $CH_2(OEt)_2$, (c) $CH_2{=}CHOEt$, (d) MeOEt. Predict their reactions with (i) acid, (ii) alkali. State the conditions of the reactions that you describe. Give reasons for your proposals and account for any similarities or differences. (Camb. Schol.)

11 How can C—O bonds be formed?

12*Record the boiling points of some alkanes, alcohols and ethers of similar relative molecular masses. Comment on the figures.

17

Aldehydes and Ketones

1 Introduction

The general formula for an aliphatic aldehyde is

$$\underset{\underset{\displaystyle H}{\big|}}{\overset{\displaystyle R}{\diagdown}} C = O$$

where R stands for a hydrogen atom or an alkyl group. The generic term, aldehyde, is derived from 'alcohol dehydrogenated'; individual members are named by replacing the -e of the -ane ending of the alkane with the same number of carbon atoms by -al. Older names are related to the carboxylic acid into which the aldehyde is oxidised e.g.

$$\underset{\underset{\displaystyle H}{\big|}}{\overset{\displaystyle H}{\diagdown}} C = O \qquad \qquad \underset{\underset{\displaystyle H}{\big|}}{\overset{\displaystyle CH_3}{\diagdown}} C = O$$

Methanal	Ethanal
(Formaldehyde)	(Acetaldehyde)

In aliphatic ketones, the H of the —CHO group of an aldehyde is replaced by an alkyl group so that the general formula is

$$\underset{\underset{\displaystyle R}{\big|}}{\overset{\displaystyle R'}{\diagdown}} C = O$$

the R and R′ standing for alkyl groups but not hydrogen atoms. The systematic names are obtained by replacing the -e of the corresponding alkane by *-one*, and numbering is used if necessary to denote the position of the $\diagup C{=}O$ group, e.g.

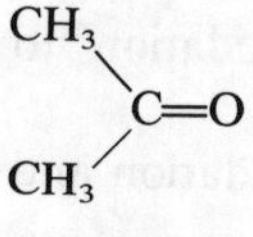

CH$_3$
$\diagdown$
C=O
$\diagup$
CH$_3$

Propanone
(Acetone)

CH$_3$—CH$_2$
$\diagdown$
C=O
$\diagup$
CH$_3$—CH$_2$

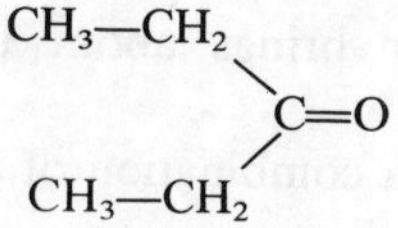

Pentan-3-one
(Diethylketone)

Aromatic aldehydes and ketones (p. 227) are exemplified by benzaldehyde, phenylethanone (acetophenone) and diphenyl methanone (benzophenone).

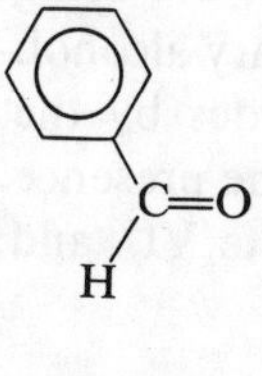

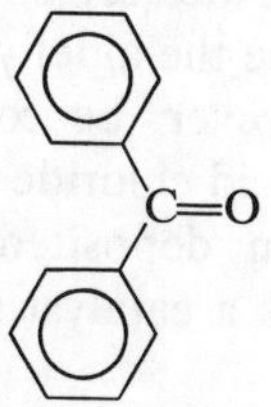

Benzaldehyde 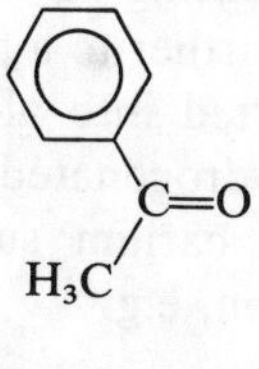   Phenylethanone   Diphenylmethanone
                      (Acetophenone)   (Benzophenone)

Aliphatic Aldehydes and Ketones

2 Methods of preparation of aldehydes

a. Oxidation of primary alcohols. Primary alcohols can be converted into aldehydes by many oxidising agents. An acidified solution of potassium dichromate(VI) is commonly used in the laboratory. The alcohol is dropped into a hot solution of the oxidising agent at the same rate as the lower boiling point aldehyde formed distils off. The removal of the aldehyde from the reaction mixture limits further oxidation to carboxylic acids.

Primary alcohols can also be oxidised industrially either by direct air-oxidation or by dehydrogenation. A mixture of methanol vapour and air, for example, gives methanal when passed over a silver catalyst at 600 °C,

$$2CH_3{-}OH + O_2 \xrightarrow{\text{Ag, 600 °C}} 2H{-}CHO + 2H_2O \quad \Delta H^\ominus(298\ K) = -131\ kJ$$

$$CH_3{-}OH \xrightarrow{\text{Ag}} H{-}CHO + H_2 \quad \Delta H^\ominus(298\ K) = +176\ kJ$$

The reaction temperature can be maintained by adjusting the amount of air and the balance between the exothermic and endothermic reactions. In the laboratory a red-hot coil of platinum wire suspended over some methanol in a beaker brings about the oxidation to methanal.

Ethanal is also manufactured by a combination of air-oxidation and dehydrogenation,

$$2C_2H_5—OH + O_2 \xrightarrow{Ag, 600°C} 2CH_3—CHO + 2H_2O$$
$$C_2H_5—OH \xrightarrow{Ag} CH_3—CHO + H_2$$

b. *Reduction of esters or acid chlorides.* Carboxylic acids cannot be effectively reduced to aldehydes for any reducing agent that will bring about the change will reduce the aldehyde further to a primary alcohol.

Acid chlorides can, however, be converted into aldehydes by the *Rosenmund reaction.* The acid chloride is hydrogenated in the presence of a catalyst of palladium deposited on barium sulphate(VI) and containing some sulphur as a catalyst poison, e.g.

$$R—COCl + H_2 \xrightarrow{Pd/BaSO_4} R—CHO + HCl$$

The sulphur limits the catalytic activity in the possible second-stage reduction of the aldehyde to an alcohol. The method is not available for making methanal as methanoyl chloride is very unstable.

c. *The Wacker process* (p. 113). Ethanal is manufactured from ethene and oxygen by reaction with an aqueous solution of palladium(II) and copper(II) chloride e.g.,

$$CH_2{=}CH_2 + PdCl_2 + H_2O \rightarrow CH_3—CHO + Pd + 2HCl$$
$$2Pd + 4HCl + O_2 \rightarrow 2PdCl_2 + 2H_2O$$

3 Methods of preparation of ketones

a. *Oxidation of secondary alcohols.* Whereas primary alcohols give aldehydes on oxidation, secondary alcohols give ketones, e.g.

$$R—\underset{\underset{H}{|}}{\overset{\overset{R'}{|}}{C}}—OH \xrightarrow{(O)} \underset{R}{\overset{R'}{}}C{=}O + H_2O$$

An acidified solution of potassium dichromate(VI) is an effective oxidising agent in the laboratory.

On an industrial scale, propanone is made by the dehydrogenation of propan-2-ol on passing it over various catalysts, e.g. copper at 350 °C, i.e.

$$(CH_3)_2CH{-}OH \rightarrow \begin{array}{c} CH_3 \\ \diagdown \\ \diagup \\ CH_3 \end{array} C{=}O + H_2$$

b. The cumene process. Propanone is obtained as a by-product in the manufacture of phenol by the cumene process (p. 198).

c. The Wacker process. Just as ethanal is made from ethene in the Wacker process so can propanone be made from propene. (p.113).

4 General properties of aldehydes and ketones

a. The carbonyl group. Both aldehydes and ketones contain the carbonyl, $\diagdown C{=}O$, group and, as a result have some properties in common. Differences arise because the carbonyl group is linked to two hydrogen atoms in methanal, to one hydrogen atom in ethanal and to two alkyl groups in aliphatic ketones.

The carbon atom in the carbonyl bond forms three sp^3 hybrid bonds; these are coplanar, σ-bonds with bond angles of about 120 °. The remaining p-orbital of the carbon atom overlaps with the p-orbital of the adjacent oxygen atom to form a π-bond. The bond between carbon and oxygen is, therefore, a double bond made up of a σ-bond and a π-bond. In this respect it resembles the $C{=}C$ bond though it is stronger, the bond energies being $C{=}O$ ($736 \, kJ \, mol^{-1}$) and $C{=}C$ ($610 \, kJ \, mol^{-1}$).

The $C{=}O$ bond is, also, much more polar than the $C{=}C$ bond because of the difference in electronegativity between carbon and oxygen. The polarity builds up a positive charge on the carbon atom, $C^{\delta+}{=}O^{\delta-}$, making it susceptible to attack by nucleophilic reagents. This contrasts with the susceptibility of the $C{=}C$ bond to attack by electrophilic reagents (p. 106).

The lower the positive charge on the carbon atom of a $C{=}O$ bond the less reactive it will be towards nucleophiles. In general, then, the reactivity falls in passing from methanal to ethanal to propanone, for the increasing +I effect of the alkyl groups lowers the positive charge on the carbon atom of the carbonyl group. The increase in the number of alkyl groups also increases the steric hindrance making it more difficult to attack the carbon atom of the carbonyl group.

$$\begin{array}{ccc}
\begin{array}{c} H \\ \diagdown \\ \\ H \diagup \end{array}\!\!C\!\!=\!\!O & \begin{array}{c} R \\ \diagdown \\ \\ H \diagup \end{array}\!\!C\!\!=\!\!O & \begin{array}{c} R \\ \diagdown \\ \\ R \diagup \end{array}\!\!C\!\!=\!\!O
\end{array}$$

→ Decrease in positive charge on C atom →
→ Increase in steric hindrance to nucleophilic attack →
→ Decrease in ease of nucleophilic attack →

Stretching of the C=O bond in aliphatic aldehydes and ketones gives strong infra-red absorptions between 1 720–1 740 and 1 702–1 725 cm^{-1} respectively. In aromatic compounds the frequencies are lowered by about 20 cm^{-1}

b. *Physical properties.* Methanal is a gas at room temperatures and it is generally used and stored as a 40 % solution in water, known as *formalin.* Other aldehydes of low relative molecular mass are volatile liquids, whilst higher ones are solids. The solubility in water decreases in ascending the homologous series.

The aqueous solutions contain an equilibrium mixture of the aldehyde and a 1,1-diol formed by reaction of the aldehyde with water, i.e.

$$R\!-\!\overset{\displaystyle H}{\underset{}{C}}\!\!=\!\!O + H_2O \rightleftharpoons R\!-\!\overset{\displaystyle H}{\underset{\displaystyle OH}{C}}\!-\!OH$$

The equilibrium lies well to the right with methanal (no $\diagup\!\!\!\diagdown$C=O bond is detectable), is evenly balanced with ethanal and moves over to the left with higher aldehydes.

Propanone is a colourless, volatile liquid with a characteristic sweet smell; it is a useful solvent. It is completely miscible with water but does not interact to form any 1,1-diol. Higher ketones become much less soluble in water.

c. *Enolisation.* In aldehydes and ketones with alkyl hydrogen atoms adjacent to the C=O bond there is a possibility of keto-enol tautomerism (p. 270), i.e.

$$-\overset{\displaystyle H}{\underset{\displaystyle |}{C}}\!-\!C\!\!=\!\!O \rightleftharpoons -C\!\!=\!\!C\!-\!OH$$

keto-form   enol-form

In simple aliphatic aldehydes and ketones the equilibrium lies so far to the left that the enol form does not normally exist* but its formation may be favoured by H^+ and/or OH^- ions as summarised below:

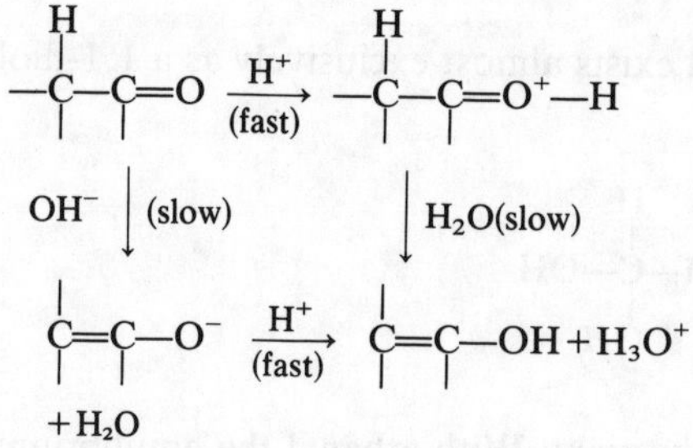

As a result, some reactions of aldehydes and ketones, e.g. aldol formation (p. 224) and halogenation (p. 223) are acid and/or base-catalysed. Such reactions can only take place via an enol form if there is an alkyl hydrogen atom adjacent to the carbonyl group. This is not so in methanal (or in benzaldehyde), which explains why they are slightly different from other aldehydes.

5 Nucleophilic attack on C=O bond

Many nucleophiles attack the positively charged carbon atom of the C=O bond, e.g.

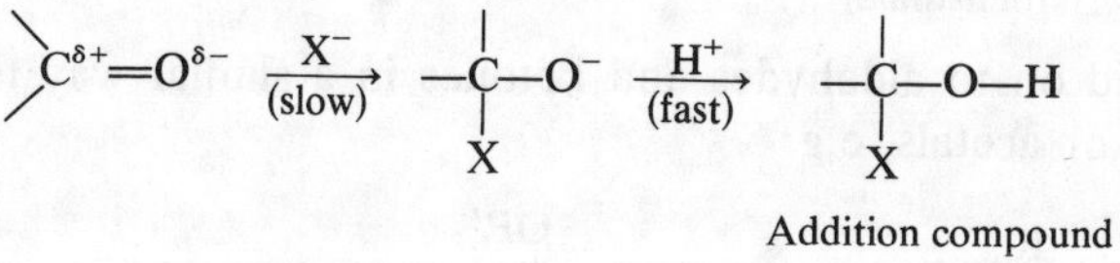

Addition compound

* Ethenyl (vinyl) alcohol,

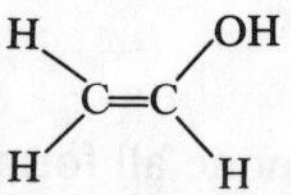

the enol-form of ethanal, does not exist and any attempts to prepare it yield ethanal. On the other hand it is the keto-form of phenol,

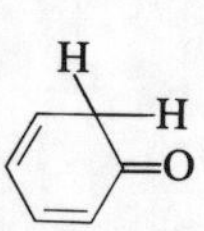

which does not exist; the enol-form is much stabler because of its aromatic character.

the overall result being the formation of an addition compound. Sometimes the addition compound then undergoes a further elimination reaction (p. 220).

a. Reaction with water. Methanal exists almost exclusively as a 1,1-diol in aqueous solution,

$$\underset{H}{\overset{H}{}}C=O \overset{OH^-}{\rightleftharpoons} H-\overset{H}{\underset{OH}{\overset{|}{C}}}-O^- \overset{H^+}{\rightleftharpoons} H-\overset{H}{\underset{OH}{\overset{|}{C}}}-OH$$

The equilibrium lies very far to the right. With ethanal the equilibrium is evenly balanced; with propanone it lies far to the left. The differences arise because of the differing degree of positive charge on the carbon atom of the carbonyl bond (p. 216).

If the positive charge is increased, as, for example, in trichloroethanal, the 1,1-diol can be isolated as a crystalline substance known as a hydrate.

$$Cl\leftarrow\overset{Cl}{\underset{Cl}{\overset{|}{C}}}\!-\!\overset{H}{C}=O \overset{H_2O}{\longrightarrow} Cl-\overset{Cl}{\underset{Cl}{\overset{|}{C}}}\!-\!\overset{H}{\underset{OH}{\overset{|}{C}}}-OH$$

Stable 1,1-diol
(or hydrate)

Alcohols can add on to aldehydes and ketones in a similar way to form hemi-acetals or acetals, e.g.

$$R-\overset{OH}{\underset{H}{\overset{|}{C}}}-OR' \xleftarrow{R'OH} \underset{H}{\overset{R}{}}C=O \xrightarrow[(-H_2O)]{2R'OH} R-\overset{OR'}{\underset{H}{\overset{|}{C}}}-OR'$$

A hemi-
acetal   An acetal

b. Reaction with CN⁻ ions. Methanal, ethanal and propanone all form cyanhydrins by the overall addition of HCN. The initial, rate-determining step is a nucleophilic attack by CN⁻ ions and this is followed by attack by H⁺ ions, e.g.

$$\underset{H}{\overset{CH_3}{}}C=O \xrightarrow[(slow)]{CN^-} CH_3-\overset{H}{\underset{CN}{\overset{|}{C}}}-O^- \xrightarrow[(fast)]{H^+} CH_3-\overset{H}{\underset{CN}{\overset{|}{C}}}-OH$$

2-hydroxypropanonitrile
(Ethanal cyanhydrin)

The reaction rate is proportional to the concentration of CN^- but it is also dependent on the pH of the solution. A high concentration of H^+ ions lowers the CN^- ion concentration because hydrogen cyanide is a weak acid,

$$H^+ + CN^- \rightleftharpoons HCN$$

The H^+ ion also catalyses the reaction by increasing the positive charge on the carbonyl carbon atom, i.e.

Alternatively, high OH^- ion concentration increases the CN^- ion concentration,

$$HCN + OH^- \rightarrow CN^- + H_2O$$

but lowers the charge on the positive carbon atom.

The maximum rate of reaction occurs, then, at a pH of about 5 and the necessary balance between the concentration of CN^- and H^+ can be achieved, in practice, by using hydrogen cyanide in alkaline solution or by gradually adding concentrated acid to potassium cyanide.

The cyanhydrins are easily converted into hydroxyacids by warming with dilute acids, e.g.

and they provide a convenient method for synthesising these acids.

c. Reaction with $NaSO_3^-$ *ions.* Aldehydes and ketones form addition compounds with sodium hydrogensulphate(IV), e.g.

The addition compounds formed are often insoluble solids which can be used, by melting point determination, for identification purposes. They revert to the aldehyde or ketone on warming with alkalis.

d. Reaction with ammonia. Ethanal forms an addition compound with ammonia,

Other aldehydes and ketones may form similar addition compounds but they are unstable and only formed, if at all, at low temperatures.

Methanal reacts differently with ammonia, giving a crystalline solid known as hexamine,

$$6HCHO + 4NH_3 \rightarrow (CH_2)_6N_4 + 6H_2O$$

e. Reaction with X–NH$_2$ *compounds.* Aldehydes and ketones react with a number of compounds containing the —NH$_2$ group with the overall elimination of water, e.g.

$$\underset{\underset{H}{|}}{\overset{\overset{R}{|}}{C}}=O + H_2N-X \rightarrow R-\overset{\overset{H}{|}}{C}=N-X + H_2O$$

When X is an alkyl or an aryl group the product is known as a *Schiff's base*. The reactions with hydroxylamine, hydrazine and substituted hydrazines, e.g.

H$_2$N—OH   H$_2$N—NH$_2$
Hydroxylamine   Hydrazine

2,4-dinitrophenyl-
hydrazine

yield crystalline solids with convenient melting points; they can be used for the identification of aldehydes.

The overall reaction is a condensation, water being eliminated from between the two molecules; the products are known as *condensation compounds*. The reactions actually take place, however, via a nucleophilic addition followed by a proton-transfer rearrangement and an elimination, e.g.

$$CH_3-\underset{\underset{NH_2OH}{|}}{\overset{\overset{H}{|}}{C}}=O \longrightarrow CH_3-\underset{\underset{^+NH_2OH}{|}}{\overset{\overset{H}{|}}{C}}-O^- \longrightarrow CH_3-\underset{\underset{NHOH}{|}}{\overset{\overset{H}{|}}{C}}-OH \overset{-H_2O}{\longrightarrow} CH_3-\underset{\underset{NOH}{\|}}{\overset{\overset{H}{|}}{C}}$$

Ethanal
oxime

The overall reactions with hydrazine and 2,4-dinitrophenylhydrazine are summarised as follows:

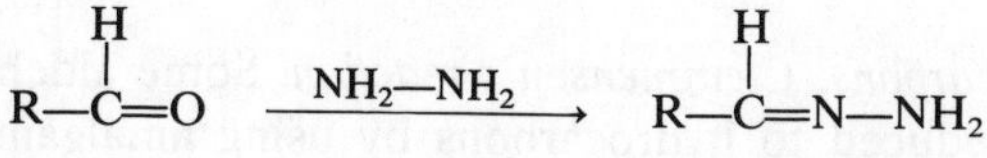

A hydrazone

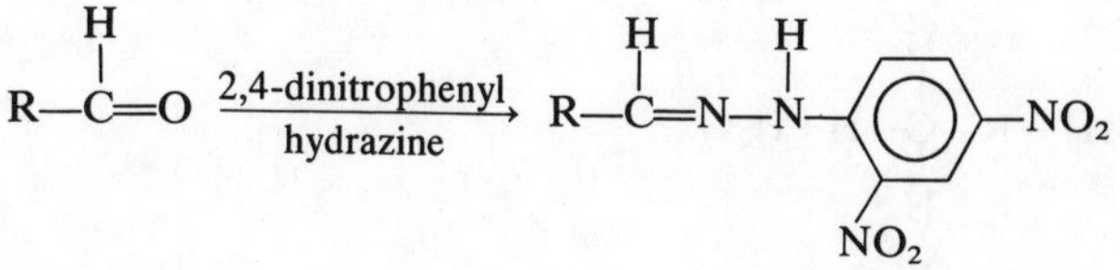

A substituted hydrazone

Similar reactions take place with primary amines (RNH_2) and semicarbazide ($H_2N—NHCONH_2$).

The reactions take place most rapidly at a pH of about 5. The presence of H^+ ions catalyses the reactions by increasing the positive charge on the carbonyl group carbon atom (p. 219), but the H^+ ion also converts the basic reagents into species which are not nucleophilic, e.g.

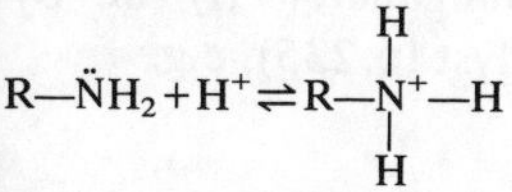

$$R—\ddot{N}H_2 + H^+ \rightleftharpoons R—\overset{H}{\underset{H}{N^+}}—H$$

6 Reduction of aldehydes and ketones

a. Reduction to alcohols. Aldehydes can be reduced to primary alcohols, and ketones to secondary alcohols, e.g.

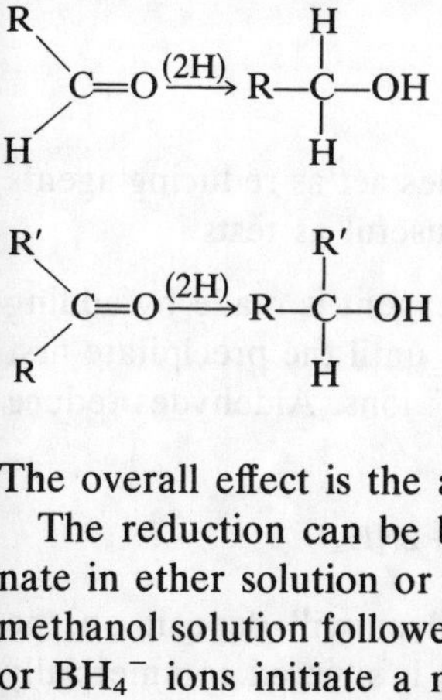

The overall effect is the addition of hydrogen to the $C{=}O$ bond.

The reduction can be brought about by sodium tetrahydridoaluminate in ether solution or by sodium tetrahydridoborate in aqueous or methanol solution followed by treatment with a dilute acid. The AlH_4^- or BH_4^- ions initiate a nucleophilic attack on the positively charged carbon atom of the $C{=}O$ bond. Alternatively, nascent hydrogen obtained from zinc and a dilute acid, or catalytic hydrogenation, can be used.

b. Reduction to hydrocarbons. Clemmensen's reaction. Some aldehydes and ketones can be reduced to hydrocarbons by using amalgamated zinc and concentrated hydrochloric acid, e.g.

$$\begin{matrix} R \\ \diagdown \\ C=O+[4H] \rightarrow R-C-H+H_2O \\ \diagup \\ H \end{matrix}$$

$$\begin{matrix} R' \\ \diagdown \\ C=O+[4H] \rightarrow R-C-H+H_2O \\ \diagup \\ R \end{matrix}$$

7 Oxidation of aldehydes

Aldehydes can be oxidised to carbon dioxide and water by combustion. They can also be oxidised to carboxylic acids by, for example, acidified solutions of potassium dichromate(VI) or manganate(VII) or by atmospheric oxidation in the presence of a catalyst (p.235), e.g.

$$R-CHO \xrightarrow{[O]} R-COOH$$

Methanoic acid may be oxidised further by strong oxidising agents for it still contains an aldehyde group, unlike other carboxylic acids, i.e.

$$H-C \underset{O}{\overset{H}{\diagup}} \xrightarrow{[O]} H-C \underset{O}{\overset{OH}{\diagup}} \xrightarrow{[O]} H_2O+CO_2$$

As they can be readily oxidised the aldehydes act as reducing agents and some reactions in which they do so are useful as tests.

a. Reduction of Tollens's reagent. Tollens's reagent is made by adding ammonia solution to silver nitrate(V) solution until the precipitate first formed just redissolves; it contains $Ag(NH_3)_2{}^+$ ions. Aldehydes reduce the complex ion to silver,

$$R-CHO+Ag(NH_3)_2{}^+ +H_2O \rightarrow R-COOH+2Ag+2NH_4{}^+$$

Under carefully controlled conditions the silver will deposit on the walls of the reaction vessel as a mirror. Glass is silvered commercially by a similar process.

b. Reduction of Fehling's solution. Fehling's solution is made by dissolving Rochelle salt ($C_4H_4O_6KNa.4H_2O$, sodium potassium 2,3-dihyroxybutanedioate or tartrate) and sodium hydroxide in water and adding the mixture to a solution of copper(II) sulphate(VI). The Rochelle salt prevents the precipitation of copper(II) hydroxide and the mixture formed is violet in colour containing complexed Cu^{2+} ions. It is reduced by aldehydes to give a red precipitate of copper(I) oxide. Under controlled conditions the reaction can be used quantitatively to estimate compounds containing the —CHO group, e.g. glucose and other reducing sugars (p. 312).

c. Recolorisation of Schiff's reagent. Schiff's reagent is made by decolorising a solution of rosaniline in water by sulphur dioxide. Aldehydes restore the magenta colour.

8 Oxidation of ketones
Ketones can only be oxidised by strong oxidising agents such as chromic(VI) acid and hot nitric(V) acid. The C—C bond between the alkyl and carbonyl groups is split, e.g.

$$CH_3 \atop CH_3 \!\!\!\!\diagdown \!\!\!\!\! C=O + [4O] \to CH_3—COOH + CO_2 + H_2O$$

Ketones are not, therefore strong enough as reducing agents to reduce Tollens's reagent, Fehling's solution or Schiff's reagent. These tests provide a simple method of distinguishing between aldehydes and ketones.

9 Halogenation of aldehydes
The alkyl group adjacent to the —CHO group in an aldehyde can be halogenated by reaction with a halogen. Methanal and benzaldehyde, however, have no such alkyl group so that they do not react.

Ethanal reacts with chlorine, for example, to give trichloroethanal (chloral),

$$CH_3—CHO + 3Cl_2 \to CCl_3—CHO + 3HCl$$

Trichloroethanal

The product is a colourless, pungent-smelling, oily liquid. It forms a monohydrate with water, $CCl_3CHO.H_2O$; this is a crystalline solid (p. 218) which has been used as a soporific.

Trichloroethanal occurs as an intermediate in the preparation of trichloromethane (chloroform), and the corresponding triodo-compound as in intermediate in the preparation of triodomethane, iodoform (p. 166).

10 Halogenation of ketones

Ketones can be halogenated in much the same way as aldehydes and the chlorination, bromination and iodination of propanone have been particularly well studied. The halogenation processes are catalysed by both acids and bases and the rates are independent of the concentration of halogen but proportional to the concentrations of propanone and H^+ or OH^-, at least for moderate concentrations.

The mechanism involves the slow conversion to the enol form (p. 217) followed by a rapid reaction with the halogen, e.g.

$$CH_3-\overset{\overset{\displaystyle O}{\|}}{C}-CH_3 \rightarrow CH_3-\overset{\overset{\displaystyle OH}{|}}{C}=CH_2 \rightarrow CH_3-\overset{\overset{\displaystyle O}{\|}}{C}-CH_2Br + HBr$$
$$\text{Enol form}$$

In the base-catalysed reaction the rate increases as each halogen atom is substituted into the propanone. It is, therefore, not possible to stop the reaction at the intermediate stages and a trihalogenopropanone is produced. The step-wise substitutions do not get quicker in acid-catalysed halogenation (the mechanism is slightly different) so that mono-, di- or tri-halogeno-products can be isolated.

11 Reactions of aldehydes with sodium hydroxide

Ethanal gives a polymer with concentrated solutions of sodium hydroxide (p. 226) but forms 3-hydroxybutanal (aldol) with a cold dilute solution,

$$2CH_3-CHO \xrightarrow{OH^-} CH_3-\overset{\overset{\displaystyle H}{|}}{\underset{\underset{\displaystyle OH}{|}}{C}}-\overset{\overset{\displaystyle H}{|}}{\underset{\underset{\displaystyle H}{|}}{C}}-\overset{\displaystyle H}{C}\diagdown_{O}$$
$$\text{3-hydroxybutanal}$$
$$\text{(Aldol)}$$

The reaction provides a convenient method of forming C—C bonds. The product is both a primary alcohol and an aldehyde; it is a colourless liquid which is dehydrated on heating to give an unsaturated aldehyde ($CH_3-CH=CH-CHO$, but-2-enal).

The reaction has been referred to as the aldol condensation but it is really a dimerisation of the ethanal involving the enol form (p. 217),

$$H-\underset{\underset{H}{|}}{\overset{\overset{H}{|}}{C}}-C=O \leftrightarrow H-\underset{}{\overset{\overset{H}{|}}{C}}=\overset{H}{C}-OH$$

Keto form   Enol form

In the presence of OH^- ions an enolate carbanion is formed

$$OH^{\frown}H-\overset{\overset{H}{|}}{\underset{\underset{H}{|}}{C}}-\overset{H}{\underset{\underset{O}{\|}}{C}} \rightarrow H_2O + {}^-CH_2-\overset{H}{\underset{}{C}}=O$$

Enolate carbanion

and as this is nucleophilic it adds on to ethanal in much the same way as CN^- and HSO_3^- (p. 219), i.e.

$$CH_3-\underset{\underset{{}^-CH_2-CHO}{|}}{\overset{\overset{H}{|}}{C}}=O \longrightarrow CH_3-\underset{\underset{CH_2-CHO}{|}}{\overset{\overset{H}{|}}{C}}-O^- \overset{H^+}{\longrightarrow} CH_3-\underset{\underset{CH_2-CHO}{|}}{\overset{\overset{H}{|}}{C}}-OH$$

Aldehydes with no alkyl hydrogen atom adjacent to the —CHO group, e.g. methanal and benzaldehyde, have no enol form and they react differently with sodium hydroxide solution. Half the aldehyde is oxidised to a carboxylic acid and half is reduced to a primary alcohol (Cannizzaro's reaction), e.g.

$$2H-CHO + NaOH \rightarrow CH_3-OH + H-C\overset{\overset{O}{\|}}{\underset{O^-Na^+}{}}$$

The reaction involves the nucleophilic addition of OH^-

$$H-\overset{H}{\underset{\underset{{}^-OH}{\uparrow}}{C}}=\overset{\frown}{O} \rightarrow H-\overset{H}{\underset{\underset{OH}{|}}{C}}-O^-$$

and the intermediate ion then reduces methanal to methanol,

$$H-\overset{H}{\underset{}{C}}=\overset{\frown}{O} \quad H-\overset{H}{\underset{\underset{OH}{|}}{C}}-O^{\frown} \rightarrow H-\overset{H}{\underset{\underset{H}{|}}{C}}-O^- + \overset{}{\underset{\underset{OH}{|}}{C}}=O$$

$$\downarrow$$

$$CH_3OH + H-C\overset{\overset{O}{\|}}{\underset{O^-}{}}$$

12 Polymerisation of aldehydes

Aldehydes polymerise in a number of different ways, but there are no similar reactions for ketones.

a. Methanal. Evaporation of an aqueous solution of methanal gives a white solid known as poly(methanal) or paraformaldehyde. It has a long chain structure

$$n\text{H—CHO} + \text{H}_2\text{O} \rightarrow \text{H—(O—CH}_2)_n\text{—OH}$$

with n having values between 8 and 100.

If poly(methanal) is heated with dilute acid, or an acidified solution of formalin is distilled, another white, solid polymer, methanal trimer (trioxymethylene) is formed; it has a cyclic structure.

Methanal trimer
(Trioxymethylene)

Ethanal trimer
(Paraldehyde)

Ethanal tetramer
(Metaldehyde)

Both polymers yield methanal on heating, but a stable form of poly(methanal), which is not affected by acids or heat, can be made by esterifying the —OH groups at the end of a chain and adding a stabiliser. The product (Delrin) is strong and can be easily moulded; it is used as a replacement for metals.

b. Ethanal. The addition of a few drops of concentrated sulphuric(VI) acid to ethanal produces a liquid polymer, ethanal trimer or paraldehyde, in a highly exothermic reaction. The polymer has a cyclic structure and has been used as a mild soporific.

If hydrogen chloride gas is passed into cooled ethanal a white, insoluble, solid polymer, ethanal tetramer or metaldehyde, is formed. It also has a cyclic structure and is used as a solid fuel (meta fuel) and as a slug killer.

With concentrated solution of sodium or potassium hydroxide, ethanal forms a brown, resinous polymer with a long chain structure.

$$\text{CH}_3\text{—(CH}=\text{CH)}_n\text{—CHO}$$

13 Uses of aldehydes and ketones

Methanal is used in making plastics and polymers, e.g. Bakelite (p. 328), urea-formaldehyde resins (p. 328) and Delrin (p. 226). Its aqueous solution is also used as a disinfectant and preservative.

Ethanal is required for the manufacture of ethanoic acid (p. 234), metaldehyde (p. 226) and trichlorethanal (p. 223).

Propanone is widely used as a solvent for dissolving ethyne (p. 118), many natural and synthetic resins and gums, cellulose nitrate(V), cellulose acetate, ethyl cellulose and celluloid. It is also used in making Perspex (p. 327).

Aromatic Aldehydes and Ketones

14 Benzaldehyde, C_6H_5—CHO

a. Preparation. Benzaldehyde a typical aromatic aldehyde can be obtained from methylbenzene either by selective oxidation (p. 134) or by dihalogenation in the side chain and subsequent hydrolysis (p. 134), e.g.

b. Physical properties. Benzaldehyde is a colourless liquid. It is also known as oil of bitter almonds as it occurs naturally in amygdalin, which is found in bitter almonds and in peach or cherry stones. The aldehyde has a characteristic odour of almonds and is used as a flavouring. It is only slightly soluble in water, but soluble in organic solvents.

c. Chemical properties. The summary on p. 230 shows that benzaldehyde resembles aliphatic aldehydes in many ways. It is particularly like methanal, both these aldehydes having no alkyl hydrogen atom adjacent to the carbonyl group (p. 217).

Benzaldehyde is, in general, less reactive than the aliphatic aldehydes for the delocalised π-electrons of the benzene ring tend to lower the positive charge on the carbon atom of the $C=O$ bond so that it is less easily attacked by nucleophiles. The benzene ring also inhibits nucleophilic attack by steric hindrance.

The main distinctive property of benzaldehyde is that it will dimerise in the presence of potassium cyanide to form a colourless solid, 2-hydroxy-1,2-diphenylethanone (benzoin),

It can also be ring substituted.

15 Aromatic ketones

Phenyl ethanone (acetophenone) and diphenylmethanone (benzophenone) are typical aromatic ketones; they are both colourless, crystalline solids, best made by an application of the Friedel–Crafts reaction in which an arene reacts with an acid chloride in the presence of aluminium chloride, as a catalyst (p.131), e.g.

Aromatic ketones have essentially the same ketonic properties of aliphatic ketones such as propanone; they can also be ring substituted.

Questions on Chapter 17

1 Describe in detail how you would prepare, from ethanol, an aqueous solution containing ethanal. What experiments would you carry out to show that the solution obtained differed from an aqueous solution of (a) methanal, (b) propanone.

2 Compare the reaction of methanal, ethanal and propanone with (a) acidified potassium dichromate(VI), (b) sodium hydrogen sulphate(IV), (c) ammonia, (d) iodine.

3 How would you bring about the following changes: (a) ethyne to ethanal, (b) ethanal to triiodomethane, (c) ethanol to aldol, (d) ethanal to butan-1-ol?

4 Write structural formulae, and suggest names, for all aldehydes and ketones of molecular formula $C_5H_{10}O$.

5 How would you obtain benzaldehyde from methylbenzene? Compare the properties of ethanal and benzaldehyde, drawing attention to any marked differences.

6 How would you convert benzaldehyde into (a) methylbenzene, (b) phenylmethanol, (c) benzaldoxime, (d) 2-hydroxy-1,2-diphenyl ethanone (benzoin)?

7 An organic liquid contains 66.6 per cent C, 11.1 per cent H and 22.3 per cent O. The compound forms an oxime containing 16.1 per cent N but does not reduce ammoniacal silver nitrate(V). Deduce the structural formula of the compound and write equations to show how it would react with (a) sodium and ethanol, (b) hydrogen and (c) hydrogen cyanide.

8 List the following in what you would expect to be their order of reactivity with nucleophiles. (a) CH_3—CHO; NO_2—CH_2—CHO; Cl—CH_2—CHO; CH_3—CH_2—CHO; (b) C_6H_5—CHO; NO_2—C_6H_4—CHO; OH—C_6H_4—CHO; (c) CH_3—CO—CH_3; $C(CH_3)_3$—CO—$C(CH_3)_3$. Account for the orders you give.

9 One mole of a hydrocarbon A of relative molecular mass 158 absorbed 4 g of hydrogen on hydrogenation over platinum at room temperature. Mild ozonolysis of A gave three compounds, B, C and D. Oxidation of B, C_2H_4O, gave ethanoic acid; C reacted with hydroxylamine to give a dioxime $C_2H_4O_2N_2$; treatment of D, C_8H_8O, with aqueous sodium bromate(I) gave tribromomethane and benzoic acid. Deduce the structure of A and explain the above reactions. (Oxf. Schol.)

10 For the substance of molecular formula C_4H_8O give the isomers which contain a carbonyl ($C{=}O$) group and those which contain a primary alcoholic group (—CH_2OH). (Other isomers are not required). What would be the reaction of each isomer with dilute acidified potassium manganate(VII) (a) in the cold, and (b) on prolonged heating? (L)

11 Give the structural formulae of the compounds obtained when propanone reacts with (a) hydroxylamine, (b) sodium hydrogen

	H—CHO	CH₃—CHO	(CH₃)₂CO	C₆H₅—CHO
By oxidation of	CH₃—OH	C₂H₅—OH	(CH₃)₂CHOH	C₆H—CH₂OH or C₆H₅—CH₃
Special methods		Reduction of CH₃COCl (Rosenmund reaction). Wacker process	Cumene process. Wacker process.	Hydrolysis of C₆H₅—CHCl₂
Oxidation	H—COOH or H₂O and CO₂ formed	CH₃—COOH formed	CH₃—COOH formed by strong O.A.'s only	C₆H₅COOH formed very easily
Reduction by NaAlH₄	CH₃—OH formed	C₂H₅—OH formed	(CH₃)₂CHOH formed	C₆H₅—CH₂OH formed
Reduction by Zn and conc. HCl (Clemmenson's reaction)	No CH₄ formed	No C₂H₆ formed	C₃H₈ formed	C₆H₅—CH₃ formed
Action as reducing agent	Reduce Tollens's reagent to Ag and Fehling's solution to Cu₂O. Recolorise Schiff's reagent		No reducing properties	Reduces Tollens's reagent. Little effect on Fehling's solution

← Preparation →

Properties				
Addition compounds	Formed with hydrogen cyanide and sodium hydrogen sulphate(VI)			Forms benzoin with KCN
Reaction with NH$_3$	Forms hexamine, (CH$_2$)$_6$N$_4$	Forms an addition compound		Form addition compounds at low temperatures
Condensation compounds	Form phenylhydrazones with phenylhydrazine, C$_6$H$_5$—NH—NH$_2$, and oximes with hydroxylamine, NH$_2$OH			
Polymerisation	Forms poly(methanal) and methanal trimer	Forms ethanal trimer, ethanal tetramer, and a resin	No polymers formed	
Reaction with NaOH	Forms CH$_3$OH and HCOONa (Cannizzaro's reaction)	Forms aldol with cold, dilute solution. Polymerises with conc. solution	No reaction	Forms C$_6$H$_5$—CH$_2$OH and C$_6$H$_5$—COO$^-$Na$^+$ (Cannizzaro's reaction)
Halogenation	Not halogenated. No alkyl H atom	Forms trichloroethanal	Forms trichloro-propanone	Halogenated in ring. No alkyl H atom
Iodoform reaction	No reaction	Forms iodoform	Forms iodoform	No reaction

Preparation and properties of methanal, ethanal, propanone and benzaldehyde

sulphite, (c) concentrated sulphuric acid, (d) lithium aluminium hydride.

To a solution of propanone (y g) in water was added an excess of bromine water followed by sodium hydroxide solution. The bromoform, $CHBr_3$, obtained weighed 4.365 g. Calculate the value of y. (O and C)

12 Compare the behaviour of compounds containing the grouping $>C=C<$ with that of compounds containing the grouping $>C=O$. (O and C)

13 How is an aqueous solution of methanal prepared? Give three reactions in which methanal and ethanal show similar behaviour and three reactions in which they differ. (O and C)

14 A chemist has produced what he believes to be pyruvic acid,

$$CH_3-\underset{\underset{O}{||}}{C}-CO_2H$$

After purifying the compound, he carries out a series of qualitative tests and finds that the compound contains no elements other than carbon, hydrogen and oxygen. Suggest (a) how the chemist might proceed to verify the identity of his compound and elucidate its structure; (b) how he might synthesise further samples of pyruvic acid starting from ethanal. (L.)

15 (a) A hydrocarbon A (b.p. −6 °C) has a molecular mass of 56 and contains 85.7 per cent of carbon. A reacts with concentrated hydrobromic acid to form a compound B which is optically active. Boiling aqueous sodium hydroxide converts B to a compound C, from which a compound D is formed by oxidation with hot, acidified potassium dichromate solution. D has b.p. 80 °C, and gives a yellow precipitate with 2,4-dinitrophenylhydrazine, but does not react with boiling Fehling's solution. Identity A, B, C and D, giving their names and structural formulae. Write equations for all the reactions mentioned above. (b) Devise a flow chart for the synthesis of 2-methylbutanoylchloride from 2-iodobutane. Include in the chart the reagents and conditions for each step in the synthesis. Write equations for all reactions. (L. Nuffield)

16*Compare the enthalpies of hydrogenation of the $C=C$ and $C=O$ bonds and explain why it is difficult to reduce a carbonyl group without also saturating any $C=C$ bond that may be present.

18

Organic Acids

1 Introduction

Aliphatic mono-carboxylic acids contain one —COOH group linked to a hydrogen atom or to an alkyl group, e.g.

$$H-C{\overset{O}{\underset{O-H}{}}} \qquad CH_3-C{\overset{O}{\underset{O-H}{}}} \qquad C_2H_5-C{\overset{O}{\underset{O-H}{}}}$$

Methanoic acid   Ethanoic acid   Propanoic acid
(Formic acid)   (Acetic acid)   (Propionic acid)

The name of the group is derived from *carb*onyl (C═O) and hydr*oxyl* (O—H). The names of the acids are obtained by replacing the —*e* of the corresponding alkane by —*oic*. Older names refer to the origin or use of the acid. Methanoic acid, for example, was first made from ants (*Formica* = ant); ethanoic acid occurs in vinegar (*acetus* = sour). Some of the higher acids occur in plant and animal fats so that the acids, as a group, were once called fatty acids.

In aromatic mono-carboxylic acids one —COOH group is linked to an aryl group (p. 244).

Dicarboxylic acids (pp. 242, 246) contain two —COOH groups substituted into an alkane or an arene.

Sulphonic acids are much less common than carboxylic acids. They contain the —$SO_2(OH)$ group linked, in the commonest examples, to an aryl group (p. 246).

In substituted acids, the hydrogen atoms in the alkyl or aryl group are substituted by such groups as —OH, —Cl, —CN and —NH_2 (p. 303).

Aliphatic Mono-carboxylic Acids

2 Methods of preparation and manufacture

a. By oxidation of primary alcohols or aldehydes, e.g.

$$CH_3-OH \rightarrow H-CHO \rightarrow H-COOH$$

Methanol   Methanal   Methanoic
acid

$$C_2H_5-OH \rightarrow CH_3-CHO \rightarrow CH_3-COOH$$

Ethanol   Ethanal   Ethanoic
acid

The conditions under which these reactions take place are described on pages 181 and 222.

b. By hydrolysis of nitriles (p. 287). This hydrolysis can be brought about both in acid and alkaline solution. In acid solution, using, say, dilute sulphuric(VI) acid, the product is the free acid, e.g.

$$2CH_3-CN + H_2SO_4 + 4H_2O \rightarrow 2CH_3-COOH + (NH_4)_2SO_4$$

Ethanonitrile   Ethanoic
acid

$$R-CN + H^+ + 2H_2O \rightarrow R-COOH + NH_4^+$$

In alkaline solution, using, say, a solution of sodium hydroxide, the sodium salt of the acid is obtained, e.g.

$$CH_3-CN + NaOH + H_2O \rightarrow CH_3-COO^-Na^+ + NH_3$$

$$R-CN + OH^- + H_2O \rightarrow R-COO^- + NH_3$$

The free acid can be obtained from the salt by treating with a dilute acid.

c. Manufacture of methanoic acid. Carbon monoxide reacts with sodium hydroxide at 200 °C and 10 atmospheres to form sodium methanoate,

$$CO(g) + NaOH(s) \rightarrow H-COO^-Na^+(s)$$

The acid can be liberated from the salt by distilling with dilute sulphuric(VI) acid.

d. Manufacture of ethanoic acid. This acid is made by the catalytic, atmospheric oxidation of ethanal or alkanes obtainable from naphtha, e.g.

$$2CH_3\text{---}CHO + O_2 \xrightarrow[60°C]{Mn(CH_3COO)_2} 2CH_3\text{---}COOH$$

$$C_nH_{2n+2} + O_2 \xrightarrow[200°C, \, 200 \, atm.]{Co(CH_3COO)_2} CH_3\text{---}COOH \, (+ \, by\text{-}products)$$
$$(n = 4\text{--}7)$$

In the latter process methanoic and propanoic acids are also formed depending on the starting material.

3 The carboxyl group

This group contains both a carbonyl (C=O) group, as in aldehydes and ketones, and a hydroxyl (O—H) group, as in alcohols. Both bonds, in isolation, are quite strongly polar ($C^{\delta+}=O^{\delta-}$ and $O^{\delta-}\text{---}H^{\delta+}$) and they interact when combined together so that the characteristics of the bonds in carboxylic acids are significantly different from the separate bonds. In particular, the carbon atom of the C=O bond in a carboxylic acid is less susceptible to attack by nucleophiles than in aldehydes and ketones (the carbon atom is less positively charged), and the O—H bond in carboxylic acids splits and releases an H^+ ion much more readily than in alcohols.

a. Carbonyl properties of carboxylic acids. In the C=O bond in an aldehyde or ketone there is some π-bonding and some polarity, $C^{\delta+}=O^{\delta-}$. In the C=O bond in a carboxylic acid there is a second oxygen atom adjacent to the bond and the available p-orbitals of this oxygen atom can overlap with the π-bond giving delocalised π-bonding over the O—C—O atoms as illustrated in Fig. 48.

This reduces the positive charge on the carbon atom and makes it less susceptible to attack by nucleophiles. Carboxylic acids will not,

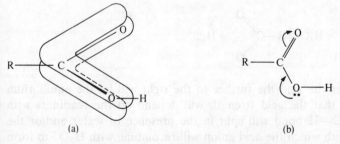

Fig. 48. Two ways of representing delocalisation in the carboxyl group. In (a), one charge cloud is above the plane of the molecule with the other below.

therefore, form addition or condensation compounds as aldehydes and ketones do. The carbon atom of the —COOH group can, however, still be attacked by some nucleophiles as explained for example, on p. 252.

b. Acidic properties. Carboxylic acids are weak acids as compared with mineral acids but they are very much stronger than alcohols (p. 177). The variation of strength in passing from one acid to another has been extensively studied and typical pK_a values are given below. The variation in strength can be accounted for very satisfactorily.

HCOOH	3.75	CH_2FCOOH	2.66	CH_3COOH	4.76
CH_3COOH	4.76	$CH_2ClCOOH$	2.86	$CH_2ClCOOH$	2.86
C_2H_5COOH	4.88	$CH_2BrCOOH$	2.89	$CHCl_2COOH$	1.29
C_3H_7COOH	4.82	CH_2ICOOH	3.16	CCl_3COOH	0.65

pK_a values for some carboxylic acids

The strength of any carboxylic acid will depend on the equilibrium position for the equilibrium,

The stronger the acid the further to the right will be the equilibrium position so that the acid strength will depend on the readiness with which the O—H bond will split in the presence of water and/or the readiness with which the acid anion will recombine with H_3O^+ to form the undissociated acid.

Alcohols are very weak acids (p. 177) because the O—H bond does not split easily and because the RO^- ion combines readily with H_3O^+, i.e.

$$R—OH + H_2O \rightleftharpoons RO^- + H_3O^+$$

The O—H bond in a carboxylic acid splits more readily than in an alcohol because the possible electromeric shifts withdraw electrons away from the H atom, as summarised in Fig. 48, and facilitate the loss of H^+ to a water molecule. This can be expressed in other words by referring to the formation of delocalised π-bonding over the O—C—O atoms as shown in Fig. 48, or the possible existence of resonance hybrids, i.e.

I   II

Structure II would not be expected to be very stable for it involves charge separation so that the resonance energy would not be high.

On the contrary, the acid anion is very highly stabilised because its two resonance hybrids, i.e.

are symmetrical and do not involve any charge separation. This stabilisation of the acid anion hinders its recombination with H_3O^+ to form undissociated acid. There are no similar resonance hybrids for RO^-.

The situation can also be described in relation to the delocalisation over the O—C—O atoms (Fig. 48). This spreads the negative charge over two oxygen atoms whereas it is concentrated on one oxygen atom in the RO^- ion.

c. Effect of R on acid strength. As R changes from H to higher and higher alkyl groups, the increasing $+I$ effect (p. 24) causes a generally decreasing acid strength. The higher the $+I$ effect the greater the drift of electrons towards the H of the O—H bond and the lower the positive charge on the H atom. Alternatively, the higher the $+I$ effect of the R group in the $RCOO^-$ ion the greater the negative charge on the oxygen atoms and the greater the readiness to recombine with H_3O^+ ions to form undissociated acid.

The converse is also true, for introduction of $-I$ groups into the alkyl group, R, of a carboxylic acid causes increased acid strength in close relation to the $-I$ effect and the position of the substituted group. A study of the figures on p. 236 shows that the acid strength increases as more and more halogen groups are introduced, that fluorination increases the acid strength more than iodination, and that the increase in acid strength falls off as the halogen is substituted further and further away from the carboxyl group.

d. Infra-red absorption. The —COOH bond causes infra-red absorption by stretching of the C=O and O—H bonds. The former gives a band at about $1\,700\,cm^{-1}$, close to the value in aldehydes and ketones (p. 216). The hydroxyl group is strongly hydrogen-bonded and gives a broad band around $3\,000\,cm^{-1}$.

4 Physical properties

Methanoic acid is a colourless liquid with a pungent smell. It is completely miscible with water and has a corrosive action on the skin. It occurs in the sting of ants, bees and other insects and the stinging nettle.

Pure (glacial) ethanoic acid is a colourless liquid which solidifies on cold days (m.p. = 17 °C) to form glass-like crystals. It has a very strong 'vinegar' smell and is completely miscible with water.

Higher members of the series are oily liquids, sparingly soluble in water, and still higher ones are wax-like solids, insoluble in water.

Many carboxylic acids, particularly the lower ones, exist as dimers in many organic solvents and, to some extent, in the vapour phase. The dimerisation occurs through hydrogen bonding (Fig. 49) which is

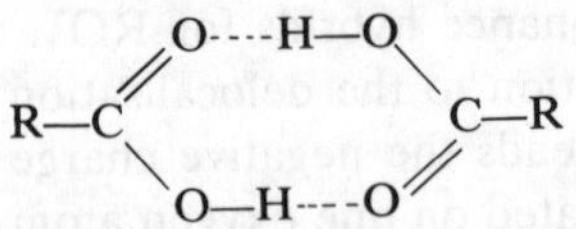

Fig. 49. Dimerisation through hydrogen bonding in carboxylic acids.

particularly strong because of the high polarity of the $C^{\delta+}$—$O^{\delta-}$ and $O^{\delta-}$—$H^{\delta+}$ bonds. The hydrogen bonding causes the acids to have higher boiling points than would be expected. Ethanoic acid and propanol, for example, both have the same relative molecular mass of 60 but the acid has a boiling point of 118 °C and the alcohol one of 97 °C.

5 Chemical reactions of carboxylic acids

The carboxylic acids have many general properties in common but methanoic acid, like the first member of other homologous series, is not typical as described in Section 6 (p. 240).

a. Formation of salts. Carboxylic acids will react with bases and carbonates to form salts. The reaction with carbonates shows that the acids are stronger than carbonic acid, unlike phenol (p. 199).

b. Formation of esters. Carboxylic acids react with alcohols to form esters; the H atom of the carboxyl group is replaced by an alkyl group, e.g.

$$H—COOH + CH_3—OH \rightleftharpoons H—COOCH_3 + H_2O$$

Methyl methanoate

$$CH_3—COOH + C_2H_5—OH \rightleftharpoons CH_3—COOC_2H_5 + H_2O$$

Ethyl ethanoate

$$R—COOH + R'—OH \rightleftharpoons R—COOR' + H_2O$$

An ester

The reactions are reversible and special conditions must be used to get the maximum yield of ester (p. 257).

c. Formation of acid chlorides. The —OH group in carboxylic acids (excepting methanoic acid) is readily replaced by a chlorine atom (as it is in alcohols) by reaction with sulphur dichloride oxide, $SOCl_2$, or phosphorus pentachloride. The products are known as acid chlorides:

$$CH_3—COOH + SOCl_2 \rightarrow CH_3—COCl + SO_2 + HCl$$

$$CH_3—COOH + PCl_5 \rightarrow CH_3—COCl + POCl_3 + HCl$$

Ethanoyl chloride

An acid chloride

Notice that the $CH_3—\overset{|}{C}=O$ group is known as *ethanoyl*; in general, RCO— groups are *acyl* or *alkanoyl* groups. Methanoyl chloride does not exist. Further details of acid chlorides are given on p. 253.

d. Chlorination. When hot and in the presence of sunlight, carboxylic acids (excepting methanoic acid) react with chlorine, to form chloro-acids. The hydrogen atoms in the alkyl group adjacent to the carboxyl group are replaced by chlorine atoms, the number of hydrogen atoms replaced depending on the amount of chlorine used, e.g.

$$CH_3—COOH + Cl_2 \rightarrow CH_2Cl—COOH + HCl$$
$$\text{Chloroethanoic acid}$$

$$CH_3—COOH + 2Cl_2 \rightarrow CHCl_2—COOH + 2HCl$$
$$\text{Dichloroethanoic acid}$$

$$CH_3—COOH + 3Cl_2 \rightarrow CCl_3—COOH + 3HCl$$
$$\text{Trichloroethanoic acid}$$

Chloromethanoic acid cannot be prepared.

The chloro-acids are of importance because of their varying strengths (p. 236) and because the halogen atoms can easily be replaced by —OH groups, giving hydroxy-acids (p. 303), or by —NH$_2$ groups, giving amino-acids (p. 304), e.g.

CH$_2$OH—COOH   CH$_2$NH$_2$—COOH
Hydroxyethanoic acid   Aminoethanoic acid
                                 (Glycine)

e. Reduction to primary alcohols. Carboxylic acids can be reduced to primary alcohols but only by specific reducing agents such as lithium tetrahydridoaluminate. An intermediate complex alkoxide is formed and the primary alcohol released from it by acid hydrolysis (p. 350), e.g.

$$R—COOH \xrightarrow{\text{LiAlH}_4} \{(RCH_2O)_4Al\}Li \xrightarrow{\text{dil. HCl}} R—CH_2OH$$

f. Coloration of neutral iron(III) chloride solution. Most carboxylic acids, when just neutralised, give characteristic colours with a neutral solution of iron(III) chloride solution, due to the presence of the hydroxyl group with an adjacent double bond (p. 203).

6 Particular properties of methanoic acid

This acid is not a typical carboxylic acid; it is a stronger acid, contains no alkyl group and is the only acid to contain the —CHO group.

It will not form methanoyl chloride or chloromethanoic acid and it does not form a true anhydride. It can be dehydrated by sulphuric(VI) acid to form carbon monoxide but this will not react with water to reform the acid.

The existence of the aldehyde group in methanoic acid means that it can be oxidised and, therefore, act as a reducing agent; it reduces potassium manganate(VII), Tollens's reagent and Fehling's solution.

Methanoic acid is also less thermally stable than other carboxylic acids, decomposing into carbon dioxide and hydrogen on heating,

$$H—COOH \rightarrow H_2 + CO_2$$

7 Uses of carboxylic acids

Methanoic acid is used in dyeing, electroplating, tanning and in coagulating rubber.

Ethanoic acid is used as a weak acid and as an inert solvent. It is stable enough to be used as the solvent for hot hydrogenations, ozonolyses and other similar processes. As a raw material it is used in manufacturing metallic ethanoates, vinyl and cellulose ethanoates and other esters.

Higher carboxylic acids are needed for making soaps and detergents.

	H—COOH	CH_3—COOH
By oxidation of By acid hydrolysis of special methods	CH_3—OH and H—CHO HCN $CO + NaOH \rightarrow H—COO^-Na^+$ $H—COO^-Na^+ \xrightarrow{H_2SO_4} H—COOH$	C_2H_5—OH and CH_3—CHO CH_3—CN Catalytic atmospheric oxidation of CH_3CHO or $C_nH_{2n+2}(n = 4-7)$
Dimerisation	Both acids exist as dimers in organic solvents	
Acidity Reaction with bases or carbonates Reaction with alcohols Reduction Neutral $FeCl_3$	Stronger acid Forms methanoates Forms esters (methanoates) Reduced to primary alcohols by $LiAlH_4$ Give red colorations	Weaker acid Forms ethanoates Forms esters (ethanoates)
Reaction with $SOCl_2$ or PCl_5 Reaction with Cl_2 Oxidation Reducing action	Methanoyl chloride does not exist Chloromethanoic acid does not exist Oxidised to $CO_2 + H_2O$ Reduces $KMnO_4$, Tollens's reagent and Fehling's solution	Forms ethanoyl chloride Forms chloroethanoic acids Not oxidised No reducing action
Effect of heat Dehydration with c.H_2SO_4	$H—COOH \rightarrow CO_2 + H_2$ Forms CO	Thermally stable Not dehydrated

Preparation and properties of methanoic and ethanoic acids

Aliphatic Dicarboxylic Acids

8 Introduction

The simplest homologous series of dicarboxylic acids have a general formula $HOOC(CH_2)_nCOOH$, e.g.

$$\begin{array}{ccc}
& \overset{\textstyle COOH}{} & \\
COOH & H_2C & H_2C\!-\!COOH \\
| & \diagdown & | \\
COOH & COOH & H_2C\!-\!COOH
\end{array}$$

| Ethanedioic acid | Propanedioic acid | Butanedioic acid |
| (Oxalic acid) | (Malonic acid) | (Succinic acid) |

The systematic names have a -*dioic* ending based on the alkane with the same number of carbon atoms.

The two carboxyl groups react in much the same way as they do in monocarboxylic acids. The acids are dibasic, forming both normal and acid salts. They are stronger than the corresponding monocarboxylic acids, the —COOH group itself exerting a −I effect (p. 24).

Esters, acid chlorides and acid amides are formed as for monocarboxylic acids, e.g.

$$\begin{array}{ccc}
\overset{\textstyle COOC_2H_5}{} & & \\
H_2C & H_2C\!-\!COCl & CONH_2 \\
\diagdown & | & | \\
COOC_2H_5 & H_2C\!-\!COCl & CONH_2
\end{array}$$

| Diethylpropanedioate | Butanedioylchloride | Ethanediamide |
| (Ethylmalonate) | (Succinylchloride) | (Oxamide) |

9 Ethanedioic acid

a. Manufacture. The acid is made by heating sodium methanoate at 400 °C,

$$2H\!-\!COO^-Na^+(s) \xrightarrow{\;400\,°C\;} H_2 + (COO)_2^{2-}(Na^+)_2$$

The sodium salt obtained is dissolved in water and then converted into insoluble calcium ethanedioate by adding calcium chloride or hydroxide. The calcium ethanedioate is filtered off and treated with sulphuric(VI) acid. Calcium sulphate(VI) precipitates and ethanedioic acid can be crystallised out from the filtrate as a dihydrate, $(COOH)_2.2H_2O$.

b. Properties. Ethanedioic acid is a poisonous solid, readily soluble in water. It occurs in rhubarb, particularly in the leaves, which are inedible.

The hydrate loses its water of crystallisation on careful heating but stronger heating causes, first, decarboxylation and, then, further decomposition,

$$(COOH)_2 \xrightarrow{\text{heat}} H-COOH + CO_2 \xrightarrow{\text{heat}} H_2 + 2CO_2$$

Decarboxylation on heating is common amongst the lower dicarboxylic acids, e.g.

$$H_2C\begin{array}{c} \diagup COOH \\ \diagdown COOH \end{array} \xrightarrow[(-CO_2)]{\text{heat}} CH_3-COOH + CO_2$$

Ethanedioic acid is also dehydrated on heating with concentrated sulphuric(VI) acid: carbon monoxide and carbon dioxide are formed,

$$(COOH)_2 \xrightarrow{-H_2O} CO + CO_2$$

Other dicarboxylic acids may form a cyclic anhydride on dehydration, e.g.

Ethanedioic acid is also a reducing agent. It reduces potassium manganate(VII) quantitatively, on warming, the reaction being used in volumetric analysis,

$$(COOH)_2 + [O] \rightarrow H_2O + 2CO_2$$
$$(COO)_2^{2-} \qquad \rightarrow 2CO_2 + 2e^-$$

but this reducing action is not typical of all dicarboxylic acids.

Aromatic Acids

10 Introduction

In aromatic monocarboxylic acids one —COOH group is substituted into a ring, e.g.

Benzoic acid
(Benzene carboxylic acid)

4-methylbenzoic acid
(4-methylbenzenecarboxylic acid)

Notice that the arene content of the name refers to the arene with one fewer carbon atoms than the acid. Benzoic acid, for example, contains *seven* carbon atoms per molecule. That is why some prefer the name benzenecarboxylic acid.

In dicarboxylic acids two carboxyl groups are substituted into the ring, e.g.

Benzene-1,2-dicarboxylic acid
(Phthalic acid)

Benzene-1,4-dicarboxylic acid
(Terephthalic acid)

In sulphonic acids —$SO_2(OH)$ groups are substituted into a ring, e.g.

Benzene sulphonic acid

11 Preparation and manufacture of benzoic acid

Benzoic acid can be made by oxidising phenylmethanol, benzaldehyde or methyl benzene, and by the hydrolysis of (trichloromethyl)benzene or benzonitrile, i.e.

$$C_6H_5—CH_2OH + [2O] \longrightarrow C_6H_5—COOH + H_2O$$

$$C_6H_5—CHO + [O] \longrightarrow C_6H_5—COOH$$

$$C_6H_5—CCl_3 \xrightarrow{\text{hydrolysis}} C_6H_5—COOH$$

$$C_6H_5—CN \xrightarrow{\text{hydrolysis}} C_6H_5—COOH$$

The main commercial source is the catalytic, atmospheric oxidation of methylbenzene,

$$2 \, C_6H_5{-}CH_3 + 3O_2 \xrightarrow{\text{Sn(IV)vanadate}} 2 \, C_6H_5{-}COOH + 2H_2O$$

2 Properties and uses of benzoic acid

Benzoic acid is a white, crystalline solid, readily purified because it is so much more soluble in hot than in cold water. It is very similar to ethanoic acid in its carboxylic acid properties. It is widely used as a food preservative.

a. Acid strength. Benzoic acid is weaker than methanoic acid which means that the C_6H_5 group is donating electrons to the carboxyl group. The C_6H_5 group has a $-I$ effect (p. 24), which would cause electron withdrawal, but this must be overcome by interaction between the delocalised π-electrons in the ring with those in the carboxyl group. Substitution of electron-withdrawing groups, e.g. NO_2 or OH, into the benzene ring causes, in general, an increase in acid strength (p. 236).

b. Acid derivatives. Benzoic acid forms salts with bases or carbonates, and esters, e.g. ethyl benzoate with alcohols. The —OH group can also be replaced by —Cl by reaction with phosphorus pentachloride or sulphur dichloride oxide, $SOCl_2$; an acid chloride, benzoyl chloride, a colourless, fuming liquid, is formed. It reacts with ammonia to form an acid amide, benzamide.

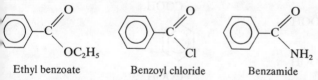

Ethyl benzoate   Benzoyl chloride   Benzamide

3 2-hydroxybenzoic (salicylic) acid

This acid occurs naturally as its methyl ester in some essential oils, e.g. oil of wintergreen.

It is a white, crystalline solid which reacts both as a phenol and a carboxylic acid. Ethanoylation with ethanoyl chloride produces aspirin.

Aspirin

14 Aromatic sulphonic acids

Benzenesulphonic acid is made by the direct sulphonation of benzene (p. 130).

It is a colourless, crystalline solid comparable in strength to the mineral acids. The greater strength, as compared with benzoic acid, arises because sulphur is more electronegative than carbon and because there are more possible resonance hybrids to stabilise the $-SO_3^-$ ion than there are for the $-CO_2^-$ ion.

Benzene sulphonic acid forms an acid chloride and an acid amide by replacing the $-OH$ group by $-Cl$ and $-NH_2$ respectively,

Benzenesulphonylchloride   Benzenesulphonamide

Its sodium salt gives sodium phenate on heating with sodium hydroxide (p. 197).

The sulphonic acids derived from alkyl substituted benzenes are important because their salts are good detergents (p. 192).

15 Aromatic dicarboxylic acids

The commonest aromatic dicarboxylic acids are benzene-1,2- or benzene-1,4-dicarboxylic acids,

Benzene-1,2-dicarboxylic acid
(Phthalic acid)

Benzene-1,4-dicarboxylic
(Terephthalic acid)

The 1,2 isomer is obtained by oxidation of 1,2-dimethyl benzene or naphthalene,

$$\xrightarrow{\text{KMnO}_4} \qquad \xleftarrow[+V_2O_5,\ 450\,°C]{\text{vapour}+\text{air}}$$

The 1,4 isomer is made by oxidation of 1,4-dimethyl benzene.

Both acids have the expected properties of compounds containing two —COOH groups. They form salts, esters, acid chlorides and acid amides. The close proximity of the two carboxyl groups in the 1,2-isomer means that it will readily lose water to give a cyclic anhydride, but the 1,4-isomer will not do this.

The 1,2-acid is important in making phenolphthalein and glyptals; the 1,4-acid in making terylene (p. 332).

Questions on Chapter 18

1 How is vinegar made? How would you obtain a sample of ethanoic acid from vinegar?

2 Methanoic acid differs from ethanoic acid because it contains an aldehyde group but not an alkyl group. Illustrate the truth of this statement.

3 How would you distinguish by chemical means between (a) ethanoic and propanoic acids, (b) methanoic and ethanoic acids, (c) methanoic acid and a mixture of it with ethanoic acid?

4 How would you attempt to show that butanoic acid is formed when butter goes rancid?

5 Explain how the —COOH group in a carboxylic acid can be replaced by (a) —H, (b) —CHO, (c) —COR.

6 What are the possible structural formulae for a carboxylic acid which forms a silver salt containing 51.6 per cent of silver?

7 Comment on the following pK_a values for various acids:

H—COOH	CH_3—COOH	CH_2F—COOH	CH_2I—COOH	C_6H_5—COOH
3.77	4.76	2.66	3.12	4.20

C_6H_5—OH	CH_3—CH_2—CH_2—COOH	HO—CH_2—COOH	$CHCl_2$—COOH
9.9	4.82	3.83	1.29

8 How would you prepare benzoic acid from methylbenzene and how would you convert benzoic acid into ethyl benzoate? How

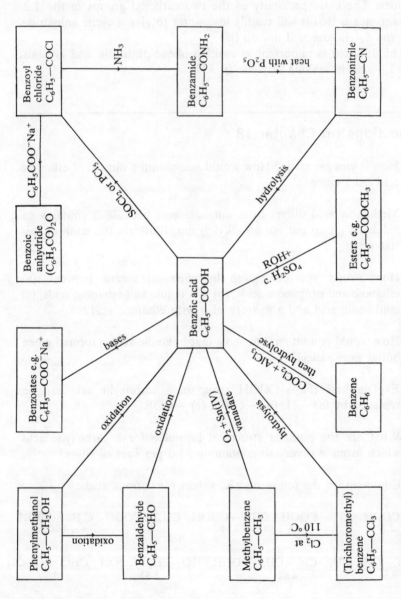

Preparation and properties of benzoic acid

would you expect benzoic acid and ethyl benzoate to differ from ethanoic acid and ethyl ethanoate?

9 Compare and contrast the chemistry of (a) methanoic acid, (b) ethanoic acid, (c) benzoic acid.

10 3-phenylpropeneoic (cinnamic) acid has a formula:

$$C_6H_5—CH=CH—COOH.$$

How would you try to prepare it and what isomeric forms would you expect?

11 The substitution of a methyl group for a hydrogen atom in simple organic compounds usually produces a profound change in their properties'. By using two properties in each case considered, show the validity of this statement when hydrogen atom(s) are substituted by methyl group(s) in the following (a) methanol, (b) methanal, (c) methanoic acid. (L)

12 Given a sample of the acid R—COOH how would you obtain specimens of (a) R—CH$_2$—OH, (b) R—CHO, (c) R—CH$_2$NH$_2$, (d) R—CO—CH$_3$? (O and C)

13 Give examples of (a) addition reactions, and (b) condensation reactions, of the carbonyl group $\diagdown$C=O. Account for the difference in reactivity of the carbonyl group in (a) acids, (b) aldehydes, (c) ketones, (d) esters. (L)

14 An organic acid, X, has a relative molecular mass of 138 and a composition C = 60.87 per cent, H = 4.35 per cent and O = 34.78 per cent. After heating with soda lime it is changed into a compound, Y, of molecular formula C$_6$H$_6$O, which is weakly acidic. What are X and Y?

15 An organic compound, A, of empirical formula C$_2$H$_2$N was reduced to a compound, B, of empirical formula C$_2$H$_6$N. On boiling A with dilute sulphuric(VI) acid an acid, C, was obtained. C was esterified with ethanol and gave an ester, D, which had a relative molecular mass of 174. Deduce the formulae for A, B, C and D. Account for the above reactions and suggest how A could be prepared.

16 An organic acid gave on analysis the following results: C = 57.70 per cent, H = 3.71 per cent and O = 38.57 per cent. The silver salt

of the acid gave 56.8 per cent of silver on ignition and the relative molecular mass of the ethyl ester was 222. What are the molecular formula and the basicity of the acid?

17 Aldehydes, ketones and carboxylic acids contain the carbonyl group ($>C=O$). This group is often said to be responsible for the addition and condensation reactions undergone by aldehydes and ketones. (a) Review briefly the major addition and condensation reactions shown by aldehydes and ketones; in each case discussed, examine the role of the carbonyl group in bringing about these reactions. (b) Carboxylic acids do not undergo any of the addition and condensation reactions shown by aldehydes and ketones, despite the fact that they contain the carbonyl group. Give explanations for this. (L)

19

Derivatives of Carboxylic Acids

1 Introduction

Carboxylic acids form a variety of important derivatives, summarised in the following table:

Name of derivative	General formula	Typical aliphatic examples	
Salts	R—C(=O)—O⁻M⁺	$HCOO^-Na^+$ Sodium methanoate	$(CH_3COO^-)_2Zn^{2+}$ Zinc ethanoate
Acid chlorides	R—C(=O)—Cl	Methanoyl chloride does not exist	CH_3—C(=O)—Cl Ethanoyl chloride
Acid anhydrides	R—C(=O)—O—C(=O)—R	Methanoic acid does not form an anhydride of this type	CH_3—C(=O)—O—C(=O)—CH_3 Ethanoic anhydride
Esters	R—C(=O)—O—R′	H—C(=O)—O—CH_3 Methyl methanoate	CH_3—C(=O)—O—C_2H_5 Ethyl ethanoate
Acid amides	R—C(=O)—NH_2	H—C(=O)—NH_2 Methanamide	CH_3—C(=O)—NH_2 Ethanamide

These derivatives can all be hydrolysed to the acid and they contain the R—C$\diagup^{O}$, *acyl* or *alkanoyl*, group. Their properties depend mainly on the influence of the group to which it is linked. The more strongly electron-withdrawing this group, the higher the positive charge on the C atom of the acyl group and the greater the reactivity so far as nucleophilic attack is concerned. −I groups would be expected to cause electron-withdrawal but this inductive effect may be off-set or overcome by interaction between unshared electrons and the π-electrons of the C=O link. The general order of reactivity is as follows

$$\underset{O^-}{\overset{O}{R-C}} \qquad \underset{NH_2}{\overset{O}{R-C}} \qquad \underset{O-R}{\overset{O}{R-C}} \qquad \underset{O-CO-R}{\overset{O}{R-C}} \qquad \underset{Cl}{\overset{O}{R-C}}$$

→ Increasing reactivity to nucleophilic attack →

It is, then, in general, easier to convert a compound from the right into one to the left than vice versa.

The derivatives have some formal similarity to aldehydes and ketones though they are significantly less susceptible to nucleophilic attack (p. 217). The following summary shows the large number of different environments in which the acyl group occurs.

$$\underset{H}{\overset{O}{H-C}} \qquad \underset{H}{\overset{O}{R-C}} \qquad \underset{R}{\overset{O}{R-C}} \qquad \underset{O-H}{\overset{O}{R-C}}$$

Aldehydes   Ketones   Carboxylic acids

$$\underset{X}{\overset{O}{R-C}} \quad (X = -Cl, -O-OC-R, -OR, -NH_2)$$

Derivatives of carboxylic acids

The change in environment shows up in the stretching frequency of the C=O bond. This changes from $1\,690\ \mathrm{cm^{-1}}$ in amides, to $1\,735$ in esters, $1\,748$ and $1\,820$ in anhydrides and $1\,815$ in acid chlorides. These changes are in line with the change in reactivity.

Salts

2 Preparation of salts

Soluble salts of carboxylic acids are made by the general methods employed in preparing inorganic salts. Most commonly the acid is neutralised by reaction with a base or carbonate and the solid is obtained by crystallisation of the resulting solution.

Insoluble salts are made by reaction between solutions of two soluble salts.

The sodium and potassium salts of higher carboxylic acids, as found in soaps, are obtained from naturally occurring fats (p. 190).

3 General properties of salts

Salts of carboxylic acids are generally soluble, crystalline solids. Their aqueous solutions give characteristic colourations with neutral iron(III) chloride solution. The salts are useful in the preparation of alkanes, acid anhydrides, esters and acid amides, e.g.

$$R{-}COO^-Na^+ + NaOH \xrightarrow{heat} R{-}H + Na_2CO_3 \quad (\text{p. 91})$$

An alkane

$$R{-}COO^-Na^+ + R'{-}COCl \rightarrow R{-}CO{-}O{-}CO{-}R' + NaCl \quad (\text{p. 256})$$

An acid anhydride

$$R{-}COO^-Ag^+ + R'{-}Cl \rightarrow R{-}COOR' + AgCl \quad (\text{p. 258})$$

An ester

$$R{-}COO^-NH_4^+ \xrightarrow{heat} R{-}CONH_2 + H_2O \quad (\text{p. 259})$$

An acid amide

Acid Chlorides

4 Preparation

Acid chlorides are made by replacing the —OH group in a carboxylic acid by reaction with sulphur dichloride oxide or phosphorus penta-chloride, e.g.

$$CH_3{-}COOH + SOCl_2 \rightarrow CH_3{-}COCl + SO_2 + HCl$$

Ethanoyl
chloride

$$R{-}COOH + PCl_5 \rightarrow R{-}COCl + POCl_3 + HCl$$

An acid
chloride

Methanoyl chloride does not exist but other aliphatic and aromatic acid chlorides are common, R standing for any alkyl or aryl group. Acid bromides and iodides also exist; they are made by using red phosphorus and bromine or iodine respectively in reaction with carboxylic acids.

5 General properties of acid chlorides

The lower acid chlorides are colourless liquids with a pungent smell; they fume in moist air as they react with water. They are very reactive towards nucleophilic reagents, the Cl atom being replaced by other groups as summarised below for ethanoyl chloride:

Reagent	Product	Name
Water, H_2O	CH_3—COOH	Ethanoic acid
Alcohols, ROH	CH_3—COOR	Ester
Phenol, C_6H_5—OH	CH_3—COOC$_6$H$_5$	Ester
Ammonia, NH_3	CH_3—CONH$_2$	Ethanamide
Primary amines, RNH_2	CH_3—CONHR	Substituted amide
Sodium salt of carboxylic acid, R—COO$^-$Na$^+$	CH_3—CO—O—OC—R	Acid anhydride
Reduction by Rosenmund reaction	CH_3—CHO	Aldehyde
Reduction by $LiAlH_4$	CH_3—CH$_2$OH	Primary alcohol
Friedel Crafts reaction, C_6H_6	CH_3—CO—C$_6$H$_5$	Phenylethanone

In most of these reactions a hydrogen or metallic atom is replaced by an ethanoyl, CH_3CO—, group. The process is known as *ethanoylation* and the original compound treated with ethanoyl chloride is said to be ethanoylated. Ethanoyl chloride is known as an *ethanoylating* agent.

The reactions are essentially nucleophilic addition reactions followed by an elimination. The C atom in the RCOCl molecule is more positively charged than in carboxylic acids or other acid derivatives (p. 252) because of the strong −I effect of the Cl atom. The C atom is, therefore, open to attack even by weak nucleophiles. Thus, with water or with a charged nucleophile, e.g. CH_3COO^-, as in anhydride formation

The initial nucleophilic attack is like that on aldehydes or ketones (p. 217) but with acid chlorides the intermediate loses Cl^- so that the overall effect is the *substitution* of Cl by X,

The intermediate formed with aldehydes and ketones would have to lose the unstable R^- or H^- ions if it were to react similarly. It is easier to add an H^+ ion leading to an *addition* product,

Benzoyl chloride is less reactive than aliphatic chlorides because the positive charge on the C atom of the COCl group is lowered by delocalisation of electrons between the ring and the COCl group. Benzoylation is, therefore, a slower process than ethanoylation; it is generally carried out in the presence of sodium hydroxide in the *Schotten–Baumann* reaction (p. 200).

Acid Anhydrides

6 Preparation of acid anhydrides

Carboxylic acids, excepting methanoic acid, form anhydrides by the elimination of one molecule of water from two molecules of acid, e.g.

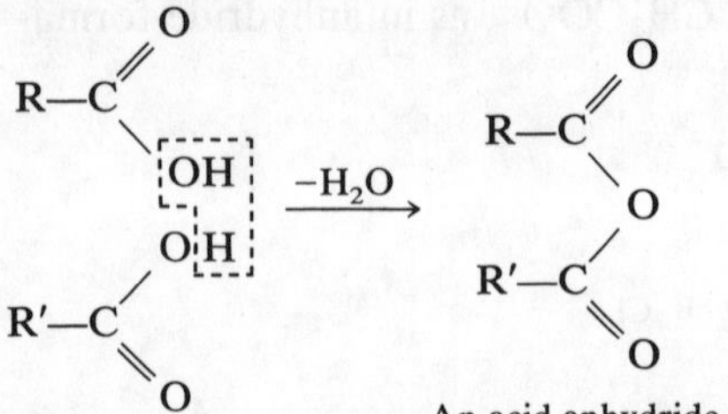

$$R-C \overset{O}{\underset{\begin{array}{c}|OH|\\O|H|\end{array}}{\diagdown}} \quad \xrightarrow{-H_2O} \quad R-C\overset{O}{\diagdown}_{O} $$

$$R'-C\diagup_{O}\qquad\qquad R'-C\diagup_{O}$$

An acid anhydride

If the two alkyl groups are alike the product is known as a *simple anhydride*; if different, as a *mixed anhydride*.

Anhydrides can be obtained by catalytic dehydration of acids but the most widely applicable method involves reaction between an acid chloride and the anhydrous sodium salt of a carboxylic acid, e.g.

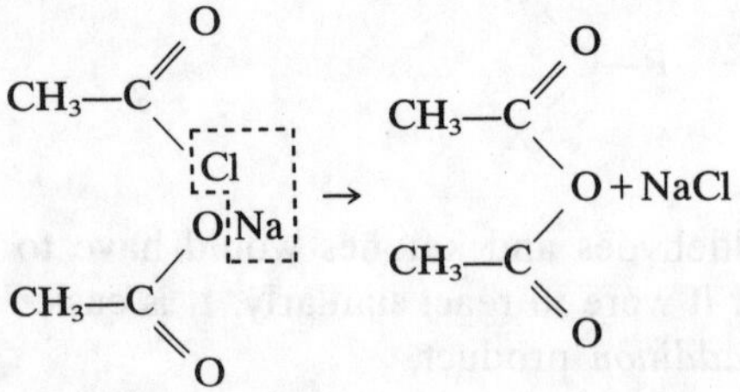

$$CH_3-C\overset{O}{\underset{\begin{array}{c}Cl\\O|Na|\end{array}}{\diagdown}} \quad \rightarrow \quad CH_3-C\overset{O}{\diagdown}_{O} + NaCl$$

$$CH_3-C\diagup_{O}\qquad\qquad CH_3-C\diagup_{O}$$

The reaction involves nucleophilic attack on the acid chloride by the carboxylate ion (p. 255).

Ethanoic anhydride, which is most commonly used, is manufactured from propanone via ethenone,

$$CH_3-CO-CH_3 \xrightarrow[800\,°C]{Ni/Cr\ alloy} CH_2=C=O+CH_4$$

Ethenone
(Ketene)

$$CH_2=C=O + CH_3-COOH \rightarrow CH_3-CO-O-CO-CH_3$$

7 General properties of acid anhydrides

Acid anhydrides react as ethanoylating agents in much the same way as acid chlorides though they are less reactive because the electron-withdrawing powers of the $CH_3-CO-O-$ group are lower than those of the Cl atom.

The anhydrides participate in all the reactions listed on p. 254 except the Rosenmund reaction and the reaction with sodium salts of carboxylic acids. Ethanoic anhydride is used on a large scale in making cellulose ethanoate (p. 333) and ethenyl (vinyl) ethanoate (p. 333).

Aromatic acid anhydrides are less reactive than the aliphatic ones.

Esters

8 Preparation of esters

a. From carboxylic acid and alcohol. The reaction between a carboxylic acid and an alcohol (esterification) is reversible, e.g.

$$R-COOH + R'-OH \rightleftharpoons R-COOR' + H_2O$$

The equilibrium constant is about 1 and it is necessary to use a large excess of alcohol to obtain a good yield of ester. A little concentrated sulphuric(VI) acid, or dry hydrogen chloride gas, is also necessary as a catalyst.

A convenient procedure for making ethyl ethanoate is to run a mixture of ethanoic acid and ethanol into a hot mixture of ethanol and concentrated sulphuric(VI) acid at the same rate as ester distils off. The distillate is then shaken with sodium carbonate (to remove acids) and with concentrated calcium chloride solution to remove ethanol. After standing over anhydrous calcium chloride, the ethyl ethanoate can be redistilled.

The acid catalyst functions by protonating the oxygen atom of the C=O bond in the carboxylic acid. This increases the positive charge on the carbon atom so that it is more readily attacked by the alcohol acting as a nucleophile. Thus

$$R-\overset{O^{\delta-}}{\underset{OH}{C^{\delta+}}} \xrightarrow{H^+} R-\overset{OH}{\underset{OH}{C^+}} \xrightarrow{R'\ddot{O}H} R-\overset{OH}{\underset{R'O^+H}{C}}-OH$$

$$R-\overset{OH}{\underset{R'O^+H}{C}}-OH \xrightarrow{-H_2O} R-C\overset{O^+H}{\underset{OR'}{}} \xrightarrow{H_2O} R-C\overset{O}{\underset{OR'}{}} + H_3O^+$$

The water eliminated from the carboxylic acid and alcohol is formed from the —OH group of the acid and the H atom of the alcohol. This has been proved by using alcohols containing ^{18}O atoms; the water resulting from their reaction with acids did not contain ^{18}O atoms.

b. From acid chlorides and acid anhydrides. Both these acid derivatives react with alcohols, more easily than carboxylic acids do, to form esters. The alcohols function as nucleophiles as in (a).

c. From the silver salt of carboxylic acids. These salts react with halogenoalkanes to form esters. The carboxylic ion attacks the halogenoalkane as a nucleophile, e.g.

$$R\text{—}COO^- Ag^+ + R'\text{—}I \rightarrow R\text{—}COOR' + AgI$$

9 General properties and uses of esters
Esters are colourless liquids or solids, sparingly soluble in water. Most of them have a characteristic fragrant odour and many of them occur naturally. They are not very reactive but can be hydrolysed to carboxylic acids and alcohols, reduced to primary alcohols and converted into amides by reaction with ammonia.

a. Hydrolysis of esters. Esters are hydrolysed very slowly by boiling with water,

$$R\text{—}C\overset{\displaystyle O}{\underset{\displaystyle OR'}{\Big\langle}} + H_2O \rightleftharpoons R\text{—}C\overset{\displaystyle O}{\underset{\displaystyle OH}{\Big\langle}} + R'\text{—}OH$$

The use of $H_2{}^{18}O$ in the reaction and the discovery that the ^{18}O occurs in the carboxylic acid and not in the alcohol shows that the bond cleavage is as indicated in the above equation.

The hydrolysis can be catalysed both by acids and bases. The acid-catalysed process is the reverse of esterification (p. 257) and, as it is reversible, an excess of acidic solution is required to give a good yield of ester.

The base-catalysed hydrolysis is a more efficient process for the carboxylate ion formed does not react appreciably with alcohols so that the reaction goes almost to completion, e.g.

$$R\text{—}COOR' + OH^- \rightarrow R\text{—}COO^- + R'\text{—}OH$$

As it is the salt of a carboxylic acid that is formed and as naturally occurring esters are converted into soaps by alkaline hydrolysis the process is known as *saponification* (p. 189).

The OH^- ion functions as a catalyst by nucleophilic attack

$$R-C^{\delta+}\underset{OR'}{\overset{O^{\delta-}}{\Big<}} \xrightarrow{OH^-} R-\underset{OR'}{\overset{OH}{\underset{|}{\overset{|}{C}}}}-O^- \rightleftharpoons R-C\underset{O}{\overset{OH}{\Big<}} + R'O^-$$

$$\downarrow$$

$$R-C\underset{O}{\overset{O^-}{\Big<}} + R'-OH$$

b. Reduction of esters. Esters can be reduced to primary alcohols more easily than carboxylic acids. Lithium tetrahydridoaluminate, sodium and ethanol, and catalytic hydrogenation are all effective, but the reduction cannot be brought about by sodium tetrahydridoborate,

$$R-COOR' \xrightarrow[\text{or } H_2 + \text{catalyst}]{\text{LiAlH}_4 \text{ or Na/C}_2\text{H}_5\text{OH}} R-CH_2OH + R'OH$$

c. Reaction with ammonia. Esters react with ammonia to form amides but less readily than acid chlorides or acid anhydrides, e.g.

$$R-COOR' + NH_3 \rightarrow R-CONH_2 + R'-OH$$

10 Uses of esters

Some esters, e.g. ethyl ethanoate, are good solvents but the more general use of esters is as perfumes and flavourings. Ethyl methanoate (rum), 3-methylbutyl ethanoate (pear) and ethyl butanoate (pineapple) are all used in this way.

Certain useful waxes are composed mainly of esters, e.g. beeswax, $C_{15}H_{31}COOC_{31}H_{63}$, and spermaceti, $C_{15}H_{31}COOC_{16}H_{33}$.

Esters of trihydric alcohols and carboxylic acids are of great importance in making soap and propane-1,2,3-triol (glycerol) (p. 189).

Acid Amides

11 Preparation of acid amides

a. By heating the ammonium salt of a carboxylic acid, e.g.

$$R-COO^-NH_4^+ \xrightarrow{\text{Heat}} R-CONH_2 + H_2O$$

Ethanamide is made by this method by boiling a mixture of ammonium carbonate and excess ethanoic acid under a reflux condenser. The amide is then distilled off.

b. From other acid derivatives. Acid chlorides, anhydrides and esters will all react with ammonia to form acid amides. The chlorides react most easily, and the esters least easily (p. 252).

12 General properties of acid amides

Apart from methanamide, which is a colourless liquid, the amides are colourless, crystalline solids. The relatively high boiling points are caused by hydrogen bonding. The lower members of the series are soluble in water.

a. Acid-base character. Acid amides are weak bases,

$$R—CONH_2 + H^+ \rightleftharpoons R—CONH_3^+$$

forming salts with strong acids, e.g. $CH_3CONH_2 \cdot HNO_3$. The basicity arises because the lone pair on the N atom can donate to a hydrogen ion. This happens less readily with the amides than with primary amines (p. 294) so that the former are weaker bases. The amide, but not the amine, is stabilised by resonance between structures I and II below,

Alternatively, the unshared pair of electrons on the N atom of an amide interacts with the π-electrons of the adjacent C=O bond as in other acid derivatives (p. 252). Neither the resonance nor the interaction are possible with primary amines.

Acid amides can also function as very weak acids

$$R—CONH_2 \rightleftharpoons R—CONH^- + H^+$$

but this is of no practical significance.

b. Hydrolysis of amides. Acid amides can be hydrolysed in acid or alkaline solution, e.g.

$$R—CONH_2 + H_2O + H^+ \rightarrow R—COOH + NH_4^+$$
$$R—CONH_2 + OH^- \rightarrow R—COO^- + NH_3$$

The mechanisms of the reactions are very similar to those for the hydrolysis of esters but amides are hydrolysed more slowly. The liberation of ammonia on warming with an alkaline solution enables an amide to be distinguished from a primary amine (p. 291).

c. Dehydration of amides. Phosphorus(V) oxide dehydrates amides to nitriles on heating, e.g.

$$R—CONH_2 \xrightarrow{-H_2O} R—C\equiv N$$

d. Reaction with nitric(III) acid. Acid amides are converted into the corresponding acid by reaction with nitric(III) acid, the reaction resembling that of primary amines (p. 296), e.g.

$$R—CONH_2 + HNO_2 \rightarrow R—COOH + N_2 + H_2O$$

e. Reaction with bromine and potassium hydroxide solution (Hofmann's reaction). The overall effect of this reaction is to convert the —CONH$_2$ group of an amide into an —NH$_2$ group, e.g.

$$R—CONH_2 + Br_2 + 4KOH \rightarrow R—NH_2 + 2KBr + K_2CO_3 + 2H_2O$$

The process can be used for descending a homologous series.

The mechanism is well understood for many of the intermediates can be isolated:

f. Reduction of amides. Most acid amides can be reduced to a primary amine by lithium tetrahydridoaluminate, e.g.

$$R—CONH_2 \xrightarrow{LiAlH_4} R—CH_2NH_2$$

Carbamide (Urea)

13 Introduction

Carbonic acid, H_2CO_3, is generally regarded as an inorganic acid but it can be thought of as hydroxymethanoic acid. It is a dibasic acid forming both acid and normal salts. It also forms typical organic acid derivatives such as esters, chlorides and amides, e.g.

$$O=C\Big\langle{}^{OH}_{OH} \qquad O=C\Big\langle{}^{OC_2H_5}_{OC_2H_5} \qquad O=C\Big\langle{}^{Cl}_{Cl} \qquad O=C\Big\langle{}^{NH_2}_{NH_2}$$

| Carbonic acid | Ethyl carbonate | Carbonyl chloride | Carbamide (Urea) |

The synthesis of carbamide (Wohler, 1827) by heating ammonium cyanate,

$$NH_4CNO \xrightarrow{\text{Heat}} CO(NH_2)_2$$

was of historical significance. Hitherto, carbamide, which is present to an extent of about 2.5 per cent in mammalian urine, had been regarded as a typically organic material and it was imagined that a mysterious *Vital Force* was in some way associated with organic, but not with inorganic, substances. To make an organic substance from purely inorganic sources was, in 1827, an important discovery.

14 Preparation of carbamide

Carbamide can be obtained by reaction between carbonyl chloride and ammonia, the carbonyl chloride functioning as a typical acid chloride. It is, however manufactured by heating ammonia and carbon dioxide at 200 °C and 200 atmospheres pressure,

$$CO_2 + 2NH_3 \xrightarrow[\text{200 atm}]{\text{200 °C}} CO(NH_2)_2 + H_2O$$

Some ammonium carbonate, $(NH_4)_2CO_3$, and ammonium carbamate, $NH_2-COO^-NH_4^+$, tend to be formed as by-products.

15 Properties of carbamide

Carbamide is a white, crystalline solid, readily soluble in water but insoluble in organic solvents. It resembles the amides of carboxylic acids in many of its properties.

a. Formation of salts. Carbamide is a stronger base than ethanamide but weaker than ethylamine. It will form salts with acids, those with ethanedioic and nitric(V) acids being the most important because they are insoluble.

b. Hydrolysis. Carbamide is hydrolysed on heating with aqueous solutions of caustic alkalies, e.g.

$$CO(NH_2)_2 + 2NaOH \rightarrow 2NH_3 + Na_2CO_3$$

This hydrolysis is also brought about by certain enzymes, e.g. urease (found in soya bean), or by soil bacteria.

c. Reaction with nitric(III) acid. Nitric(III) acid converts the —NH_2 group of acid amides into an —OH group (p. 261) and it functions in the same way with carbamide,

$$CO(NH_2)_2 + 2HNO_2 \rightarrow 2N_2 + 2H_2O + H_2CO_3$$

d. Effect of heat. Urea melts at 132 °C but decomposes on stronger heating with the loss of ammonia. The products include a substance known as biuret, which forms as a white solid when the hot mixture is cooled,

$$2CO(NH_2)_2 \xrightarrow{\text{heat}} NH_2—CO—NH—CO—NH_2 + NH_3$$
Biuret

Biuret dissolves in sodium hydroxide solution and gives a purple colouration when a few drops of copper(II) sulphate(VI) are added to the alkaline solution. This serves as a simple test (*the biuret test*) for carbamide, and also for proteins, peptides (p. 306) and other substances containing —CO—NH— groups.

16 Uses of carbamide

a. In making plastics. Carbamide reacts with methanal to form plastics (p. 328).

b. As a fertiliser. Carbamide contains 46.7 per cent by weight of nitrogen and is a good nitrogenous fertiliser. Soil bacteria convert the carbamide into, first, ammonia and carbon dioxide and, then, nitrates(V), which plants can assimilate. Many proteins in animal and vegetable manure are first broken down into carbamide in the soil.

c. In making derivatives of carbamide. Many substituted carbamides and carbamide derivatives are important medicinally. The barbiturates, derivatives of barbituric acid, made from carbamide and ethylpropanedioate, are best known e.g.

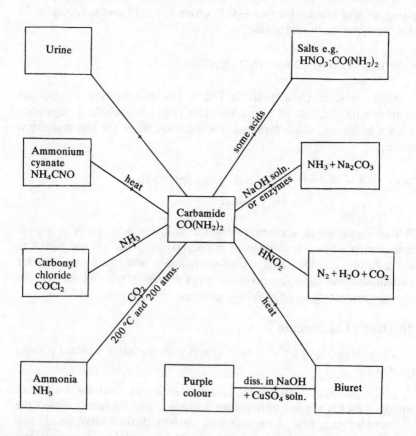

Barbitone
(barbital, veronal)

Phenobarbitone
(luminal)

Preparation and properties of carbamide

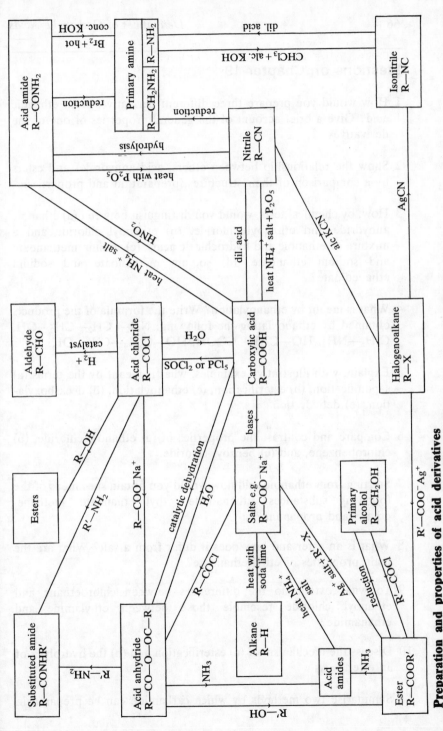

Preparation and properties of acid derivatives

Questions on Chapter 19

1 How would you prepare three different derivatives from ethanoic acid? Give a brief account of the general properties of one of the derivatives.

2 Show the relationship between ethers, acid anhydrides and esters by a comparison of their structures, preparation and properties.

3 How, by chemical tests, would you distinguish between (a) ethanoic anhydride and ethanoyl chloride, (b) ethanoyl chloride and a mixture of ethanoic and hydrochloric acids, (c) sodium methanoate and sodium ethanoate, (d) sodium methanoate and sodium ethanedioate?

4 What is meant by ethanoylation? Write the formula of the products obtained by ethanoylating the following: $NH_2—CH_2—CH_2—OH$, $C_6H_5—NH_2$, $HO—C_6H_4—COOH$, $HO—C_6H_4—CH_2—OH$.

5 Explain, with illustrative examples, what is meant by the terms (a) saponification, (b) esterification, (c) ethanoylation, (d) decarboxylation, (e) dehydration.

6 Compare and contrast the properties of (a) ethanoyl chloride, (b) chlorobenzene, and (c) benzoyl chloride.

7 Starting from ethanoic acid how would you obtain specimens of the following substances: ethanamide, ethyl ethanoate, methane, ethane and propanone?

8 What is an ester and how does it differ from a salt? What are the main properties of ethyl ethanoate?

9 To what extent do the differences between chloroethane and ethanoyl chloride resemble those between ethylamine and ethanamide?

10 Discuss the mechanism of (a) esterification and (b) the hydrolysis of an ester.

11 Summarise two methods by which carbamide can be prepared in

the laboratory. How does it react with (a) nitric(III) acid, and (b) on heating? How can carbamide be estimated in aqueous solution?

12 How can carbamide be extracted from urine? How does it get into the urine? How would you (a) detect the presence of and (b) estimate the amount of nitrogen in carbamide?

13 How would you distinguish between aqueous solutions of (a) carbamide and ammonium carbonate, (b) methylamine and ethanamide, (c) methanoic and ethanedioic acids?

14 Describe the preparation of benzoyl chloride from benzoic acid, How does benzoyl chloride react with (a) water, (b) ammonia, (c) sodium hydroxide and (d) glycine?

15 A substance, A, of molecular formula $C_3H_4OCl_2$ reacted with cold water to give a compound, B, $C_3H_5O_2Cl$. A on treatment with ethanol gave a liquid C, $C_5H_9O_2Cl$. When A was boiled with water, a compound D, $C_3H_6O_3$ was obtained. D was optically active and could be ethanoylated. Deduce the nature of the compounds A, B, C and D, and account for the reactions.

16 Name and give the structural formulae of the aliphatic acids and esters which have an empirical formula C_2H_4O and a molecular weight of 88. Outline the chemical tests that you would apply to enable you to distinguish between each isomer. (L)

17 1.76 g of a mixture of a monobasic organic acid and its ethyl ester required 10 cm³ of M NaOH solution to neutralise the acid and an equal volume of the same solution to saponify the ester. Calculate the relative molecular mass of the acid and the amount of ester present in the mixture.

18 The structures of ethers, esters and anhydrides are similar in that each consists of two alkyl and/or acyl groups attached to an oxygen atom. Compare and contrast in tabular form the methods of preparation, physical properties and chemical reactions of an ether, an ester and an acid anhydride. (L)

19 Give the full structural formulae of all the esters having the molecular formula $C_4H_8O_2$. How would you distinguish chemically between these isomeric esters? Give equations, reagents and conditions for the reactions you mention. (Camb. Local)

20 An aromatic compound P, $C_{10}H_{12}O_2$, was refluxed with sodium hydroxide solution and gave, on distillation, a liquid Q, C_2H_6O. The residue, on acidification, gave R, $C_8H_8O_2$, which, on dry distillation with soda lime, gave S, C_7H_8. Write possible structural formulae for P, Q, R and S and explain the reactions. Give one chemical test in each case to confirm the identity of Q and S. What is the general reaction of P and its isomers with concentrated ammonia solution? (Camb. Local)

21 A compound A has an empirical formula $C_2H_2Cl_2O$ and a molecular weight of 113. It fumes in air, and reacts with water to give B, empirical formula $C_2H_3ClO_2$. The immediate reaction of A with aqueous sodium hydroxide is vigorous, and prolonged action of this reagent gives C, empirical formula $C_2H_3O_3Na$. Identify A, B and C and explain the above reactions. Write equations for the reactions you would expect between the following, giving the name of the organic product in each case. (a) A and aqueous ammonia. (b) B and chlorine. (c) C and acidified potassium permanganate solution. (L)

20

Isomerism

1 Structural isomerism (p. 13).

Structural isomers may belong to the same or different homologous series as seen in the following classification of types.

a. Chain isomerism. This depends on the way in which a chain of carbon atoms are linked together, e.g.

$$CH_3—CH_2—CH_2—CH_2—CH_3$$
Pentane

$$CH_2—\overset{\overset{\displaystyle CH_3}{|}}{CH_2}—CH_2—CH_3$$
2-methylbutane

$$CH_3—\overset{\overset{\displaystyle CH_3}{|}}{\underset{\underset{\displaystyle CH_3}{|}}{C}}—CH_3$$
2,2-dimethyl propane

b. Position isomerism. Isomers may differ from each other in the position of the same functional group, e.g.

$$CH_3—CH_2—CH_2—OH$$
Propan-1-ol

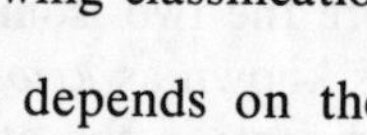

$$CH_3—\overset{\overset{\displaystyle OH}{|}}{CH}—CH_3$$
Propan-2-ol

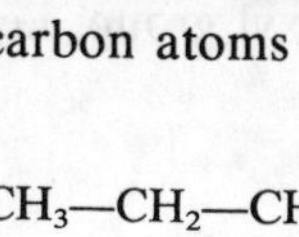

Chloro-2-methyl benzene

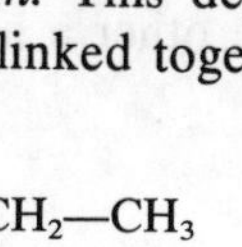

Chloro-3-methyl benzene

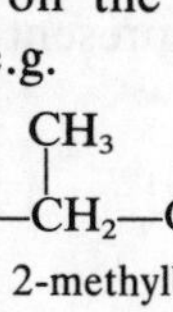

Chloro-4-methyl benzene

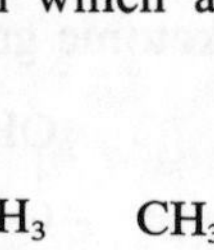

(Chloromethyl) benzene

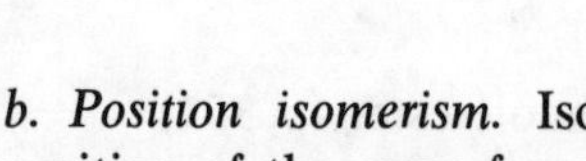

$$CH_3—CH_2—\overset{\overset{\displaystyle O}{||}}{C}—CH_2—CH_3$$
Pentan-3-one

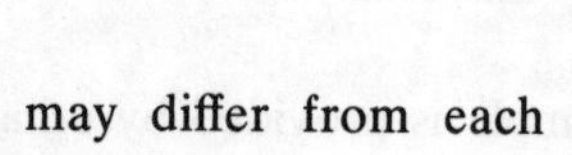

$$CH_3—\overset{\overset{\displaystyle O}{||}}{C}—CH_2—CH_2—CH_3$$
Pentan-2-one

c. Functional group isomerism. Isomerism can arise between different homologous series which contain different functional groups, e.g.

$$CH_3—O—CH_3 \qquad CH_3—CH_2—OH$$

Methoxymethane   Ethanol

d. Tautomerism. Dynamic isomerism. This type of isomerism involves the existence of two structural isomers which are mutually interconvertible and form a mixture which is in dynamic equilibrium. It may, or may not, be possible to separate the two isomers from the mixture.

Hydrogen cyanide is thought to be tautomeric,

$$H—C{\equiv}N \;\rightleftharpoons\; H—N{\equiv}C$$

but the two possible isomers have not been isolated.

Commoner examples where the two isomers can be isolated at low temperatures involve what is known as *keto-enol* tautomerism. In one isomer a ketonic group is present; in the other a hydroxyl group, i.e.

Keto-form   Enol-form

A typical example is provided by ethanoethanoic ester (p. 351)

Ethyl-3-oxobutanoate   Ethyl-3-hydroxybut-2-enoate

Aldehydes and ketones also exhibit keto-enol tautomerism (p. 216).

Geometric Isomerism

2 Free rotation about single bonds

Two groups linked by a single bond can, normally, rotate about the bond and take up an infinite number of positions in relation to each other. The two extreme forms in ethane, C_2H_6, are known as the eclipsed and staggered conformations (Fig. 50).

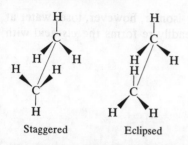

Staggered   Eclipsed

Fig. 50. Two possible extreme conformations of a C_2H_6 molecule. The energy difference between the two forms is so low that they have not been isolated.

The staggered form is slightly more stable than the eclipsed form as the hydrogen atoms in it are as far away from each other as they can possibly be. But the energy difference between the forms (about 12 kJ mol^{-1}) is very small and it has not been possible to isolate them. Ethane molecules exist in any of the infinite number of possible conformations.

If the H atoms in ethane are replaced by larger atoms as, for example, in $CHBr_2$—$CHBr_2$, free rotation about the C—C bond will be impeded because the large bromine atoms will 'get in the way'; the phenomenon is an example of what is known as steric hindrance. The eclipsed and staggered forms of $CHBr_2$—$CHBr_2$ have been isolated at low temperatures.

3 Geometric isomers

The presence of a $\diagup C{=}C \diagdown$ link or a ring of atoms in a molecule can 'lock' the molecule in position by preventing free rotation and this causes geometric isomerism.

a. Butenedioic acids. There are two isomers of butenedioic acid, that with like groups on the same side of the $\diagup C{=}C \diagdown$ bond being called the *cis*-form and the other the *trans*-form, i.e.

H   H   COOH

C=C   C=C

HOOC   COOH   HOOC   H

cis-butenedioic acid   *trans*-butenedioic acid

(maleic acid)   (fumaric acid)

The two isomers differ in physical properties, particularly in dipole moment. In most respects they are chemically similar for they are both unsaturated dicarboxylic acids. The *cis*-isomer, however, loses water at 150 °C to form an anhydride which readily re-forms the *cis*-acid with water,

$$
\begin{array}{ccc}
H & COOH \\
\ \ \diagdown C \diagup \\
\ \ \ \| \\
\ \ \diagup C \diagdown \\
H & COOH
\end{array}
\quad
\underset{H_2O}{\overset{150\,°C}{\rightleftharpoons}}
\quad
\begin{array}{ccc}
 & & O \\
 & & \| \\
H & & C \\
\ \ \diagdown C \diagup \ \ \diagdown \\
\ \ \ \| \qquad\quad O \\
\ \ \diagup C \diagdown \ \ \diagup \\
H & & C \\
 & & \| \\
 & & O
\end{array}
$$

The trans-isomer does not lose water at 150 °C but it can be converted into the cis-anhydride at higher temperatures, presumably via a conversion into the cis-acid.

b. Oximes (p. 220). The $\diagup C{=}N{-}$ link in oximes prevents free rotation and causes isomerism, e.g.

$$
\begin{array}{cc}
H_3C & OH \\
\ \ \ \diagdown C {=} N \diagup \\
\ \ \diagup \\
H
\end{array}
\qquad
\begin{array}{cc}
H_3C & \\
\ \ \ \diagdown C {=} N \diagdown \\
\ \ \diagup \qquad OH \\
H
\end{array}
$$

c. Ring compounds. A ring can make a molecule rigid and cause geometric isomerism. 1,2-dimethylcyclopropane, for example, exists in two isomeric forms, i.e.

trans-isomer   *cis*-isomer

The cyclopropane ring is planar. As only three carbon atoms are involved there is no alternative though there is very considerable distortion in the C—C—C bond angles (60 °) as compared with the normal value of 109.5 °. Cyclobutane, C_4H_8, and cyclopentane, C_5H_{10}, are both non-planar rings.

The cyclohexane ring exhibits what can be regarded as a type of geometric isomerism. There are two theoretically possible isomers; they are known as the '*boat*' and the '*chair*' conformations and they both have bond angles of 109.5 ° (Fig. 51). It has not been possible to

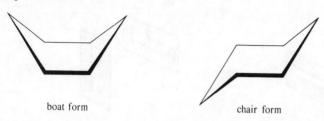

boat form   chair form

Fig. 51. Boat and chair conformations of cyclohexane, C_6H_{12}.

separate the two isomers for the energy difference between them is small, so that rapid inter-conversion between the forms is possible at room temperature. Some 99 per cent of the molecules in cyclohexane at room temperature are in the 'chair' form for this is the more stable form as the hydrogen atoms in it are further away from each other and, therefore, interact less.

Optical Isomerism

4 Polarisation of light

In an ordinary beam of light, vibrations are taking place in all directions at right-angles to the direction of the beam. If such a beam of light is passed through a *polariser*, e.g. a tourmaline crystal, a piece of polaroid (made from a periodide of quinine sulphate(VI)), or crystals of Iceland Spar cut and mounted in what is known as a *Nicol prism*, the vibrations in the emergent beam of light are found to be all in the same plane. Light of this nature is said to be *plane polarised*. A polariser is only able to transmit vibrations in one particular plane. It acts as though it is a slit, transmitting vibrations which are parallel to the slit, but stopping all other vibrations. If a beam of light which has been plane polarised by passing through a polariser is viewed through a second polariser (known as an *analyser*), what is seen depends on the relative positions of the polariser and the analyser. If their axes are parallel all light coming from the polariser will pass through the analyser; this is like having two parallel slits (Fig. 52a). But if the analyser is now rotated, more and more light will be cut out until, when the polariser and analyser have their axes at right-angles to each other, i.e. are crossed, no light will pass through the analyser. This is like having two slits at right-angles to each other (Fig. 52b); what can get through one cannot get through the other.

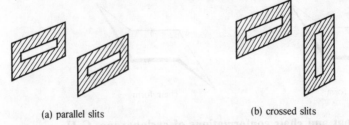

(a) parallel slits   (b) crossed slits

Fig. 52. Parallel and crossed slits to represent possible positions of polariser and analyser.

5 Rotation of plane of polarisation

If a polariser and an analyser are set up so that they are parallel, they will transmit the maximum amount of plane polarised light. But if certain crystals, e.g. quartz, or liquids, e.g. butan-2-ol, or solutions, e.g. aqueous solutions of 2-hydroxypropanoic (lactic) acid or 2-aminopropanoic acid (alanine), are placed between the polariser and analyser, then the analyser will have to be rotated either to the right or the left to restore maximum illumination.

This is bcause the substance placed between the polariser and analyser has rotated the plane of polarisation of the light coming from the polariser. To get maximum illumination, the analyser has to be rotated accordingly. Substances which have the power of rotating the plane of polarised light are said to be *optically active.*

Experimental measurements are made on a polarimeter, which consists, essentially, of one Nicol prism acting as a polariser with another acting as an analyser. Between the two there is room for placing the optically active substance under examination (Fig. 53).

Some optically active substances rotate the plane of polarisation to the right, i.e. clockwise as viewed through the analyser; they are said to be *dextro-rotatory* and they are given a postive angle of optical

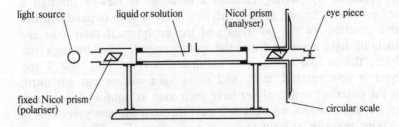

light source   liquid or solution   Nicol prism   eye piece
                                               (analyser)

fixed Nicol prism   circular scale
(polariser)

Fig. 53. A simple polarimeter.

rotation. Others rotate the plane of polarisation to the left; they are *laevo-rotatory* and are given a negative angle of optical rotation.

So that the rotating power of different substances can be compared it is necessary to specify the conditions under which this is done. In SI units, the angle of optical rotation given by a 1 metre column of a solution containing 1 kg m^{-3} is used; it is called the *specific optical rotatory power* of the substance concerned. Other units have also been used and the corresponding molar optical rotatory power for a solution containing 1 mol m^{-3} is also used.

6 Optical activity and asymmetry

Examples of solids, liquids and solutions which were optically active were first investigated in the early half of the nineteenth century. The results obtained were startling and, at the time, unexplained.

One of the most important advances in understanding the phenomena was made by Pasteur in 1848. He found that two different types of crystals of the sodium ammonium salt of racemic acid (p. 279) could be obtained. These crystals were both asymmetric and differed in that one crystal was the mirror-image (*enantiomorph*) of the other just as the left hand is the mirror image of the right (Fig. 54).* Pasteur separated the two crystals by hand-picking, dissolved them separately in water, and then discovered that one solution was dextro- whilst the other was laevo-rotatory. The original solution containing equal amounts of both types of crystals had been optically inactive.

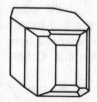 

Fig. 54. The two different types of sodium ammonium salt crystals obtained by Pasteur from racemic acid. The crystals are asymmetric and are enantiomorphic, i.e. they are mirror-images of each other.

This discovery indicated a connection between optical activity and asymmetry but it was difficult, initially, to see how a liquid or a solution could be in any way asymmetric. The problem was solved by the realisation that it was asymmetry *within a molecule* that mattered

---

* The term 'chiral' is sometimes used; it refers to left-handedness or right-handedness and is virtually synonymous with 'dissymmetric'.

and by the suggestion of Le Bel and van't Hoff (1874) that four monovalent groups might be arranged tetrahedrally around a central carbon atom.

There was a contemporary realisation that CX_4 molecules were not flat, planar molecules because it had not been possible to separate two isomers of compounds with general formula CX_2Y_2. Only one form of such a compound would be expected if the molecule was tetrahedral; two forms would be expected (Fig. 55) if it was flat. But it was the way in which the tetrahedral molecular shape could account for optical isomerism that was of greater importance.

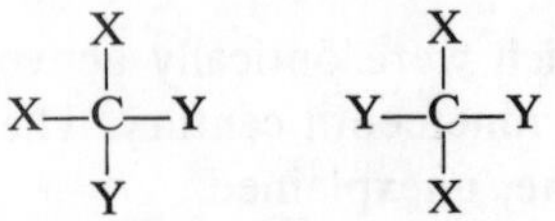

$$
\begin{array}{cc}
\text{X} & \text{X} \\
| & | \\
\text{X--C--Y} \quad & \quad \text{Y--C--Y} \\
| & | \\
\text{Y} & \text{X}
\end{array}
$$

Fig. 55. The two isomers of CX_2Y_2 that would be expected if the molecule was planar.

7 Molecules with one asymmetric carbon atom

A carbon atom linked to four different groups is known as an *asymmetric carbon atom;* it is sometimes denoted by an asterisk. Any molecule containing one asymmetric carbon atom is asymmetric as a whole and there will be two optically active forms and one optically inactive form of the substance.

The two optically active forms represent the two enantiomorphic molecular structures. One of the isomers is dextro- and the other laevo-rotatory; they are written as the (+)-form and the (−)-form.* The two optically active forms differ in their effect on polarised light and in the shape of their crystals, which are enantiomorphic. Otherwise, they have identical physical properties.

The optically inactive form is simply an equimolecular mixture of the two optically active forms; it is written as the (±)-form. Its physical properties, e.g. melting point, solubility and density, are not the same as those of the optically active isomers.

The possible isomers of 2-hydroxypropanoic (lactic) acid are shown in Fig. 56.

The optically inactive form is referred to as a *racemic mixture* (the term was introduced in dealing with racemic acid) and it is said to be *externally compensated,* the optical inactivity being due to the equal but opposite rotatory effects of the two components of the mixture. A

---

* The older usage of *d-* for (+)- and *l-* for (−)- still persists but is less satisfactory because of the confusion in the usage of *D-* and *L-* (p. 282) with *d-* and *l-*.

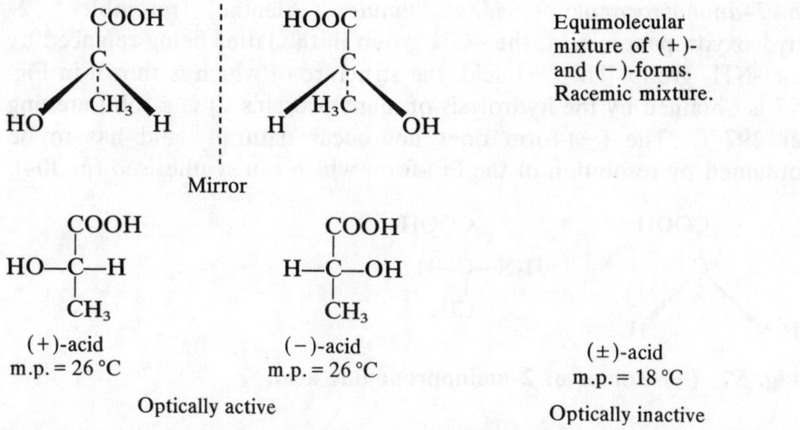

COOH
|
C
/ | \
HO   CH₃   H

HOOC
|
C
/ | \
H   H₃C   OH

Equimolecular
mixture of (+)-
and (−)-forms.
Racemic mixture.

Mirror

COOH
|
HO—C—H
|
CH₃

(+)-acid
m.p. = 26 °C

Optically active

COOH
|
H—C—OH
|
CH₃

(−)-acid
m.p. = 26 °C

(±)-acid
m.p. = 18 °C

Optically inactive

Fig. 56. The isomers of 2-hydroxypropanoic acid. In the upper line three-dimensional structures are portrayed. In the lower line a conventional projection method is used. The carbon atom is regarded as being in the plane of the paper. The two vertical bonds pass into the plane of paper and the two horizontal ones come out of the plane.

racemic mixture can be separated into its optically active components by methods described on p. 281; this separation process is known as *resolution*.

Laboratory synthesis of any optically active substance from optically inactive materials generally gives a racemic mixture for there is just as much chance of the (+)-form being formed as the (−)-form. One optically active substance can be converted into another and only one optically active isomer is often encountered in biosynthesis.

a. 2-hydroxypropanoic (lactic) acid. The three possible forms of this acid are shown in Fig. 56.

The (+)-form can be extracted from animal muscle but the (−)-form does not occur naturally and must be obtained by resolution of the racemic mixture. This can be made by a number of synthetic processes, e.g.

$$CH_3—CH(Cl)—COOH \xrightarrow{H_2O} CH_3—CH(OH)—COOH$$

$$CH_3—CH(NH_2)—COOH \xrightarrow{HNO_2} CH_3—CH(OH)—COOH$$

$$CH_3—CHO \xrightarrow{HCN} CH_3—CH(OH)—CN \xrightarrow[\text{acid}]{\text{dil}} CH_3—CH(OH)—COOH$$

b. 2-aminopropanoic acid. Alanine. Alanine resembles 2-hydroxypropanoic acid, the –OH group in the latter being replaced by an –NH$_2$ group. The (+)-acid, the structure of which is shown in Fig. 57 is obtained by the hydrolysis of many proteins. It is a solid melting at 297 °C. The (−)-form does not occur naturally and has to be obtained by resolution of the (±)-form which can synthesised (p. 304).

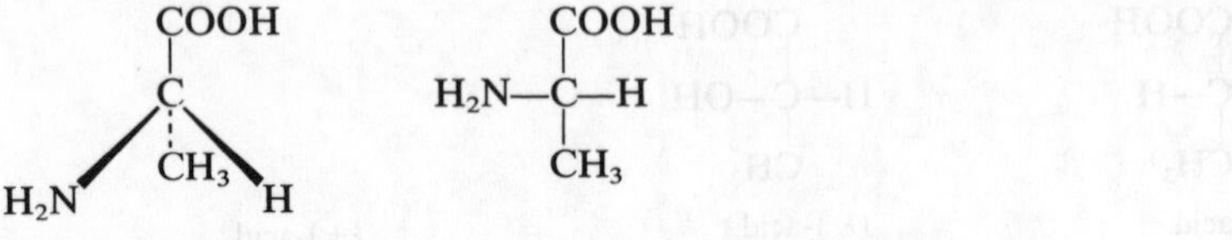

Fig. 57. (+)-isomer of 2-aminopropanoic acid.

c. Butan-2-ol. There are four structurally isomeric butanols as shown below

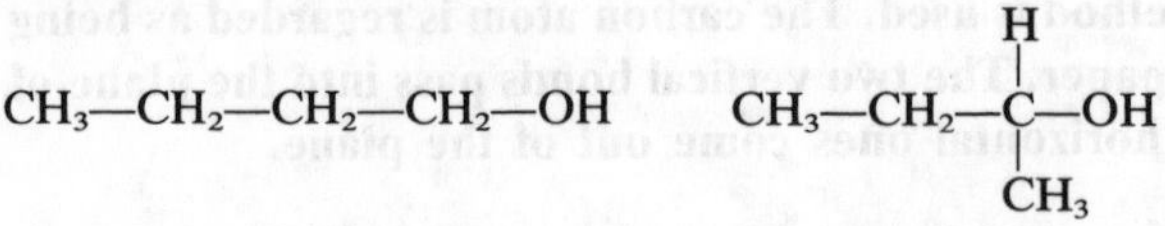

Butan-1-ol   Butan-2-ol

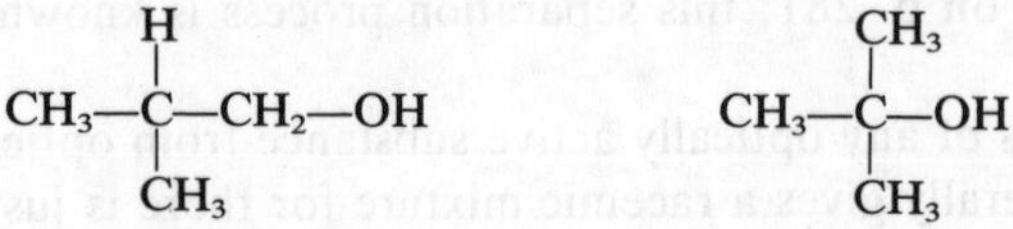

2-methylpropan-1-ol   2-methylpropan-2-ol

Butan-2-ol is the only one with an asymmetric carbon atom so that it also exists in three isomeric forms; there is a (+)-form, a (−)-form and a (±)-form.

8 Molecules with two asymmetric carbon atoms

A compound with n asymmetric carbon atoms has 2^n possible optical isomers, as in (b) below, but this number may be lowered if the asymmetric carbon atoms concerned are linked to the same three groups, as in (a).

a. 2,3-dihydroxybutanedioic (tartaric) acid. This acid exists in four forms, two of them optically active and two inactive. The two optically active forms are enantiomorphs, one being dextro- and the other laevo-rotatory (Fig. 58).

One of the optically inactive forms is a racemic mixture; it is known

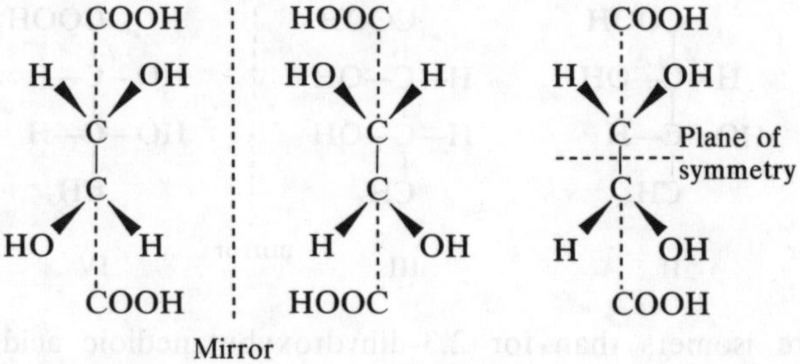

Equimolecular mixture of (+)- and (−)-acids. Racemic mixture.

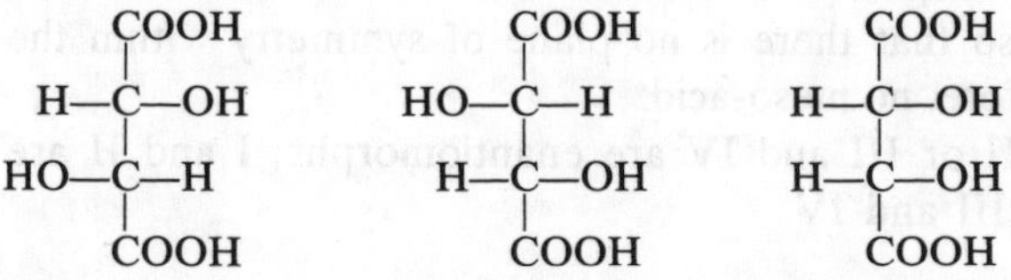

(+)-acid	(−)-acid	Meso-acid	(±)-acid
m.p. = 170 °C	m.p. = 170 °C	m.p. = 143 °C	m.p. = 206 °C
Optically active		Optically inactive	
		Internally compensated	Externally compensated

Fig. 58. The various forms of 2,3-dihydroxybutanedioic (tartaric) acid.

as the (±)-acid or as racemic acid, the name originating because the acid was extracted from grapes (*racemus* = a bunch of grapes). It is optically inactive because the effect of the (+)-acid is counteracted by the (−)-acid; it is externally compensated.

The second optically inactive form is known as meso-2,3-dihydroxybutanedioic acid. Although the molecule contains two asymmetric carbon atoms the molecule as a whole is symmetrical, for there is a plane of symmetry as shown (Fig. 58). This arises because both carbon atoms are linked to the same three groups. The acid is said to be *internally compensated* for, in effect, the rotatory power of the top half of the molecule is counteracted by that of the bottom half.

The two optically active forms, which are enantiomorphic, have the same physical properties excepting their crystal shapes and their effect on polarised light properties as can be seen from their melting points quoted in Fig. 58. Stereoisomers which are not enantiomorphs, e.g. the meso-acid and either the (+)-or (−)-acid, are known as *diastereoisomers.*

b. 2,3-dihydroxybutanoic acid. This substance has four optically active isomers as shown and each pair can form a racemic mixture.

$$\begin{array}{ccc}
\text{COOH} & \text{COOH} & \text{COOH} & \text{COOH} \\
\text{HO—C—H} & \text{H—C—OH} & \text{H—C—OH} & \text{HO—C—H} \\
\text{H—C—OH} & \text{HO—C—H} & \text{H—C—OH} & \text{HO—C—H} \\
\text{CH}_3 & \text{CH}_3 & \text{CH}_3 & \text{CH}_3
\end{array}$$

I   *mirror*   II   III   *mirror*   IV

There are more isomers than for 2,3-dihydroxybutanedioic acid because the two asymmetric carbon atoms are not linked to the same three other groups so that there is no plane of symmetry within the molecule and, therefore, no meso-acid.

Structures I and II or III and IV are enantiomorphs; I and II are diastereoisomers of III and IV.

9 Other optically active compounds

It is the asymmetry of a molecule as a whole that leads to optical activity and this can come about without the presence of an asymmetric C atom. Any atom which can form 4 tetrahedral covalent bonds will act as a centre of asymmetry and quaternary ammonium compounds of the type $(NR^1R^2R^3R^4)^+X^-$ (p. 295) can be optically active.

Optical activity also occurs in the following types of compound, typical enantiomorphic forms being shown in each case with the dotted line representing a mirror.

a. Allenes (p. 114)

$$\begin{array}{cc}
\text{CH}_3 \quad\quad \text{C}_3\text{H}_7 & \text{H}_7\text{C}_3 \quad\quad \text{H}_3\text{C} \\
\text{C=C=C} & \text{C=C=C} \\
\text{H} \quad\quad \text{C}_2\text{H}_5 & \text{H}_5\text{C}_2 \quad\quad \text{H}
\end{array}$$

b. Spiro compounds, e.g.

$$\begin{array}{cc}
\text{H} \quad \text{CH}_2 \quad \text{CH}_2 \quad \text{H} & \text{H} \quad \text{CH}_2 \quad \text{CH}_2 \quad \text{H} \\
\text{C} \quad\quad \text{C} \quad\quad \text{C} & \text{C} \quad\quad \text{C} \quad\quad \text{C} \\
\text{NH}_2 \quad \text{CH}_2 \quad \text{CH}_2 \quad \text{NH}_2 & \text{NH}_2 \quad \text{CH}_2 \quad \text{CH}_2 \quad \text{NH}_2
\end{array}$$

c. Substituted biphenyls. If the substituents in the ortho positions are large enough they will prevent free rotation about the bond linking the two rings by steric hindrance, e.g.

10 Resolution of racemic mixtures

The separation of a racemic mixture into its two optically active components is known as resolution. The processes available are not easy to carry out successfully.

a. Biochemical method. When certain bacteria or moulds are added to a solution of a racemic mixture they may bring about the decomposition of one of the optically active forms more rapidly than the other. *Penicillium glaucum*, for example, decomposes $(+)$-2,3-dihydroxybutanedioic acid more readily than the $(-)$-isomer so that the latter can be obtained from the residue remaining after treatment with the mould.

b. Compound formation. If a racemic mixture can react with some other optically active substance the resulting compound can often be separated. A racemic mixture of an acid, A, will for example, contain $(+)$-A and $(-)$-A. On reaction with an optically active base, $(-)$-B, two salts will be formed, i.e. $(+)$-A.$(-)$-B and $(-)$-A.$(-)$-B. These two salts are not enantiomorphic; they are diastereoisomers (p. 279). They have different physical properties and they can be separated by fractional crystallisation or, possibly, by chromatography. Once the two salts have been separated $(+)$-A and $(-)$-A can be obtained from them by treatment with acid.

For a racemic mixture which is not acidic it is not possible to use salt formation, but some other kind of compound formation may be possible.

c. Crystal picking. If a solution of a racemic mixture is crystallised it might produce crystals of the racemate or it might produce crystals of the $(+)$- and $(-)$-forms. If conditions can be found in which separate crystals of the two optically active forms can be made they can be separated by hand picking with the aid of a lens because the crystals will be enantiomorphic. The method is very laborious and of very limited appplication; it was, however, used by Pasteur in his separation of sodium ammonium racemate (p. 275).

11 Racemisation and inversion

Racemisation, the opposite of resolution, is the conversion of an optically active stereoisomer into a racemic mixture. Inversion is the conversion of a $(+)$-into a $(-)$-isomer, or vice versa.

Both changes may occur in reactions in which a bond to one of the asymmetric carbon atoms in the molecule is split. In the S_N1 reaction between an optically active tertiary bromoalkane and OH^- ions, for example, the initial bond splitting leads to a planar carbonium ion which can be attacked by OH^- on either side of the plane. If attacked equally on both sides a racemic mixture will result; if attacked on the side opposite to the original Br atom inversion will result.

Under some conditions complete racemisation or complete inversion may occur but, more commonly, racemisation is linked with some inversion.

12 Absolute and relative configuration

It is easy, experimentally, to find which of two optically active isomers is the (+)- and which the (−)-form, but it was not, originally, easy to decide which of the two enantiomophic structures represented which form.

Initially an arbitrary choice was made, (+)-glyceraldehyde being allocated the structure as shown. This was then known as the D-configuration, the particular isomer being given the full name of D-(+)-glyceraldehyde. The D- refers to the structure; the (+)- to the sign of the experimentally measured angle of rotation.

Other optical isomers which could be related to the D-glyceraldehyde structure were denoted as D-forms irrespective of whether they were (+)- or (−)-forms. It was, for example, found that (−)-2-hydroxypropanoic acid was a D-form.

The (+)-isomer of 2-hydroxypropanoic acid was, correspondingly an L-form.

This arbitrary choice was eventually found to have been correct for, in 1951, it became possible by a special X-ray technique to establish the absolute configuration of the sodium rubidium salt of (+)-2,3-dihydroxybutanedioic (tartaric) acid.

$$\underset{\substack{| \\ \text{CH}_2\text{OH} \\ \text{D-(+)-glyceraldehyde}}}{\overset{\text{CHO}}{\underset{|}{\text{C}}}}_{H \quad \text{OH}}$$

CHO
|
H—C—OH
|
CH₂OH

COOH
|
H—C—OH
|
CH₃

D-(+)-glyceraldehyde   D-(−)-2-hydroxy-
                              propanoic acid

Isomerism

	Optical isomerism	(a) Molecules with one C* atom, e.g. 2-hydroxypropanoic (lactic) acid. 2-aminopropanoic acid (alanine). Butan-2-ol.
		(b) Molecules with two C* atoms, e.g. 2,3-dihydroxybutanedioic (tartaric) acid. 2,3-dihydroxybutanoic acid
	Geometric isomerism	(a) Dependent on the presence of a double bond, e.g. *cis-* and *trans-*butenedioic acids *cis-* and *trans-*ethanal oximes
		(b) Dependent on a ring, e.g. *cis-* and *trans-*1,2,dimethylcyclopropanes

Stereoisomerism Isomers containing the same number of each kind of atom linked in the same way but differing in the way in which the atoms are arranged in space.

Isomerism The existence of two or more compounds having the same molecular formula.

Structural isomerism Isomers containing the same number of each kind of atom but differing in regard to which atom is linked to which.

	Chain isomerism	e.g. Butane and 2-methylpropane. Pentane, 2-methylbutane and 2,2-dimethyl propane.
	Position isomerism	e.g. Propan-1-ol and propan-2-ol. Ring and side-chain chlorinated methyl benzene. Pentan-3-one and pentan-2-one.
	Functional isomerism	e.g. Methoxymethane and ethanol.
	Tautomerism	e.g. Hydrogen cyanide. Keto-enol tautomerism; ethyl-3-oxobutanoate and ethyl-3-hydroxybut-2-enoate.

Questions on Chapter 20

1 Discuss the evidence in favour of a tetrahedral arrangement of carbon's four valency bonds.

2 Explain the terms optical activity, external compensation, asymmetric carbon atom, racemisation and resolution.

3 In what isomeric forms can the following compounds occur: (a) C_3H_8O, (b) $C_2H_4(OH)COOH$, (c) $C_6H_4Cl_2$?

4 Write down the structural formulae of all optically active hydrocarbons of formula C_8H_{18}.

5 What does the term unsaturation mean in organic chemistry? Indicate, briefly, (a) how it may be detected experimentally, and (b) how it can cause isomerism in compounds.

6 How may the existence of the numbers of isomerides of the following molecular formulae be explained: (a) two of formula C_2H_3N; (b) two of formula $C_2H_4Br_2$; (c) three of formula C_3H_8O. (O and C)

7 How many planar patterns can be made from (a) 3, (b) 4, (c) 5 and (d) 6 equal squares? Which of the patterns are asymmetric?

8 Discuss the isomerism displayed by the following: (a) the dichloroethanes, (b) $NO_2—C_6H_4—OH$, (c) C_5H_{12}, (d) C_3H_9N, (e) $COOH—(CHOH)_2—COOH$.

9 An optically active compound, A, with molecular formula $C_3H_6O_3$ effervesces with sodium carbonate. On oxidation, a compound, B, is formed which gives a phenylhydrazone, C, containing 60.6 per cent carbon, 5.6 per cent hydrogen and 15.7 per cent nitrogen. What are the structures of A, B and C? Give one method by means of which A could be synthesised.

10 Explain clearly what is meant by the terms optical isomerism, optical activity, racemate, and resolution. Draw clear structural diagrams for each of the compounds below to show whether it would or would not exhibit optical isomerism; (a) CH_2ClCH_2OH,

(b) CHClBr—CH$_3$,   (c) CH$_3$—CH=CH—CH$_2$—CH$_3$,
(d) CH$_3$—CH$_2$—CH(NH$_2$)CO$_2$H, (e) (CH$_3$)$_2$CH=CH—CH$_2$.
Outline the method you might adopt in order to resolve a pair of optical isomers which contain a carboxyl group. (C. Local)

11 Discuss the stereochemistry of the following: (a) CH$_3$—CH(OH)—CH$_2$OH; (b) CH$_3$—CH(OH)—CH$_3$, (c) CH$_3$—CH=C(CH$_3$)$_2$, (d) C$_6$H$_5$—CH(OH)—CN, (e) CH$_3$—CH=CH—CO$_2$H, (f) C$_6$H$_5$—CH=CH—CH(CH$_3$)—CO$_2$H. (O and C)

12 But-2-ene is an alkene (CH$_3$—CH=CH—CH$_3$) which can exist in *cis*- and *trans*-forms. When *cis*-but-2-ene is treated with bromine in glacial ethanoic acid as solvent the product is the racemic (±)-form of 2,3-dibromobutane. From *trans*-but-2-ene, the *meso* form of 2,3-dibromobutane is obtained. Write out the structural formulae of the four compounds involved. What deductions can you make about the way in which bromine has added? (Oxf. Schol.)

13 A neutral compound A, C$_{10}$H$_{18}$O$_4$, dissolved slowly in boiling sodium hydroxide solution to give ethanol and, after acidification, a solid B, C$_6$H$_{10}$O$_4$. The solid B decomposed smoothly at its melting point (118 °C) to give a liquid acid C, C$_5$H$_{10}$O$_2$. The acid C was optically inactive, but could be resolved. Suggest structures for A, B and C and explain the reactions involved. (Oxf. Schol.)

14 Classify and give examples of the various types of isomerism which occur among organic compounds. (SUJB)

15 A compound A, C$_8$H$_8$Cl$_2$, yields on hydrolysis B, C$_8$H$_8$O, which on mild reduction yields the optically active compound C, C$_8$H$_{10}$O. When C reacts with iodine and sodium hydroxide solution, iodoform is produced, together with D, C$_7$H$_5$O$_2$Na. With hydrochloric acid, D gives a solid which is markedly more soluble in hot water than in cold. Identify A, B, C and D, and elucidate the reactions indicated. (L.)

16 Can the following compounds be optically active? (a) CH$_2$=C=CBr$_2$, (b) BrHC=C=C=CHBr, (c) spiro (2.2) pentane, C$_5$H$_8$.

17 Using the compounds shown below as your examples, explain the

meaning of the terms *racemic* and *meso*.

$$CH_3-CHCl-CHCl-CH_3 \qquad CH_3-CHCl-CH_2-CH_3$$

Assign possible structures to compounds **A**, **B** and **C** below, giving your reasoning. The only one of these compounds which can be resolved is **B**.

$$\textbf{C, } C_6H_{14}O \xleftarrow[\text{(2) H}^+/\text{H}_2\text{O}]{\text{(1) CH}_3\text{MgI}} \textbf{A, } C_5H_{10}O \xrightarrow[\text{(2) H}^+/\text{H}_2\text{O}]{\text{(1) LiAlH}_4} \textbf{B, } C_5H_{12}O$$

$$\Big\downarrow \text{NH}_2\text{OH}$$

Crystalline
solid   (Oxf. Schol.)

21

Nitriles, Nitro-compounds and Amines

1 Nomenclature

Nitriles have a general formula R—C≡N, and ethanonitrile, CH₃—C≡N, is typical; it owes its name to the fact that it contains *two* carbon atoms. Hydrogen cyanide (prussic acid), HCN, can be regarded as methanonitrile but it is not a typical nitrile as it contains no alkyl group.

Nitriles may also be regarded as cyanoalkanes, and they are isomeric with isocyanoalkanes, R—N≡C.

2 Preparation of nitriles

Nitriles are prepared by heating ammonium salts of carboxylic acids, or acid amides, with phosphorus(V) oxide,

$$R—COO^-NH_4^+ \xrightarrow[\text{heat}]{-H_2O} R—CONH_2 \xrightarrow[P_2O_5]{-H_2O} R—C≡N.$$

or by reaction between an alcoholic solution of potassium cyanide and an iodoalkane (p. 161),

$$R—I + KCN \rightarrow R—C≡N + KI$$

3 General properties of nitriles

The lower members are water soluble liquids with relatively high boiling points. These are caused by association brought about by the high polarity of the $C^{\delta+}≡N^{\delta-}$ bond. This polarity renders the bond open to attack by both nucleophiles and electrophiles. Most of the reactions are, in fact, addition reactions.

a. Hydrolysis. Nitriles are hydrolysed in both acid and alkaline solution. Acid amides are first formed but they are then hydrolysed to the carboxylic acid or its salt (p. 234). The mechanisms are summarised as follows:

$$R-C^{\delta+}\!\equiv\!N^{\delta-}$$

$$\xrightarrow{H^+} R-C^+\!=\!NH \xrightarrow{H_2O} R-\overset{\displaystyle OH}{\underset{}{C}}=NH+H^+$$

$$\xrightarrow{OH} R-\underset{\displaystyle OH}{C}=N^- \xrightarrow{H_2O} R-\underset{\displaystyle OH}{C}=NH+OH^-$$

$$R-\overset{\displaystyle O}{C}\diagdown_{NH_2}$$

b. Reduction. Hydrogen can be added on to the C≡N bond to convert nitriles into primary amines. Sodium and ethanol, catalytic hydrogenation and lithium tetrahydridoaluminate will all bring this about,

$$R-C\!\equiv\!N \xrightarrow{[4H]} R-CH_2-NH_2$$

4 Isocyanoalkanes

These compounds, R—N≡C, are isomeric with the nitriles. They are made from alcoholic solutions of iodoalkanes and silver cyanide, e.g.

$$R-I+AgCN \rightarrow R-N\!\equiv\!C+AgI$$

or by heating primary amines with trichloromethane and potassium hydroxide solution, e.g.

$$R-NH_2+CHCl_3+3KOH \rightarrow R-N\!\equiv\!C+3KCl+3H_2O$$

Simple isocyanoalkanes are evil-smelling, poisonous liquids. They are hydrolysed by dilute acids, e.g.

$$CH_3-N\!\equiv\!C+2H_2O \rightarrow CH_3-NH_2+H-COOH$$

but not by alkalis. They give secondary amines on reduction whereas nitriles give primary amines.

Nitro-compounds

5 The nitro-group

Nitro-compounds contain —NO$_2$ groups linked to alkyl or aryl groups. The aromatic compounds are of greater importance than the aliphatic ones.

The nitro group can be represented either in terms of resonance hybrids or of delocalisation:

$$-N^+\!\!\overset{\displaystyle O}{\underset{\displaystyle O^-}{}} \;\leftrightarrow\; -N^+\!\!\overset{\displaystyle O^-}{\underset{\displaystyle O}{}} \;\leftrightarrow\; -N\!\!\overset{\displaystyle O}{\underset{\displaystyle O}{}}$$

The group is highly polar and exerts a strong electron-withdrawing power. So strong, in fact, that nitroalkanes exhibit acidic properties. These arise because the weakening of an adjacent C—H bond due to the electron-withdrawal facilitates the loss of H^+ and the formation of a carbanion, e.g.

$$\underset{\displaystyle H}{\overset{\displaystyle H}{H-C-NO_2}} \;\rightleftharpoons\; \underset{\displaystyle H}{\overset{\displaystyle H}{H-C^--NO_2}} + H^+$$

The pK_a value for nitromethane is 10.2 and this compares with a value of 4.76 for ethanoic acid and an estimated value of about 60 for methane. The nitration of an aromatic acid also increases the acid strength very markedly (pages 199 and 236).

Nitro-compounds are isomeric with *alkyl nitrates* (III) which contain the —O—N=O group; they are esters of nitric(III) acid.

6 Aliphatic nitro-compounds

a. Preparation. Nitroalkanes are not easy to prepare for side reactions tend to take place in the methods used. Direct nitration of alkanes by heating with nitric(V) acid at 450 °C is possible but for higher alkanes the alkanes tend to crack,

$$R—H + HNO_3 \xrightarrow{450\,°C} R—NO_2 + H_2O$$

Halogenoalkanes will also react with nitrates(III) to form nitro-compounds but nitrates(III) are formed as by-products. Silver nitrate(III) gives a rather better yield of the nitro-compound than sodium nitrate(III),

$$R—X + AgNO_2 \rightarrow R—NO_2 + AgX$$

b. Properties and uses. The lower aliphatic nitro-compounds are liquids which are useful as solvents. They can be reduced by catalytic hydrogenation or by lithium tetrahydridoaluminate to primary amines,

$$R—NO_2 \xrightarrow{[4H]} R—NH_2 + 2H_2O$$

Care must be taken as the reaction is highly exothermic.

Because of the acidic nature caused by the nitro group they will dissolve in alkaline solutions to form salts.

7 Preparation of aromatic nitro-compounds

Aromatic nitro-compounds can be obtained by direct nitration. The process generally involves electrophilic substitution by the nitronium ion, NO_2^+ (p. 129). The choice of conditions depends on the reactivity of the aromatic compound being nitrated and on whether mono-, di- or tri-substitution is wanted.

a. Preparation of nitrobenzene (p. 129). Benzene is added slowly to a mixture of concentrated nitric(V) and sulphuric(VI) acids in a flask. The temperature is kept below 50 °C to limit the formation of dinitrobenzene. When the reagents are all mixed, the mixture is heated under a reflux condenser, with shaking, for about thirty minutes. The contents of the flask are then poured into a large amount of water to separate out the nitrobenzene as a heavy oil. The nitrobenzene layer is separated from the aqueous layer, washed with water and dilute sodium carbonate solution, dried over anhydrous calcium chloride, and finally distilled using an air condenser.

b. Preparation of 1,3-*dinitrobenzene.* The nitro group is *meta*-directing (p. 148) and 1,3-dinitrobenzene can be made by heating nitrobenzene with a mixture of concentrated sulphuric(VI) acid and fuming nitric(V) acid on a boiling water bath for an hour. On pouring the mixture into excess cold water the 1,3-dinitrobenzene separates out as a pale yellow crystalline solid. The crystals may be purified by recrystallisation from methylated spirits.

c. Preparation of 1,3,5-*trinitrobenzene.* It is not easy to introduce a third nitro group into the benzene ring but it can be achieved by prolonged treatment with a mixture of concentrated sulphuric(VI) acid and fuming nitric(V) acid. The product is a colourless, explosive solid.

d. Nitration of methylbenzene (p. 133). Methylbenzene is much more easily nitrated than benzene due to the activating influence of the —CH_3 group (p. 146). Direct nitration with a mixture of concentrated sulphuric(VI) and nitric(V) acids gives a mixture of methyl-2-nitrobenzene and methyl-4-nitrobenzene. These mono-nitro products can be further nitrated to methyl-2,4,6-trinitrobenzene (T.N.T.).

8 Properties and uses of aromatic nitro-compounds

Nitrobenzene is a pale yellow liquid smelling, like benzaldehyde, of bitter almonds; it is used as a source of phenylamine (p. 293) and as a solvent. Most other aromatic nitrocompounds are solids.

Compounds containing sufficient nitro groups are explosive and methyl-2,4,6-trinitrobenzene (T.N.T.) is particularly well known.

Careful reduction of nitrobenzene in acid solution, using tin or iron and hydrochloric acid, provides the main source of phenylamine,

$$\langle\bigcirc\rangle\text{—NO}_2 + [6H] \rightarrow \langle\bigcirc\rangle\text{—NH}_2 + 2H_2O$$

Under different conditions, other reduction products, e.g. N-phenyl-hydroxylamine, C_6H_5—NHOH; phenylazobenzene, C_6H_5—N=N—C_6H_5; N,N'-diphenylhydrazine, C_6H_5—NH—NH—C_6H_5, can be obtained.

Amines

9 Introduction

Amines are derived from ammonia by replacement of one or more of the hydrogen atoms by alkyl and/or aryl groups.

a. Primary amines. If only one hydrogen atom is replaced a primary amine is formed. Typical examples are

$$CH_3\text{—NH}_2 \qquad \langle\bigcirc\rangle\text{—NH}_2 \qquad CH_3\text{—}\overset{\overset{\displaystyle NH_2}{|}}{CH}\text{—}CH_3$$

Methyl-amine Phenyl-amine 2-aminopropane

These primary amines contain the —NH_2 (*amino*) group. The simple ones are named by adding an -*amine* ending to the alkyl or aryl group concerned; in more complicated cases the prefix *amino-* is used.

b. Secondary amines. Replacement of two hydrogen atoms in the ammonia molecule forms secondary amines which contain the $\searrow$N—H group, e.g.

$$\begin{array}{c}CH_3\\ \diagdown\\ CH_3\diagup\end{array}N\text{—H} \qquad \begin{array}{c}\langle\bigcirc\rangle\\ \diagdown\\ \langle\bigcirc\rangle\diagup\end{array}N\text{—H} \qquad \begin{array}{c}\langle\bigcirc\rangle\\ \diagdown\\ CH_3\diagup\end{array}N\text{—H}$$

Dimethyl-amine Diphenyl-amine N-methyl-phenylamine

c. Tertiary amines. These are formed by replacing all the hydrogen atoms in ammonia, e.g.

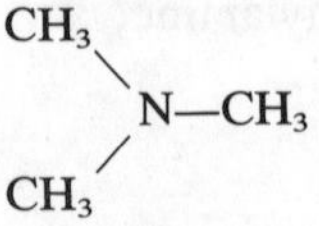

Trimethyl-
amine

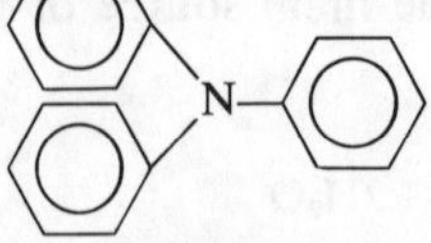

Triphenyl-
amine

d. Quaternary ammonium salts or bases. These are closely related to amines (p. 295) and may be regarded as derived from ammonium salts or ammonium hydroxide by replacing all four H atoms by alkyl and/or aryl groups, e.g.

$NH_4^+Cl^-$ $[N(CH_3)_4]^+Cl^-$ $[N(CH_3)_4]^+OH^-$

Ammonium
chloride

Tetramethyl-
ammonium chloride

Tetramethyl-
ammonium hydroxide

10 Preparation of amines
The methods used for making both aliphatic and aromatic amines are essentially the same.

a. From ammonia. The H atoms in ammonia can be replaced by alkyl groups by heating an alcoholic solution of ammonia with a halogeno-alkane in a sealed tube. The method is not very satisfactory for a mixture of primary, secondary and tertiary amines with a quaternary salt is produced, e.g.

$$NH_3 \xrightarrow{CH_3I} CH_3-NH_2 \xrightarrow{CH_3I} (CH_3)_2=NH \xrightarrow{CH_3I} (CH_3)_3\equiv N \xrightarrow{CH_3I} (CH_3)_4N^+I^-$$

Aromatic amines cannot normally be made in this way because the halogenarenes (p. 164) are less reactive than the halogenoalkanes.

b. From acid amides. Primary amines can be made from acid amides either by the *Hofmann reaction* (p. 261) with bromine and potassium hydroxide solution, e.g.

$$R-CONH_2 \xrightarrow[\text{KOH soln.}]{Br_2+} R-NH_2$$

or by reduction with lithium tetrahydridoaluminate, e.g.

$$R-CONH_2 \xrightarrow{LiAlH_4} R-CH_2-NH_2$$

The first method gives an amine containing one less C atom than the second method.

c. Reduction of nitro-compounds. Nitro-compounds can be reduced to primary amines by hydrogen from metals and acids or by catalytic hydrogenation. The method is particularly important in making phenylamine from nitrobenzene.

In the laboratory, tin and concentrated hydrochloric acid are used to reduce the nitrobenzene. The tin and acid produce hydrogen and tin(II) chloride, which are both reducing agents. In the course of the reaction the tin(II) chloride is oxidised to tin(IV) chloride,

Nitrobenzene   Phenylamine

Concentrated hydrochloric acid is added slowly to a mixture of granulated tin and nitrobenzene in a flask fitted with a reflux condenser. The flask is shaken frequently and, if the reaction produces too much heat, it is cooled. When all the acid has been added, the mixture is heated on a water bath until the odour of the nitrobenzene is no longer present. The flask is then cooled.

The phenylamine formed reacts with the tin(IV) chloride and hydrochloric acid present to form a complex hexachlorostannate(IV),

$$2C_6H_5-NH_2 + SnCl_4 + 2HCl \rightarrow [C_6H_5-NH_3]_2SnCl_6$$

This complex is decomposed by adding a concentrated solution of sodium hydroxide until the mixture is strongly alkaline. The phenylamine separates out as an oil which can be removed from the mixture by steam distillation.

The mixture of phenylamine and water obtained is treated with sodium chloride to lower the solubility of phenylamine in the aqueous layer. The phenylamine is then extracted with ether and the ether is distilled off. The phenylamine may be further purified by distillation using an air condenser.

Industrially, iron filings and concentrated hydrochloric acid or catalytic hydrogenation are used to reduce the nitrobenzene.

d. By reduction of nitriles. Nitriles (p. 288) can be reduced to primary amines by sodium and ethanol, catalytic hydrogenation or lithium tetrahydridoaluminate, e.g.

$$R-C\equiv N + [4H] \rightarrow R-CH_2-NH_2$$

11 Amines as bases

Ammonia is basic because it can use the lone pair of electrons on the N atom to bond to H^+ ions; it is, therefore, a proton acceptor,

$$:NH_3 + H^+ \rightleftharpoons NH_4^+$$

In aqueous solution the equilibrium

$$:NH_3 + H_2O \rightleftharpoons NH_4^+ + OH^-$$

is set up with the position of equilibrium well to the left as ammonia is only a weak base ($pK_b = 4.75$).

Amines behave in a similar way,

$$R_3N: + H_2O \rightleftharpoons R_3N^+ - H + OH^-$$

Anything that increases the negative charge on the N atom of the amine, or increases the stability of the cation it forms with H^+, will increase the basic strength of the amine. Aliphatic amines are, in general, slightly stronger than ammonia; aromatic amines considerably weaker. pK_b for methylamine, for instance, is 3.36; for phenylamine it is 9.38.

a. Aliphatic amines. Alkyl groups have $+I$ inductive effects which would be expected to increase the negative charge on the nitrogen atom of an amine. On these grounds the order of basic strength would be

Tertiary > Secondary > Primary > Ammonia

It is, however, found that the tertiary amine is often out of line, being weaker than expected. This is caused by the fact that the number of hydrogen atoms decreases in passing from the ion of a primary amine to that of a tertiary amine, i.e.

$$NH_4^+ \qquad R-NH_3^+ \qquad R_2-NH_2^+ \qquad R_3-NH^+$$

The more hydrogen atoms there are the more highly solvated the ion will be through hydrogen bonding to water molecules. This solvation increases the stability of the ion. $(CH_3)_3NH^+$ is, therefore, less stable than $CH_3NH_3^+$ and more ready to recombine with OH^- to form unionised amine.

The quaternary ammonium bases, e.g. $(CH_3)_4N^+OH^-$, are comparable in strength to caustic alkalis. The N atom of the cation has no hydrogen atoms; it can neither form hydrogen bonds nor recombine with OH^- ions to give unionised amine.

b. Aromatic amines. In phenylamine, which is weaker than ammonia, the unshared electron pair on the N atom interacts with the delocalised π-electrons in the ring (as in chlorobenzene and phenol) so that the electron pair is less available for bonding to H^+.

In general, substitution of electron donating groups into the ring of phenylamine increases the basic strength; introduction of electron-withdrawing groups decreases it.

12 General properties of amines

The lower aliphatic amines are gases or very volatile liquids which have a fish-like, ammoniacal smell; they are soluble in water giving alkaline solutions. Phenylamine is an oily liquid, colourless when freshly distilled but generally yellow or brown in an old bottle due to atmospheric oxidation. It has a characteristic smell and is very slightly soluble in water.

The essential difference between primary, secondary and tertiary amines lies in the number of reactive hydrogen atoms; primary amines have two, secondary amines one, and tertiary amines none.

a. Formation of salts. Primary, secondary and tertiary amines form salts in the same way as ammonia does, e.g.

$$NH_3 + HCl \rightarrow NH_4^+Cl^-$$
$$CH_3{-}NH_2 + HCl \rightarrow [CH_3NH_3]^+Cl^-$$

$$\langle \bigcirc \rangle{-}NH_2 + HCl \rightarrow [C_6H_5{-}NH_3]^+Cl^-$$

$$(CH_3)_3N + HCl \rightarrow [(CH_3)_3NH]^+Cl^-$$

The salts are crystalline solids, generally soluble in water. They liberate the free amine on warming with an alkali.

Tertiary amines react with halogenoalkanes to form quaternary ammonium salts, e.g.

$$(C_2H_5)_3N + C_2H_5I \rightarrow [(C_2H_5)_4N]^+I^-$$

and these salts give solutions of quaternary bases on shaking with a suspension of silver oxide in water,

$$2(R_4N)^+I^- + Ag_2O + H_2O \rightarrow 2(R_4N)^+OH^- + 2AgI$$

b. Reaction with nitric(III) acid. Primary amines react with nitric(III) acid (obtained by mixing a solution of sodium nitrate(III) and a dilute acid) to form, in the first place, a diazonium salt, e.g.

$$R—NH_2 + HNO_2 + HCl → R—N^+≡NCl^- + 2H_2O$$

Alkyl diazonium salts decompose instantly to liberate nitrogen and to form a carbonium ion, R^+, e.g.

$$CH_3—N^+≡NCl^- → CH_3^+ + N_2 + Cl^-$$

These alkyl carbonium ions can then undergo a variety of reactions, e.g.

$$CH_3^+ + H_2O → CH_3—OH + H^+$$
$$CH_3—CH_2—CH_2^+ → CH_3—CH=CH_2 + H^+$$

so that a mixture of products is obtained.

Aromatic diazonium salts are much more stable and they can be made and kept at temperatures close to 0 °C, e.g.

$$⟨◯⟩—NH_2 + HNO_2 + HCl → ⟨◯⟩—N^+≡NCl^- + 2H_2O$$

Benzenediazonium
chloride

These aromatic diazonium salts can be converted into a number of other chemicals and they are important synthetic reagents, as described on p. 344.

The reactions of both aliphatic and aromatic secondary and tertiary amines with nitric(III) acid are more complicated.

c. Ethanoylation and benzoylation. Primary and secondary amines can be ethanoylated or benzoylated using ethanoyl chloride or anhydride, and benzoyl chloride respectively. Hydrogen atoms are replaced by $CH_3CO—$ or $C_6H_5CO—$ groups, e.g.

$$CH_3—N\genfrac{}{}{0pt}{}{H}{H} + CH_3COCl → CH_3—N—\overset{O}{\overset{\|}{C}}—CH_3 + HCl$$

N-methylethanamide

$$CH_3—N\genfrac{}{}{0pt}{}{CH_3}{H} + CH_3COCl → CH_3—N—\overset{O}{\overset{\|}{C}}—CH_3 + HCl$$

N,N-dimethylethanamide

$$\text{(ring)}-N\overset{H}{\underset{H}{\diagup}} + \text{(ring)}-COCl \rightarrow \text{(ring)}-\overset{H}{N}-\overset{O}{\overset{\|}{C}}-\text{(ring)} + HCl$$

N-phenylbenzamide

The benzoylation is slower than the ethanoylation (p. 255) and is best carried out in the presence of sodium hydroxide solution (Schotten–Baumann reaction). The free amine can be liberated from the ethanoylated or benzoylated products by warming with an alkaline solution.

Tertiary amines cannot be ethanoylated or benzoylated for they do not contain any replaceable hydrogen atom.

d. Formation of isocyano-compound. Primary amines react on warming with an alcoholic solution of potassium hydroxide and trichloromethane to form isocyanoalkanes or arenes (p. 288), e.g.

$$CH_3—NH_2 + CHCl_3 + 3KOH \rightarrow CH_3—N{\equiv}C + 3KCl + 3H_2O$$

$$\text{(ring)}—NH_2 + CHCl_3 + 3KOH \rightarrow \text{(ring)}—N{\equiv}C + 3KCl + 3H_2O$$

The products have a very nasty smell and are poisonous; their formation can be used as a test to distinguish primary from secondary and tertiary amines, which do not form isocyanoalkanes.

13 Particular properties of phenylamine
The amino group is *o–p* directing and activating (p. 146) so that phenylamine can be substituted in the ring quite easily. The problem, in fact, is how to limit the substitution to the mono-product. Phenylamine can also be oxidised in a number of ways.

a. Nitration. The direct nitration of phenylamine by nitric(V) and sulphuric(VI) acids is not very satisfactory. Some tri-nitro-product is formed but this is mixed with by-products caused by oxidation by the nitric(V) acid. The difficulty can be overcome by ethanoylating the phenylamine before nitration. The product is not oxidised by nitric(V) acid and the —NH—CO—CH₃ groups though still *o–p* directing, is much less activating than the —NH₂ group. Nitration can, therefore, be carried out in a more controlled way either to the mono- or the tri-nitro stage. After nitration, the original —NH₂ group can be reformed by warming with an alkaline solution.

b. Halogenation. Phenylamine is so susceptible to substitution that either chlorine or bromine water react in the cold to form white precipitates, almost quantitatively, of the tri-halogeno-products, e.g.

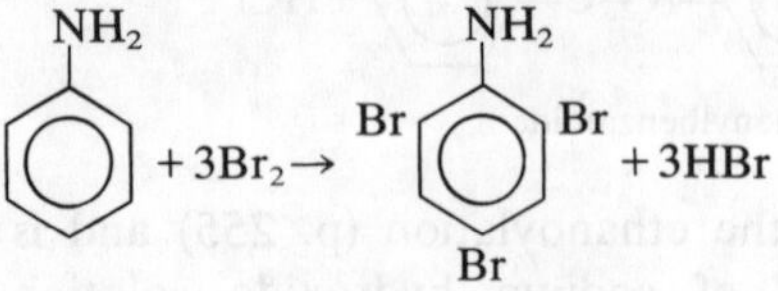

2,4,6-tribromo-
phenylamine

To obtain mono-halogenated products it is necessary to ethanoylate the phenylamine first, as in (a).

c. Sulphonation. Phenylamine can be sulphonated by heating with fuming sulphuric acid at 180 °C. The main product is 4-aminobenzene-sulphonic acid (sulphanilic acid),

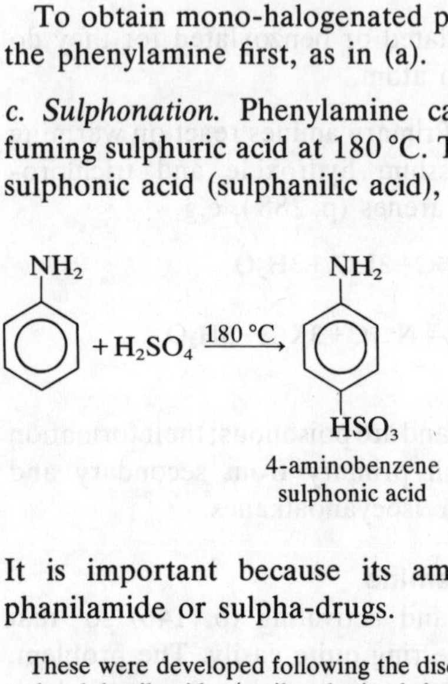

4-aminobenzene
sulphonic acid

It is important because its amide leads to the widely used sulphanilamide or sulpha-drugs.

These were developed following the discovery, in the 1930's, that a red dye, prontosil, and sulphanilamide (easily obtained from the dye) were effective in treating such infections as pneumonia, meningitis and bacillary dysentry. Modification of the sulphanilamide structure led to still more useful products; sulphapyridine (M and B693), sulphadiazine, sulphamethazine and sulphamerazine are the commonest examples:

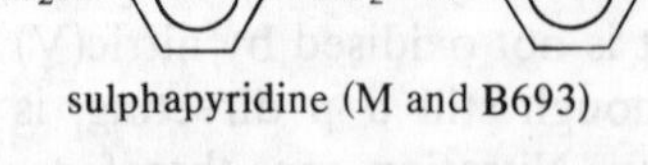

sulphapyridine (M and B693)

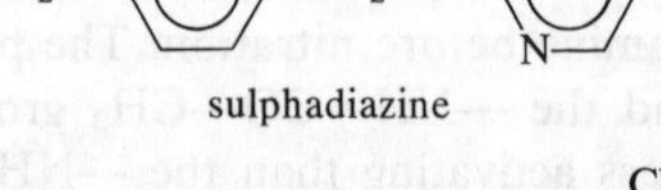

sulphadiazine

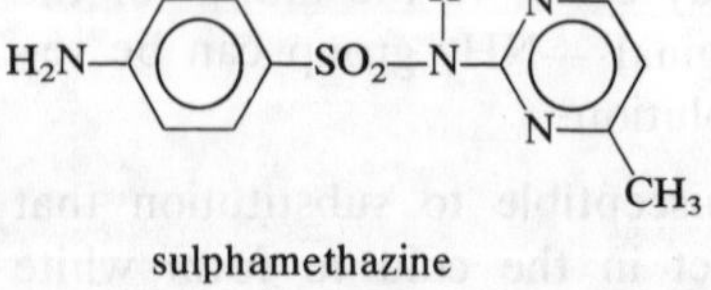

sulphamethazine

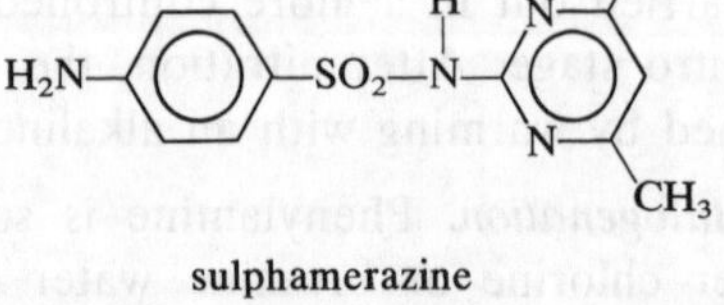

sulphamerazine

d. Oxidation. Phenylamine is easily oxidised, the nature of the product depending on the oxidising agent used. Typical examples are aniline black and cyclohexadiene-1,4-dione (benzoquinone). The former is a complex black substance used as a dye; it is made by oxidation with sodium chlorate(V) and sodium dichromate(VI). The latter is a yellow solid made by using acidified sodium dichromate(VI) as the oxidising agent.

		CH_3—NH_2	C_6H_5—NH_2
Preparation	Reaction of NH_3 with	CH_3—I	Halogenoarenes do not react readily with NH_3
	Reaction of Br_2 and KOH (Hofmann reaction) with	CH_3—$CONH_2$	C_6H_5—$CONH_2$
	Reduction, by acid and metal, of	CH_3—NO_2 HCN	C_6H_5—NO_2
Properties	Basic strength	Stronger than NH_3	Weaker than NH_3
	With acids	Form salts, e.g. $CH_3NH_3{}^+Cl$ and $C_6H_5NH_3{}^+Cl^-$	
	With HNO_2	Forms diazo salt which immediately decomposes to liberate N_2	Forms diazo salt, e.g. C_6H_5—$N^+{\equiv}NCl^-$ which is stable at low temperatures
	With CH_3—COCl (Ethanoylation)	Forms CH_3—N(H)—C(O)—CH_3	Forms C_6H_5—N(H)—C(O)—CH_3
	With C_6H_5—COCl and NaOH (Benzoylation by Schotten–Baumann reaction)	Forms CH_3—N(H)—C(O)—C_6H_5	Forms C_6H_5—N(H)—C(O)—C_6H_5
	With alc. solution of KOH and $CHCl_3$	Forms CH_3—$N{\equiv}C$	Forms C_6H_5—$N{\equiv}C$
	Oxidation	Not oxidised	Various products
	Ring substitution	Not possible	Can be nitrated, halogenated and sulphonated

Preparation and properties of methylamine and phenylamine

Questions on Chapter 21

1 Give the structural formulae of the two isomers with molecular formula C_2H_3N and show how their structures influence their properties.

2 Outline one method for detecting nitrogen in an organic compound and one method for estimating it. Summarise the main properties of nitriles.

3 Given carbon, hydrogen, nitrogen and oxygen together with normal laboratory apparatus and reagents (other than those containing any of these four elements) devise a synthesis of ethanonitrile. What product(s) are obtained on (a) the hydrolysis, and (b) the reduction of ethanonitrile? Outline the procedure in each case and comment on any suggested mechanisms involved in these processes. (L)

4 Write balanced equations to illustrate the reduction of (a) a nitroalkane, (b) a nitrile, (c) an oxime, (d) an acid amide.

5 Give the structures of the nitro groups and the nitrate(III) and nitrate(V) radicals. Show, in each case, the arrangements of electrons.

6 Discuss the part played by nitrogen-containing compounds in the explosives industry.

7 Discuss the mechanism of nitration processes.

8 Compare and contrast the properties of the —NH_2 group in methylamine, ethanamide and aminoethanoic acid.

9 How does the —NH_2 group in R—NH_2 (a) resemble, (b) differ from that in R—$CONH_2$?

10 Enumerate the isomerides of molecular formula C_3H_9N and indicate how, in principle, they might be distinguished from one another. How might a specimen of one be prepared?

11 Explain, with examples, the meaning of the terms (a) tertiary amine, (b) basic, (c) quaternary ammonium salt.

12 How can ethylamine be prepared from (a) methylamine, (b) ethanol, (c) propanoic acid? From ethylamine how can (a) ethanoisonitrile, (b) C_6H_5—CO—NH—C_2H_5 be prepared?

13 What are the chief properties of the —NH_2 group in organic compounds and how do they differ between aliphatic and aromatic compounds?

14 Comment on the following pK_b values:

$NH_3(4.75)$ $CH_3NH_2(3.36)$ $(CH_3)_2NH(3.23)$ $(CH_3)_3N(4.20)$
               $C_2H_5NH_2(3.33)$ $(C_2H_5)_2NH(3.07)$ $(C_2H_5)_3N(3.12)$

15 Many nitrogen compounds are classified as bases. What is the definition of a base? Decide which of the following compounds cannot be classified as bases, explaining your reasons. (a) $(CH_3CH_2)_3N$, (b) $CH_3CH_2NO_2$, (c) $C_6H_5NHNH_2$, (d) CH_3CN, (e) $C_6H_5CONH_2$. Suggest syntheses for three of these compounds.

16 Compare and contrast the reactions of the two compounds in each of the following pairs. (Not more than three different reactions are required for each pair).
(a) R—CH_2OH and R—CH_2NH_2 (R = alkyl)
(b) R—CO—OH and R—CO—NH_2 (R = alkyl)
(c) Ar—OH and Ar—NH_2 (Ar = aromatic nucleus).
Give two reactions of benzenediazonium chloride in which nitrogen is eliminated and one in which nitrogen is retained. (JMB)

17 An aliphatic amine containing only C, H and N gave a hydrochloride which, on combustion gave the following results: (a) 0.324 g gave 0.4472 g of CO_2 and 0.3051 g of water, (b) 0.5632 g gave 66 cm^3 of nitrogen at s.t.p. Suggest possible formulae for the amine.

18 A mono-acid base containing 61 per cent C, 15.3 per cent H and 23.7 per cent N forms a hydrochloride, 1 g of which requires 105.7 cm^3 of 0.1 M silver nitrate(V) solution for titration. What structural formulae are possible for it? How may they be distinguished?

19 Sketch the apparatus you would employ for the purification of phenylamine by steam distillation and indicate the advantages of the process. Calculate what fraction of the total mass of distillate would be phenylamine if the distillation was carried out at 98 °C under a pressure of 101.325 kN m^{-2}? (The vapour pressure of water at 98 °C is 94.260 kN m^{-2}.)

20 Give examples which illustrate the difference in reactivity of an amino (—NH$_2$) group when attached to (a) an alkyl group, (b) an aromatic nucleus, (c) a carbonyl group. How do you account for the difference in reactivity? (L)

21 (a) Give examples of addition and substitution reactions which can be classified as nucleophilic and electrophilic. Explain the reasons why you classify each example in the particular way you do. (b) Explain (i) the difference in acidity between phenol and ethanol, (ii) the difference in basicity between methylamine (aminoethane) and ammonia. (SUJB)

22 A compound A, C$_7$H$_9$O$_2$N, dissolves in water. With cold sodium hydroxide solution, an oil B, C$_6$H$_7$N, is formed which yields a white precipitate, C, with bromine and also burns with a smoky flame. On acidification of the aqueous alkaline layer, a clear solution, D, is formed which decolorizes potassium manganate(VII) solution. Comment on these observations and propose a structure for A. Identify B, C and D. How, and under what conditions, would you expect the organic compound in D to react with (a) sulphuric(VI) acid (b) silver oxide? State one confirmatory test for the oil, B. (SUJB)

23*Tabulate the relative molecular masses and the boiling points of butane, nitromethane and propanone and comment on the figures.

24*What is the enthalpy change in the reaction

$$CH_3—NO_2 \rightarrow \tfrac{1}{2}N_2 + CO_2 + 1.5H_2$$

Why is it that many nitro-compounds are explosive? Why is TNT used as an explosive rather than trinitrobenzene? Why is ammonium nitrate(V) sometimes added to TNT to improve its efficiency?

25*Use bond energy values to estimate the enthalpy of reaction for the reaction between hydrogen and nitromethane.

22

Amino Acids and Proteins

1 Substituted carboxylic acids

In all the carboxylic acids, excepting methanoic acid, the hydrogen atoms in the alkyl groups can be replaced by other monovalent groups to give substituted acids. The groups mainly concerned are the —Cl group (giving *chloro*-acids), the —OH group (giving *hydroxy*-acids) the —NH$_2$ group (giving *amino*-acids) and the —CN group (giving *cyano*-acids).

Typical mono-substituted acids derived from ethanoic acid are

H | H—C—COOH | Cl	H | H—C—COOH | OH	H | H—C—COOH | NH$_2$
Chloroethanoic acid	Hydroxyethanoic acid (Glycollic acid)	Aminoethanoic acid (Glycine)

With chlorine, in particular, it is also possible to replace more than one hydrogen atom so that dichloro- and trichloroethanoic acids are well known (p. 240).

With higher carboxylic acids there are many different hydrogen atoms that can be substituted. Numbers are used to indicate the particular atom concerned e.g.

H H | | H—C——C—COOH | | H NH$_2$	H H | | H—C——C—COOH | | NH$_2$ H
2-aminopropanoic acid (Alanine)	3-aminopropanoic acid (β-aminopropanoic acid)

The carbon atoms in the chain may also be referred to as α-, β-, γ-, etc. starting with the one next to the COOH group; the corresponding acids are known as α-, β-, γ-acids.

2 Aminoethanoic acid (glycine)

The amino acids are the most important substituted acids for they are the component parts of proteins. Aminoethanoic acid provides a simple example.

a. Preparation. The acid can be obtained by the acid hydrolysis of proteins such as glue or gelatine, or from ethanoic acid

$$CH_3-COOH \xrightarrow[\text{and sunlight}]{Cl_2+P} H-\underset{\underset{Cl}{|}}{\overset{\overset{H}{|}}{C}}-COOH \xrightarrow[NH_3]{\text{Excess}} H-\underset{\underset{NH_2}{|}}{\overset{\overset{H}{|}}{C}}-COOH$$

b. Properties. Aminoethanoic acid is a white, crystalline solid readily soluble in water. The aqueous solution is neutral to litmus, but the substance can act as an acid, on account of the —COOH group, or as a base, on account of the —NH$_2$ group. Salts can, therefore, be formed with bases, e.g. $[CH_2(NH_2)COO]^-Na^+$, or with acids, e.g. $[CH_2(NH_3)COOH]^+Cl^-$.

In neutral solution, the aminoethanoic acid exists as a *zwitterion*,

$$\underset{H}{\overset{H}{\diagdown}}N-CH_2-C\overset{\diagup O}{\underset{\diagdown OH}{}} \rightleftharpoons H^+-\underset{\underset{H}{|}}{\overset{\overset{H}{|}}{N}}-CH_2-C\overset{\diagup O}{\underset{\diagdown O^-}{}}$$

and this polar structure accounts for the surprisingly high melting point (235 °C), the solubility in water and insolubility in organic solvents.

In acid solution the positively charged ion, $[CH_2(NH_3)COOH]^+$, is present and this migrates to the cathode in electrophoresis; in alkaline solution the negatively charged ion, $[CH_2(NH_2)COO]^-$, is present and this migrates to the anode. The pH of the solution, 5.97, in which no migration occurs is known as the *isoelectric point*; it is a useful physical constant for the substance.

Aminoethanoic acid functions as a carboxylic acid in forming esters and other acid derivatives; it functions as an amine in liberating nitrogen with nitric(III) acid and in its ability to be ethanoylated or benzoylated.

Proteins

3 Introduction

A group of complex substances known as proteins occurs very widely in all plants and animals. The substances are synthesised, within a plant, from such simple compounds as carbon dioxide, nitrates(V) and water. When plants are eaten by animals, the plant proteins are decomposed in the process of digestion and the animal then resynthesises its own animal protein requirements from the decomposition products. This biosynthesis of proteins is a quite remarkable process controlled by chemicals in chromosomes which are themselves made up of proteins bonded to nucleic acids (nucleoproteins), and by enzymes which are also proteins or protein-containing substances.

Simple proteins exist in two main types. *Fibrous proteins* are insoluble and function mainly as structural materials; *globular proteins* dissolve in water to give colloidal solutions and have varied biochemical functions. Simple proteins are also found linked with other substances in what are known as *conjugated proteins*, e.g. nucleoproteins, enzymes and hormones.

Typical examples of both simple and conjugated proteins are listed below; they include structural materials, hormones, enzymes and plant viruses.

Protein	Nature and function
Collagen	A simple protein found as a structural material in connective tissue, skin and cartilage. Gives gelatin on boiling with water.
Keratin	A simple protein found as a structural material in skin, hair, nails and feathers.
Insulin	A simple protein which functions as a hormone, governing sugar metabolism.
Urease	A simple protein extracted from soya beans. Acts as an enzyme catalyst in the hydrolysis of carbamide (urea).
Haemoglobin	A conjugated protein made up of the protein, globin, combined with haem. The oxygen carrier in blood.
Tobacco mosaic virus	A conjugated protein occurring as a plant virus in infected tobacco plants.
Nucleoproteins	Conjugated proteins made up of nucleic acids and proteins. Components of cell nuclei.

4 Hydrolysis of proteins

Although different proteins vary widely in physical properties and function they can all be hydrolysed into a mixture of amino acids. The hydrolysis can be brought about by acids, alkalis or enzymes.

About twenty different amino-acids have so far been isolated as products of protein hydrolysis. Insulin, for example, yields sixteen different amino-acids on hydrolysis, some of them in larger quantities than others. The acids concerned are α-amino acids which can be regarded as substituted aminoethanoic acids, for example,

2 aminopentane dioic acid
(Glutamic acid)

Tyrosine

Lysine

Cysteine

All the amino acids contain at least one —NH_2 and one —$COOH$ group. Those which contain additional —$COOH$ groups are known as 'acidic' amino-acids, e.g. glutamic acid; those with additional —NH_2 groups are 'basic' amino-acids e.g. lysine. Apart from aminoethanoic acid all the other amino-acids obtained from proteins are optically active (p. 278) and are *L*-forms.

The amino-acids produced by the hydrolysis of any protein can be separated and identified by two-way paper chromatography (p. 61) or by the use of ion exchangers. It was the development of these methods of separation that made the investigation of proteins so much easier.

Amino-acids can be differentiated one from another because they have different iso-electric points (p. 304), different angles of optical rotation (p. 275) and different R_f values (p. 61).

5 The peptide link

The amino-acids of which a protein is made are joined by what are known as peptide links. These are formed by the elimination of water between the —$COOH$ group of one acid and the —NH_2 group of an adjacent one, i.e.

The peptide link

Aminoethanoic acid, for example, will react with its acid chloride to give a product commonly known as glycylglycine,

$$H_2N-\underset{\underset{H}{|}}{\overset{\overset{H}{|}}{C}}-\underset{\underset{Cl}{|}}{\overset{\overset{O}{\|}}{C}} \quad + \quad \underset{\underset{H}{|}}{\overset{\overset{H}{|}}{N}}-\underset{\underset{H}{|}}{\overset{\overset{H}{|}}{C}}-COOH \rightarrow H_2N-CH_2-\overset{\overset{O}{\|}}{C}-\overset{\overset{H}{|}}{N}-CH_2-COOH$$

Glycylglycine

This contains a single peptide link and is the simplest example of what is known as a *dipeptide*. It, like other dipeptides made from other pairs of amino-acids, has a free —COOH and a free —NH₂ group so that it can build up with other amino acids into *tripeptides* and, ultimately, *polypeptides*.

The 16 different amino-acids in insulin make up a total sequence of 51 amino-acid units giving a relative molecular mass of 5 700. In other proteins the number of units may be about 6 000 and the relative molecular mass of the order of 10^6. When the number of amino-acid units is less than about 40 the product is known as a polypeptide rather than a protein but the choice is rather arbitrary.

6 The structure of proteins

With about 20 amino-acids that can be used as units there is an enormous number of ways in which sequences of the acids can be linked together, and this explains why there is such a wide variety of protein materials. A long chain protein can be represented by the following general formula

$$H_2N-\underset{\underset{R_1}{|}}{\overset{\overset{H}{|}}{C}}-\overset{\overset{O}{\|}}{C}-\underset{}{\overset{\overset{H}{|}}{N}}-\underset{\underset{R_2}{|}}{\overset{\overset{H}{|}}{C}}-\overset{\overset{O}{\|}}{C}-\underset{}{\overset{\overset{H}{|}}{N}}-\underset{\underset{R_3}{|}}{\overset{\overset{H}{|}}{C}}\cdots-\underset{\underset{R_n}{|}}{\overset{\overset{H}{|}}{C}}-COOH$$

and the order in which amino-acids occur in such a formulation is known as the *primary structure* of the protein. The *secondary structure* refers to the detailed configuration of the chain; the tertiary structure involves the way in which protein molecules are arranged in relation to each other.

a. The primary structure. Discovering the primary structure of a protein, e.g. insulin, is like trying to find the one particular order in which 16 different letters might be arranged in a sequence of 51 letters.

In principle the protein is partially hydrolysed using dilute acids or by using enzymes which will only split the peptide links between certain specific types of amino-acid. The relatively small hydrolysis products are then investigated by N-terminal or C-terminal analysis.

The particular amino acid at one end of any peptide chain can be ascertained by treating the peptide with 2,4-dinitrofluorobenzene (DNFB). This reacts (p. 148) with the N-terminal amino-acid which contains a free —NH_2 group and thus labels it. Two amino-acids can link together in two ways. The nature of the hydrolysis products after treatment with DNFB shows the way in which the two amino acids were originally linked. For example,

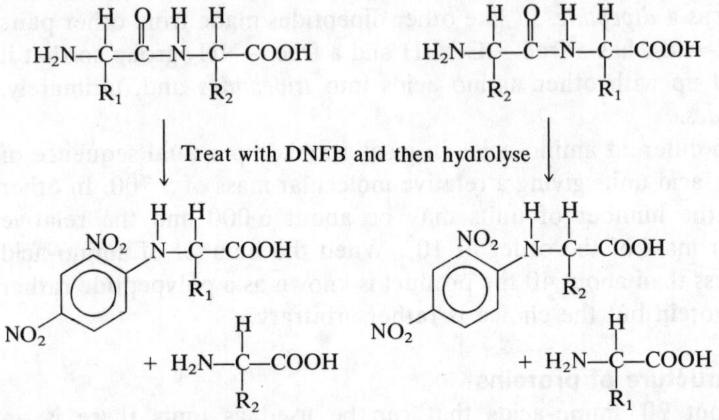

$$H_2N-\underset{R_1}{\overset{H}{C}}-\overset{O}{C}-\underset{}{\overset{H}{N}}-\underset{R_2}{\overset{H}{C}}-COOH \qquad\qquad H_2N-\underset{R_2}{\overset{H}{C}}-\overset{O}{C}-\underset{}{\overset{H}{N}}-\underset{R_1}{\overset{H}{C}}-COOH$$

Treat with DNFB and then hydrolyse

$$NO_2\text{-}C_6H_3(NO_2)-\underset{}{\overset{H}{N}}-\underset{R_1}{\overset{H}{C}}-COOH \qquad\qquad NO_2\text{-}C_6H_3(NO_2)-\underset{}{\overset{H}{N}}-\underset{R_2}{\overset{H}{C}}-COOH$$

$$+ \; H_2N-\underset{R_2}{\overset{H}{C}}-COOH \qquad\qquad\qquad + \; H_2N-\underset{R_1}{\overset{H}{C}}-COOH$$

Similarly, in C-terminal analysis, it may be possible to find an enzyme which will split off only the amino-acid at the —COOH end of the chain so that this can be identified.

The primary structure of ox insulin was elucidated by Sanger in 1955 (Fig. 59) and since then, the structures of even larger molecules, e.g. the enzymes lysozyme (present in tears) and ribonuclease (present in the pancreas) have been worked out.

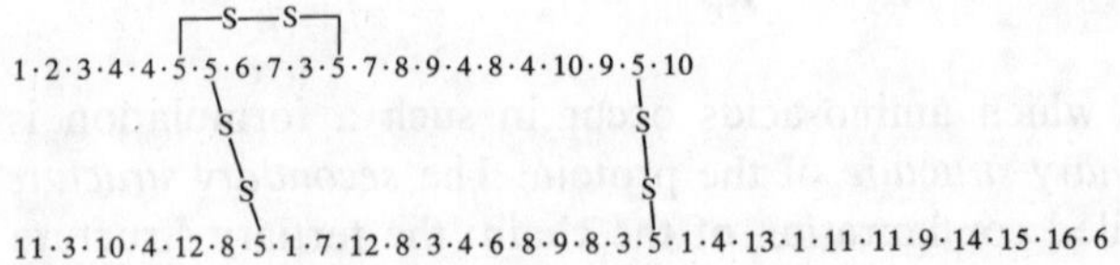

1·2·3·4·4·5·5·6·7·3·5·7·8·9·4·8·4·10·9·5·10 (with S—S cross linkage above)

11·3·10·4·12·8·5·1·7·12·8·3·4·6·8·9·8·3·5·1·4·13·1·11·11·9·14·15·16·6 (with S linkages)

Fig. 59. The arrangement of amino-acid units in ox insulin showing the -S-S-cross linkages between a chain of 21 and a chain of 30 units. The key is as follows: 1, glycine; 2, isoleucine; 3, valine; 4, glutamic acid; 5, cysteine; 6, alanine; 7, serine; 8, leucine; 9, tyrosine; 10, aspartic acid; 11, phenylalanine; 12, histidine; 13, arginine; 14, threonine; 15, proline; 16, lysine. Insulin samples from sheep, horses, pigs and whales have slightly different structures.

b. The secondary structure. The long chain of amino-acid units in a protein molecule may be coiled or folded or puckered and the actual configuration, investigated by X-ray analysis, is known as the secondary structure. One of the commonest structures is the α-helix structure resembling an extended spiral spring. Such a structure is stabilised by the existence of N—H $\cdots$ O hydrogen bonds between the N—H group of each amino-acid unit and the fourth C=O group following it along the chain. The pattern repeats itself every five turns (Fig. 60).

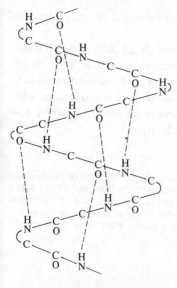

Fig. 60. Diagrammatic representation of the hydrogen bonding in the α-helix structure of a protein molecule.

The breaking down of such a secondary structure only involves the breaking of weak hydrogen bonds. It might be brought about by heat or acids or ultra-violet radiation. The sequence of amino-acids, i.e. the primary structure, is not changed but the secondary structure is; the protein is said to be *denatured* for it loses its original characteristics as, for example, in the boiling of an egg.

The secondary structure of a protein may also involve the linking together of two amino-acid chains through —S—S— bonds originating from the sulphur atoms of cysteine molecules in each of the two chains. Such cross-linking occurs in the structure of insulin first discovered by Sanger in 1954. One chain contains 21 amino-acid units; the other, 30 (Fig. 59).

c. The tertiary structure. This involves the way in which helical or extended chain molecules are arranged in relation to each other. The structure is investigated by X-ray analysis. A number of helices can twist together like the wires in an electric cable; such a tertiary structure is found in nucleic acids. Extended-chain molecules can pack together side by side or as pleated-sheets. In all cases, hydrogen bonding between adjacent molecules stabilises the structure.

7 Tests for proteins

There are many tests which serve to detect proteins, polypeptides and amino-acids.

Some give a purple colour when a few drops of copper(II) sulphate(VI) are added to a solution in sodium hydroxide (the biuret reaction, p. 263); some give a yellow colour with concentrated nitric(V) acid (the xanthoproteic reaction); some give a red colour with Millon's reagent (made by dissolving mercury in nitric(V) acid); and some give a blue colour on heating with ninhydrin.

8 The synthesis of proteins

Little is known about the synthesis of proteins in living matter but some relatively small polypeptides have been synthesised in the laboratory. Between 1900 and 1915 Fischer synthesised some polypeptides containing up to 19 amino acid units but only about two different acids were involved. Since then more complex polypeptides, e.g. the hormones oxytocin and vasopressin (secreted by the pituitary gland) have been made by du Vigneaud (Fig. 61).

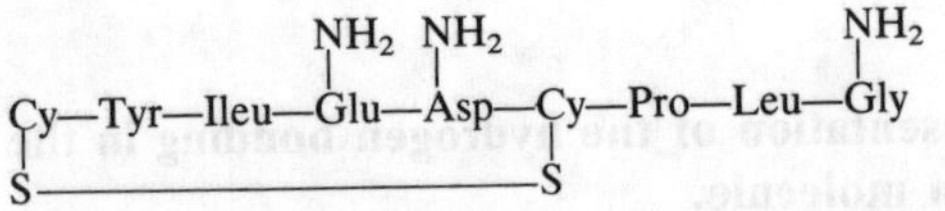

Fig. 61. The structure of oxytocin. Vasopressin has the same ring structure but a modified side-chain. Cy-S = Cystine; Tyr = Tyrosine; Ileu = Isoleucine; Glu = Glutamic acid; Asp = Aspartic acid; Pro = Proline; Leu = Leucine; Gly = Glycine.

Questions on Chapter 22

1 Explain and illustrate the behaviour of aminoethanoic acid as (a) an acid, (b) a base.

2 Describe the preparation of aminoethanoic acid. How does it react

with (a) soda-lime, (b) sodium hydroxide solution, (c) nitric(III) acid, (d) hydrochloric acid, (e) copper carbonate?

3 An aliphatic compound, A, contained C = 21.2 per cent, H = 1.8 per cent and Cl = 62.8 per cent. On being treated with water in the cold it gave a compound, B, containing C = 25.4 per cent, H = 3.15 per cent, Cl = 37.6 per cent. When B was heated with sodium carbonate solution and then acidified an acid, C, of molecular formula $C_2H_4O_3$ was isolated. What were A, B and C? Account for the above reactions.

4 A compound, A, is optically active and on analysis is found to contain 40 per cent C, 7.9 per cent H and 15.7 per cent N. On treatment with nitric(III) acid, nitrogen is evolved and an acid, B, of molecular formula $C_3H_6O_3$ is formed. Both A and B can be ethanoylated. What are A and B? Give two methods for synthesising A.

5 Outline the methods which are available for the separation of a mixture of amino acids. A pentapeptide which does not react with 2,4-dinitrofluorobenzene, on complete acid hydrolysis yields 3 moles of glycine, 1 mole of alanine and 1 mole of phenylalanine. Among the products of partial hydrolysis are found H—Ala—Gly—OH (alanylglycine) and H—Gly—Ala—OH (glycylalanine). What is the structure of the polypeptide? (Camb. Schol.)

6 A hexapeptide gave, upon complete hydrolysis, two moles of glycine(Gly), and one mole each of alanine(Ala), leucine(Leu), serine(Ser) and tyrosine(Tyr). Partial hydrolysis under milder conditions gave amongst other products, two dipeptides, glycylserine(Gly.Ser) and tyrosylglycine(Tyr.Gly), and a tripeptide, glycylalanylleucine(Gly.Ala.Leu). Treatment of the hexapeptide with 2,4-dinitrofluorobenzene and alkali, followed by hydrolysis of the N-dinitrophenylpeptide, gave N-dinitrophenylglycine. Give structures for the hexapeptide consistent with these observations, and indicate how you could distinguish between them. (Note—structures for the amino-acid residues are not required. Use the abbreviations indicated in parentheses). (Oxf. Schol.)

7 The reactivity of the amino (—NH₂) group is largely dependent on the nature of the group to which it is attached. Give examples which illustrate the difference in reactivity of the amino group when attached to (a) an alkyl group: (b) an aromatic nucleus: (c) a carbonyl group. How may this difference in reactivity be explained? (L.)

23

Carbohydrates

1 Introduction

The carbohydrates comprise a large group of substances many of which occur naturally and are of great importance. The name carbohydrate is derived from the fact that many of the group have a molecular formula which can be written as $C_x(H_2O)_y$; such a formula does not, however, indicate the correct structural arrangement.

a. Monosaccharides. Carbohydrates containing six or less carbon atoms per molecule are known as monosaccharides. The names always end in -ose and monosaccharides are divided into *aldoses* (those containing an aldehyde, —CHO, group), and *ketoses* (those containing a keto, $\geq$C=O, group). An aldose containing five carbon atoms per molecule is known as an aldopentose; a ketose with six carbon atoms is a ketohexose; and so on.

Glucose, an aldohexose, and fructose, a ketohexose, are the most important examples. They both have a molecular formula of $C_6H_{12}O_6$, but this represents many isomers.

b. Disaccharides. Carbohydrates containing twelve carbon atoms per molecule and having molecular formula $C_{12}H_{22}O_{11}$, are known as disaccharides. They may be regarded as built up by the elimination of water from two hexose molecules, i.e. $2C_6H_{12}O_6 - H_2O$. On hydrolysis they split up into two hexoses. Sucrose (a non-reducing sugar) and maltose and lactose (reducing sugars) are the most important disaccharides.

c. Polysaccharides. These may be regarded as being built up from many hexose molecules linked together in long-chains, water being eliminated between each pair of hexose molecules. They have a general formula

$(C_6H_{10}O_5)_n$, i.e. $nC_6H_{12}O_6 - nH_2O$,

with *n* having a value of about 330 in starch and about 600 in cellulose. On hydrolysis, the polysaccharides split up into disaccharides and/or monosaccharides.

d. Summary. The commonest carbohydrates are summarised as follows.

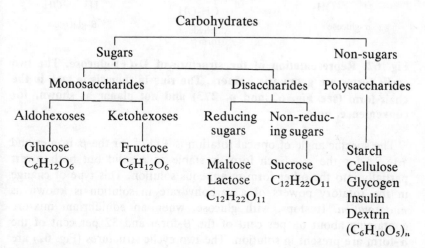

2 Glucose

There are sixteen possible isomeric aldoses with molecular formula $C_6H_{12}O_6$ made up of eight optically active pairs corresponding with the existence of four asymmetric carbon atoms in the molecule (p. 278). $(-)$-glucose does not occur naturally and the name glucose refers to the $(+)$-isomer. As it is dextrorotatory it may also be called *dextrose*.

a. Occurrence and manufacture. Glucose occurs naturally in grapes and most sweet fruits, in honey and in small quantities in many plants.

It is manufactured by heating starch with dilute sulphuric(VI) acid under pressure,

$$(C_6H_{10}O_5)_n + nH_2O \rightarrow nC_6H_{12}O_6$$
$$\text{Starch} \qquad\qquad \text{Glucose}$$

b. Structure of glucose. Solid $(+)$-glucose exists in two isomeric forms, known as α-$(+)$- and β-$(+)$-glucose. They are geometric isomers the difference in their structures depending on the position of —H and —OH groups above or below the plane of the chair-form (p. 273) of a six-membered ring (Figs. 62 and 63).

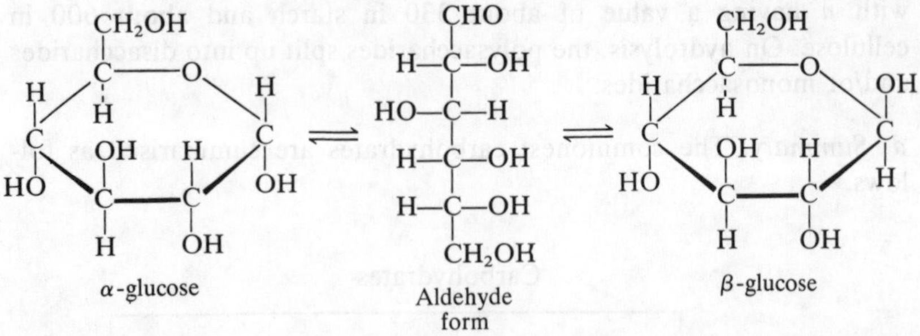

α-glucose   Aldehyde   β-glucose
                     form

Fig. 62. Representation of the structure of D-(+)-glucose. The two cyclic forms are geometric isomers. The ring in them is in fact, in the chair-form (see Fig. 63 and p. 273) and not planar as shown, for convenience, here.

The specific angle of optical rotation is +112° for the β-isomer and +18.7° for the α-. Each form is stable as a solid but both revert partially into the other form in aqueous solution. This type of change in the rotatory powers of a carbohydrate in solution is known as *mutarotation*. It stops, with glucose, when an equilibrium mixture containing about 68 per cent of the β-form and 32 per cent of the α-form are present in solution. The two cyclic structures (Fig. 62) are also in equilibrium with a small amount of a straight-chain aldehyde.

The relationship of this structure to that of D-glyceraldehyde (p. 283) shows that the common form of glucose is D-(+)-*glucose*.

c. Properties of glucose. Glucose is a white, crystalline solid, readily soluble in water but only sparingly soluble in ethanol. It tastes sweet but less so than sucrose. Its chemical properties depend on the existence of one —CHO group, one —CH₂OH group and four ⟩CHOH groups.

The —CHO group can be oxidised to —COOH by mild oxidising agents such as bromine water; it can be reduced to —CH₂OH by sodium amalgam and water; it will form cyanhydrins with hydrogen cyanide; and it forms condensation compounds with phenylhydrazine

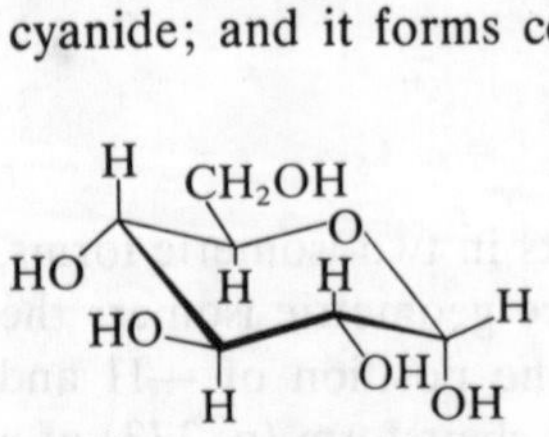

Fig. 63. The chair-form structure of α-glucose.

and hydroxylamine. The aldehydic nature of glucose also means that it is a reducing sugar. In particular, it will reduce Fehling's solution to copper(I) oxide and Tollens's reagent (p. 222) to silver.

Stronger oxidising agents than bromine water oxidise both the —CHO and —CH₂OH groups to —COOH.

The five —OH groups in the molecule can be ethanoylated with ethanoyl chloride or ethanoic anhydride; the product is a penta-ethanoylated glucose.

A solution of glucose is readily fermented by the enzyme, zymase, in yeast, i.e.

$$C_6H_{12}O_6 \xrightarrow{\text{zymase}} 2C_2H_5\text{—OH} + 2CO_2$$

d. Glucose as a foodstuff. Glucose is manufactured for use as a foodstuff and it is particularly well absorbed by the body. It is formed as a hydrolysis product of sucrose and starches and it undergoes a complex, step-wise oxidation which is exothermic and which provides energy for other bodily processes,

$$C_6H_{12}O_6 + 6O_2 \rightarrow 6CO_2 + 6H_2O \quad \Delta H^{\ominus}(298\ \text{K}) = -2\,850\ \text{kJ mol}^{-1}$$

Excess glucose is converted into a polysaccharide, glycogen (p. 320), in the liver and this is stored in the liver and muscles as a reserve supply of energy. When necessary, some of the glycogen is reconverted into glucose. This normal metabolic process is impeded or prevented in diabetics because their pancreas glands do not secrete enough insulin (p. 305). One of the first signs of this is the build up of glucose concentration in the urine, but the disease can be well treated by taking insulin and controlling the diet.

3 Fructose

Fructose is isomeric with glucose but it is a ketohexose. These ketohexoses have three asymmetric carbon atoms in the molecule so that there are four optically active pairs making eight isomers in all. (−)-fructose is the commonest isomer as it occurs naturally; (+)-fructose does not occur naturally.

The common form of fructose is, therefore, laevorotatory (and, sometimes, called laevulose), whereas that of glucose is dextrorotatory (dextrose). Both are *D*-structures.

a. Occurrence and manufacture. (−)-fructose occurs naturally with glucose in honey and many fruits. It is manufactured by the acid hydrolysis of the polysaccharide inulin (p. 320) which occurs in dahlia tubers and Jerusalem artichokes,

$$(C_6H_{10}O_5)_n + nH_2O \rightarrow nC_6H_{12}O_6$$

Inulin   Fructose

b. Structure (Fig. 64). (−)-fructose, like (+)-glucose, exists in α- and β-forms with six-membered ring-structures and in a straight-chain ketonic form. Like glucose, it exhibits mutarotation.

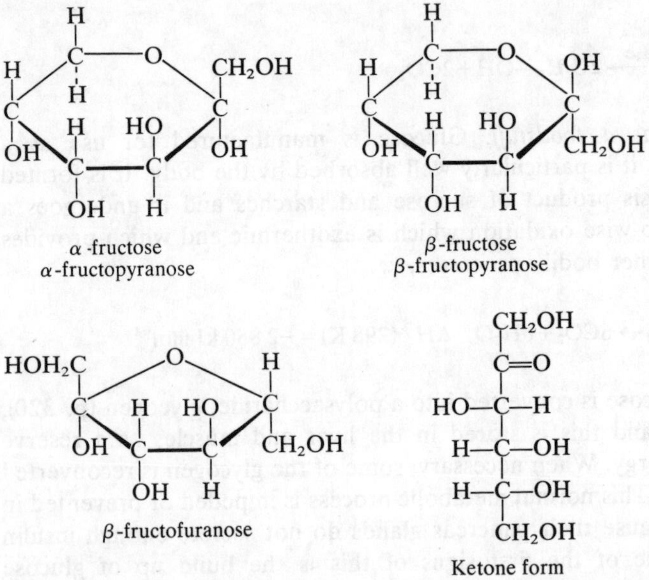

α -fructose
α-fructopyranose

β-fructose
β-fructopyranose

β-fructofuranose

Ketone form

Fig. 64. Representation of the structure of D-(−)-fructose.

The six-membered ring structures are known as α-fructopyranose and β-fructopyranose because they are formally related to the six-membered ring structure of pyran. When linked with other sugars, as in sucrose, fructose can also have a five-membered ring structure; this is known as β-fructofuranose because of its relationship to the structure of furan (p. 6).

c. Properties. Fructose is a colourless, crystalline solid, more soluble in ethanol and water than glucose. Many of its chemical properties are similar to those of glucose.

It acts as a reducing agent in the same way, despite the fact that it is a ketohexose and not an aldohexose; it is reduced to sorbitol by sodium amalgam and water; it reacts with hydrogen cyanide to form a cyanhydrin, with ethanoic anhydride to form a penta-ethanoyl derivative, and with phenylhydrazine; and it can be fermented by yeast.

Fructose differs from glucose in its optical rotatory powers, its solubility, and in the fact that it cannot be oxidised by bromine water. It is, however, oxidised by nitric(V) acid but, unlike, glucose, the fructose molecule splits up.

4 Other monosaccharides

There are four aldopentoses, known as arabinose, xylose, ribose and lyxose, each occurring in two optically active forms.

The eight aldohexoses, each known in two optically active forms are called glucose, mannose, galactose, allose, altrose, gulose, talose and idose.

The four pairs of optically active ketohexoses are fructose, sorbose, tagatose and psicose.

5 Sucrose

Sucrose is the commonest of all sugars and when the word sugar is used in everyday language it is usually sucrose that is meant. It is very widely distributed in nature and is extracted in very large quantities from sugar cane or sugar beet.

a. Properties. Sucrose is a white, crystalline solid, soluble in water but almost insoluble in ethanol. It is one of the sweetest of all sugars.

Sucrose is not a reducing sugar and will not react with Fehling's solution or Tollens's reagent. Nor will it react with hydrogen cyanide, phenylhydrazine or hydroxylamine. These facts indicate that there is no $>C=O$ group in the sucrose molecule.

Sucrose can, however, be ethanoylated to give an octaethanoyl derivative which indicates the presence of eight —OH groups. It is also oxidised by nitric(V) acid to ethanedioic acid.

Sucrose is dextro-rotatory in solution but it is hydrolysed into an equimolecular mixture of $(+)$-glucose and $(-)$-fructose. The latter has a greater rotatory power than the former so that the mixture is laevo-rotatory; it is known as *invert sugar* and its formation from sucrose is known as the inversion of sucrose. This inversion is catalysed by H^+ ions, and can also be brought about by enzymes, e.g. invertase in yeast or amylase in saliva,

$$C_{12}H_{22}O_{11} + H_2O \rightarrow \underbrace{C_6H_{12}O_6 + C_6H_{12}O_6}_{\text{Invert Sugar}}$$

$(+)$-sucrose $\quad$ $(+)$-glucose $\quad$ $(-)$-fructose

b. Structure. The absence of $>C=O$ groups in sucrose, the presence of eight —OH groups, and the hydrolysis into (+)-glucose and (−)-fructose are all accounted for by allotting sucrose the structure as shown in Fig. 65. It will be seen that a molecule of α-glucose is linked to one of β-fructofuranose through the elimination of one molecule of water.

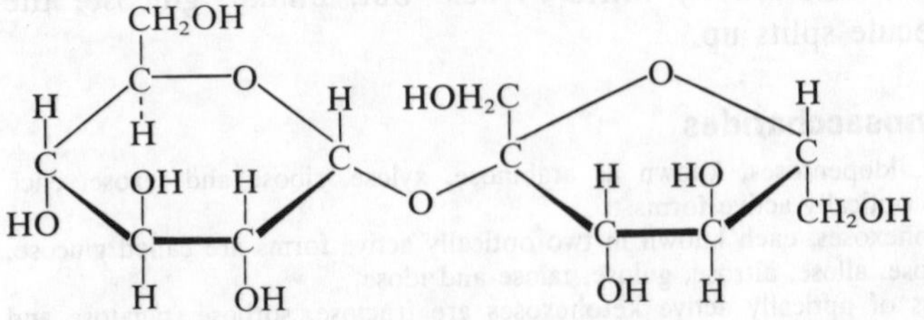

Fig. 65. Simplified structure of sucrose. In reality the α-glucose (left-hand) part of the molecule is in the chair-form (Fig. 63) and not planar as shown here.

6 Maltose and lactose

Maltose and lactose are the only disaccharides, other than sucrose, that occur naturally.

Maltose (malt-sugar) is obtained when starch is hydrolysed by the enzyme, diastase, present in malt and in saliva. It is made up of two glucose molecules joined together with the elimination of water.

Lactose (milk-sugar) occurs naturally in mammalian milk and is manufactured from whey, a by-product of cheese-making. It is oxidised by certain bacteria, normally present in the air, to 2-hydroxypropanoic (lactic) acid and butanoic acid. It is this oxidation that normally causes the souring of milk. Glucose and galactose rings are linked together in the lactose molecule.

7 Starch

Starch occurs as white granules in almost all plants. It is used by plants as a reserve food supply and it provides a very important component of animal's diets. Starch is extracted from such plants as rice, barley, maize, wheat and potatoes by grinding and washing. A suspension of starch in water is obtained and impurities are removed by filtration through a sieve which allows the starch through but retains impurities such as cellulose. The suspension of purified starch is then washed and dried. It may, also, be separated into its water-soluble component, *amylose*, and its water-insoluble component, *amylopectin*.

Starch is used in laundry work for stiffening purposes, in sizing cloth and paper, in making pastes and adhesives, in cooking, and in the manufacture of many chemicals, e.g. glucose (p. 313) and ethanol (p. 175).

a. Properties. Starch is a white, amorphous powder, partially soluble in water. On heating with water, the starch granules swell and burst forming a viscous, opalescent solution which, on cooling, sets to a jelly and can be used as a paste.

Starch decomposes into dextrin and other products on heating; it gives a characteristic blue-black colour with iodine solution; and it can be hydrolysed.

Enzymes, e.g. diastase in malt or amylase in saliva, hydrolyse it to maltose; hot mineral acids hydrolyse it, first, to maltose but then to glucose; partial hydrolysis yields dextrin (p. 320).

b. Structure. Starch is a polysaccharide with molecular formula $(C_6H_{10}O_5)_n$. The hydrolysis products indicate that it consists of glucose units linked together with the elimination of water (Fig. 66). The structures of starches varies with their source but most can be split up into water-soluble amylose and insoluble amylopectin.

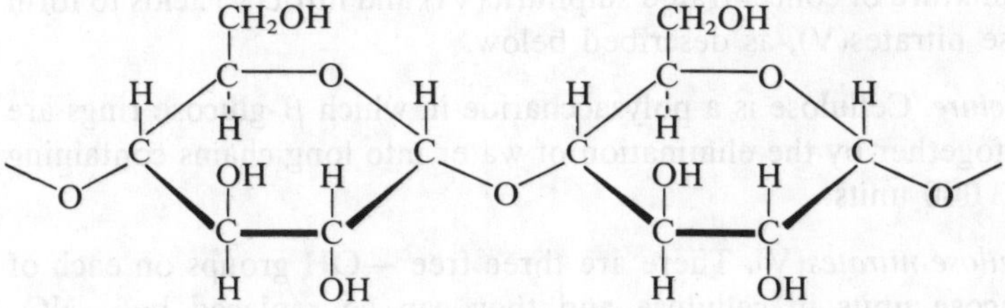

Fig. 66. Part of a starch molecule (simplified). In reality the two rings are not planar as shown here. The α-glucose units are linked in long chains in amylose but the chains are branched in amylopectin.

Both amylose and amylopectin are made up of α-glucose units linked together. In amylose the linking is in long straight chains with n having values between 1 000 and 4 000 and in amylopectin n has values between 600 and 6 000 and the structure consists of highly branched chains.

8 Cellulose

Cellulose is the main constituent of the cell-walls of plants, e.g. cotton, flax, hemp and jute and is very widely distributed. It also occurs, together with lignins and resins, in wood. The lignins and resins can be dissolved out by treatment with a mixture of sodium hydroxide, sodium sulphide and sulphur or with calcium hydrogensulphate(IV) solution. Much of the remaining cellulose, often referred to as wood pulp, is made into paper.

Pure cellulose is generally obtained either from wood or from cotton linters. These are the short fibres remaining on the cotton seed after the removal of the staple cotton. Much of this cellulose is converted into rayon (p. 333).

a. Properties. Cellulose is a white solid, insoluble in water and most organic solvents, but soluble in an ammoniacal solution of copper(II) hydroxide (Schweitzer's solution) and in a mixture of sodium hydroxide and carbon disulphide.

Cellulose is not hydrolysed so easily as starch; it cannot, for example, be digested by humans as starch can. It is, however, hydrolysed on heating with mineral acids under pressure. The tetra-, tri- and disaccharide units obtainable from cellulose can all be made, but the final hydrolysis product is glucose.

Cellulose reacts with ethanoic anhydride to form cellulose ethanoate (acetate), used in making fibres and plastic sheets and films (p. 333), and with a mixture of concentrated sulphuric(VI) and nitric(V) acids to form cellulose nitrates(V), as described below.

b. Structure. Cellulose is a polysaccharide in which β-glucose rings are linked together by the elimination of water into long chains containing 2 000–3 000 units.

c. Cellulose nitrates(V). There are three free —OH groups on each of the glucose units in cellulose and they can be replaced by —NO$_3$ groups either partially or fully by treatment with mixtures of nitric(V) and sulphuric(VI) acids.

If all the —OH groups are replaced the product is known as *gun cotton;* it is approximately cellulose trinitrate(V). Gun cotton, made from cotton waste and waste paper, is a propellant explosive used, mixed with nitroglycerol, in smokeless powders such as *cordite.*

A mixture of cellulose di- and tri-nitrates(V), made by using less concentrated acids, is known as *pyroxylin.* Solutions of it in solvents such as butanol, known as *collodion,* have been used as the base of many lacquers, but less inflammable materials are now replacing collodion. Pyroxylin itself is converted into *celluloid* by treating it with ethanol and then heating with camphor or similar compounds.

9 Other polysaccharides

Polysaccharides other than cellulose also play an important part in many natural processes.

Glycogen (p. 314) acts as the reserve food supply in animals in much the same way as starch does in plants; it is hydrolysed to glucose. *Inulin* also occurs naturally (p. 316); it is hydrolysed to fructose. *Dextrin,* used in making gums and adhesives, is the product of the partial hydrolysis of starch.

Pectins (from plant and fruit juices), *gums* (from trees), and *alginates* (from sea-weed) are also closely related to the polysaccharides.

Questions on Chapter 23

1 Distinguish, with examples, between the terms aldopentose, aldohexose, aldotetrose, ketopentose.

2 How is glucose obtained? How would you distinguish it from fructose? How does glucose react with (*a*) oxidising agents, (*b*) phenyl hydrazine?

3 Give an account of the experimental evidence for the assumption that the molecule of glucose has (*a*) a straight chain of six carbon atoms, (*b*) five —OH groups, (*c*) an aldehyde group. How could glucose be converted into fructose?

4 Explain the typical reactions of glucose using a straight-chain formula. Why is the straight-chain formula not completely satisfactory?

5 (*a*) Outline the processes involved in preparing crystalline cane sugar from either sugar cane or sugar beet. (*b*) Explain what happens when cane-sugar is heated, (i) alone, (ii) with a dilute acid. (*c*) Describe briefly how a specimen of dry ethanol could be obtained from cane-sugar.

6 How does (i) glucose, (ii) fructose react with (*a*) phenyl hydrazine, (*b*) ethanoic anhydride, (*c*) oxidising agents, (*d*) reducing agents?

7 How would you distinguish between samples of glucose, fructose, sucrose, starch and cellulose?

8 Explain the similarities and differences in the structures of (a) amylose and amylopectin, and (b) cellulose and starch.

9 Write notes on (a) mutarotation, (b) inversion, (c) enzymes, (d) specific rotation, (e) anomers, (f) epimers.

10 Discuss the ways in which (a) carbohydrates, (b) proteins can be hydrolysed.

11 Describe the uses of carbohydrates.

12*Use bond energy values to calculate the enthalpy of combustion of glucose.

24

Plastics

The term plastic, coined about 1930, covers a very wide field. It is generally taken to include both natural materials, e.g. rubber, many synthetic or semi-synthetic products, and substances, both natural and synthetic, sometimes called *resins*. The compounds concerned have very large molecules made up of recurring structural units; they are known as polymers (Gk. poly = many, meros = parts).

1 Types of polymers

Polymers are formed by the linking together of many smaller units known as monomers, the overall process being known as polymerisation (p. 112).

a. Addition polymerisation. If a mononomer is unsaturated its molecules can, in effect, form addition compounds with each other giving a polymer with the same empirical formula as the monomer. Ethene, and substituted ethenes, for example, form addition polymers,

$$2n\,CH_2{=}CH_2 \longrightarrow +\!(CH_2{-}CH_2{-}CH_2{-}CH_2)\!+_n$$
Ethene Poly(ethene) (Polythene)

Phenylethene (Styrene) Poly(phenylethene) (Polystyrene)

The products are simple *linear polymers*. When two different monomers, e.g. ethene and propene, are involved the product is a linear *copolymer*.

b. Condensation polymerisation. Condensation polymerisation occurs by the elimination of small molecules such as H_2O or NH_3 between two different molecules each of which has at least two functional groups which can participate in the condensation. Nylon-6,6, for example, is made from hexanedioic (adipic) acid and hexane-1,6-diamine (hexamethylene diamine) by elimination of H_2O molecules,

$$+\,HOOC—(CH_2)_4—COOH + H_2N—(CH_2)_6—NH_2 + HOOC—(CH_2)_4—COOH +$$

$$\downarrow -H_2O$$

$$—OC—(CH_2)_4—CO—NH—(CH_2)_6—NH—OC—(CH_2)_4—CO—$$
Nylon

The product is a linear copolymer.

c. Cross linking. Linear polymers or copolymers have only weak van der Waals' forces between their long chains. Polymers of this type, or those with only a few, weak cross links between the chains, are either *elastomers* (rubber-like substances) or *thermoplastic polymers* (p. 326). The latter soften when heated and harden when cooled as very little energy is required to break down the forces between the chains; they can be re-softened and re-hardened over and over again. Differences in the strength of the attractive forces between the chains mean that not all thermoplastic polymers are elastomers or vice versa.

Thermosetting polymers (p. 327) are those which become highly cross-linked on heating either in planar, two-dimensional sheets or in three-dimensional networks. Once hardened they cannot be softened by re-heating as the energy required to break down the cross-linkages is high enough to decompose the whole structure.

d. Atactic and isotactic polymers. If the chains in a polymer are arranged completely haphazardly the polymer will be amorphous but in most polymers there are at least regions within the polymer known as crystallites where the chains are orientated relative to one another so that the polymer has some crystalline nature.

The general properties of any polymer depend on the degree of crystalline character. A polymer with a high degree will have a higher tensile strength, a higher and sharper softening point, and a lower elasticity than one with a low degree.

The extent of crystallisation can be controlled to some extent both by the method of manufacture (p. 326) and by subsequent treatment. A monomer such as phenylethene (styrene), which is not symmetrical, can polymerise with all the phenyl groups on one side of the polymer chain (an *isotactic polymer*) or randomly on one side or the other (an *atactic polymer*). The more regular arrangement in the isotactic polymer enables the chains to pack together more tightly so that the structure is more crystalline than that of the corresponding atactic polymer.

Increased crystalline character can also be achieved by *cold-drawing* which is particularly important for Nylon and other fibres. The polymer is stretched and this has the effect of drawing the chains together into a better crystalline alignment and strengthening the material.

2 The mechanism of polymerisation

Synthetic polymers are manufactured on a very wide scale and many varied products are made to meet many varied needs. The same monomer can be polymerised into different chain lengths or into atactic or isotactic polymers by different methods of manufacture. Alternatively, two monomers can be used to give copolymers. All the processes must be carefully controlled and pure raw materials are required as the processes are affected by impurities.

Originally most polymerisations were brought about at very high temperatures and pressures but the development of a range of organometallic catalysts by Ziegler and Natta (the 1963 Noble prize winners) enabled the processes to be operated at lower temperatures and pressures and greatly extended the possibilities.

a. Free radical polymerisation. This type of polymerisation has to be initiated by the generation of a free radical, $X \cdot$, which can then lead to the build-up of a chain reaction, e.g.

$$X \cdot + CH_2 = CH_2 \rightarrow X - CH_2 - \dot{C}H_2$$

$$X - CH_2 - \dot{C}H_2 + CH_2 = CH_2 \rightarrow X - CH_2 - CH_2 - CH_2 - \dot{C}H_2 \text{ etc.}$$

The chain build up is limited by combination of the free radicals between themselves or with an impurity.

The initial free radicals are provided by organic peroxides (p. 111) or azo-compounds.

b. Anion polymerisation. This involves the formation of a carbanion of the monomer by attack of a nucleophile, the carbanion then acting as a nucleophile in attacking other molecules of the monomer, e.g.

$$Y^- + \begin{array}{c} H \\ | \\ C=C \\ | \\ H \end{array} \begin{array}{c} H \\ | \\ R \end{array} \rightarrow Y-\begin{array}{c} H \\ | \\ C \\ | \\ H \end{array}-\begin{array}{c} H \\ | \\ C^- \\ | \\ R \end{array}$$

$$Y-\begin{array}{c} H \\ | \\ C \\ | \\ H \end{array}-\begin{array}{c} H \\ | \\ C^+ \\ | \\ R \end{array} + \begin{array}{c} H \\ | \\ C=C \\ | \\ H \end{array}\begin{array}{c} H \\ | \\ R \end{array} \rightarrow Y-\begin{array}{c} H \\ | \\ C \\ | \\ H \end{array}-\begin{array}{c} H \\ | \\ C \\ | \\ R \end{array}-\begin{array}{c} H \\ | \\ C \\ | \\ H \end{array}-\begin{array}{c} H \\ | \\ C^- \\ | \\ R \end{array} \text{ etc.}$$

The build-up of the chain can be terminated by adding chemicals such as H^+, NH_3, O_2, CO_2 which react with the carbanion.

Ethene itself is not very susceptible to polymerisation by an anion mechanism for it is not readily attacked by a nucleophile (p. 106). The mechanism is most likely when R is a strongly electron-attracting group, e.g. CN.

c. Cation polymerisation. This involves formation of a carbonium ion of the monomer by attack of a proton from a suitable acid, e.g. $HClO_4$, the carbonium ion then acting as an electrophile in attacking other monomer molecules, e.g.

$$H^+ + \begin{array}{c} H \\ | \\ C=C \\ | \\ H \end{array}\begin{array}{c} H \\ | \\ R \end{array} \rightarrow H-\begin{array}{c} H \\ | \\ C \\ | \\ H \end{array}-\begin{array}{c} H \\ | \\ C^+ \\ | \\ R \end{array}$$

$$H-\begin{array}{c} H \\ | \\ C \\ | \\ H \end{array}-\begin{array}{c} H \\ | \\ C^+ \\ | \\ R \end{array} + \begin{array}{c} H \\ | \\ C=C \\ | \\ H \end{array}\begin{array}{c} H \\ | \\ R \end{array} \rightarrow H-\begin{array}{c} H \\ | \\ C \\ | \\ H \end{array}-\begin{array}{c} H \\ | \\ C \\ | \\ R \end{array}-\begin{array}{c} H \\ | \\ C \\ | \\ H \end{array}-\begin{array}{c} H \\ | \\ C^+ \\ | \\ R \end{array} \text{ etc.}$$

The build-up of the chain is terminated by loss of a proton.

This mechanism is most important when R is an electron-donating group, e.g. an alkyl group.

d. Use of Ziegler catalysts (p. 326). The Ziegler catalysts are heterogeneous catalysts and the detailed mechanism of their function is not known. A surface effect is probably involved perhaps with the formation of an intermediate compound, e.g.

$$-\overset{|}{Ti}- + CH_2=CH_2 \rightarrow -\overset{|}{Ti}-CH_2-CH_2-$$

$$-\overset{|}{Ti}-CH_2-CH_2- + CH_2=CH_2 \rightarrow -\overset{|}{Ti}-CH_2-CH_2-CH_2-CH_2-$$

3 Thermoplastic polymers

Thermoplastics are generally used in the form of moulded shapes, fibres, pipes, sheets or films. They may also be used as coatings for other materials.

If the original material is too brittle its properties can be changed by adding plasticisers, e.g. esters of phosphoric(V) or benezene dicarboxylic (phthalic) acids. Alternatively, a cheap filter can be added if a lowering of elasticity is acceptable.

Thermoplastics are generally very good insulators and are resistant to many chemicals. Most are translucent or opaque, though they can be coloured by adding pigments. Those which are transparent, e.g. Perspex, find particular uses.

a. Poly(ethene) or polythene. Low-density poly(ethene) is made by subjecting ethene to a pressure of 1 500–2 000 atmospheres at about 200 °C. Small amounts of oxygen or organic peroxides act catalytically by initiating a free-radical mechanism (p. 324).

High density poly(ethene) is made by the Ziegler or Phillips processes, which do not involve free-radical mechanisms and, therefore, give a slightly different product. In the Ziegler process ethene, at about 5 atmospheres and 60 °C is passed into an aromatic hydrocarbon solvent containing titanium(IV) chloride and aluminium triethyl, $Al(C_2H_5)_3$, as a catalyst. When the ethene has polymerised dilute acid is added to decompose the catalyst and the polymer is filtered off.

In the Phillips process, ethene at 30 atmospheres and 160 °C is passed into an aromatic hydrocarbon solvent containing aluminium and molybdenum oxides as catalysts; small amounts of hydrogen are added periodically.

The low-density polythene is less crystalline (more atactic) than the high-density type. This means that the low-density type is less strong, more elastic and has a lower softening temperature (110 °C) than the high-density type (130 °C). Both types are good electrical insulators and are resistant to acids and alkalis. They are used in electrical equipment, and in making containers, pipes, films and sheets.

b. Other examples. The discovery, by Ziegler and Natta, of catalysts which would enable polymerisation to take place at relatively low temperatures via ionic rather than free radical mechanisms quickly led to a great extension in the number of monomers that could be polymerised and the range of products that was available. The following table summarises some of the more important polymers obtained from unsaturated monomers, which may be regarded as substituted alkenes.

If the monomer is X the systematic name for the polymer is poly(X) but many of them are much better known by trade names or initials or by the old-fashioned name of the monomer.

Monomer	Name of polymer	Typical uses
Ethene, $CH_2{=}CH_2$	Polythene Alkathene	Electrical equipment. Containers. Piping. Sheets.
Chloroethene (Vinyl chloride) $CH_2{=}CHCl$	P.V.C.	Piping. Sheets.
Tetrafluoroethene, $CF_2{=}CF_2$	P.T.F.E. Teflon. Fluon	Surface coatings. Gaskets. Insulation.
Propene, $CH_2{=}CH{-}CH_3$	Poly(propene) Propathene	Fibres and mouldings. Bottle crates.
Phenylethene (Styrene), $C_6H_5{-}CH{=}CH_2$	Polystyrene	Mouldings. Foam.
Ethenylethanoate (Vinyl acetate), $CH_2{=}CH{-}O\!-\!\underset{\displaystyle O}{C}{-}CH_3$	Polyvinyl acetate	Adhesives
Methyl 2-methyl-propenoate,	Perspex Plexiglas	Mouldings. 'Safety' glass

$$CH_2{=}\underset{\displaystyle CH_3}{C}{-}\overset{\displaystyle O}{C}{-}O{-}CH_3$$

4 Thermosetting plastics

These contain highly crosslinked structure (p. 323) and they cannot be softened once they have hardened. They are generally used as moulding powders, the powder (mixed with fillers and pigments if required) being heated and pressed into shape in making such articles as electric plugs and switches, telephones, wireless and television cabinets, lavatory seats, bottle caps, ash trays and plastic tableware. The plastics are also used, in the liquid form or in solution, as adhesives and bonding resins. Typical uses are in making ply-wood, fibre boarding, laminated plastics and fibre-glass products.

a. Products from methanal. Methanal will form thermosetting plastics with phenol, urea (carbamide) and melamine.

The phenol-methanal products were developed by Baekeland in the first decade of this century. They are known as *Bakelite* and are made by heating the reagents in alkaline solution. Water is eliminated between two molecules of phenol and one of methanal, e.g.

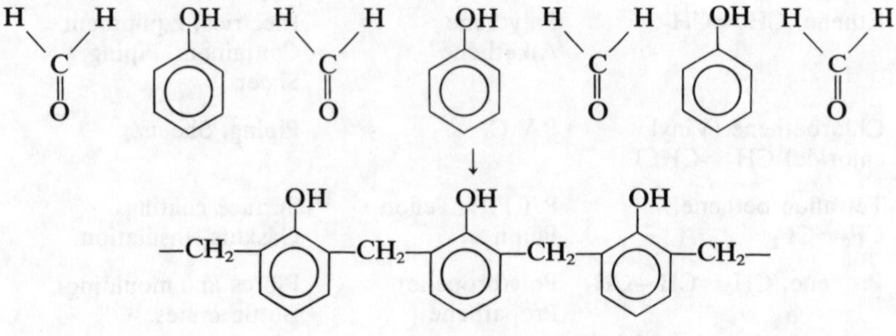

All the *o*- or *p*-positions in the phenol molecule can be used so that complex networks can build up as well as the straight chain arrangement shown above.

Bakelite is a good electrical insulator and has been widely used in making electrical goods but it can only be obtained in dark colours such as brown or black.

Methanal also condenses with carbamide (urea) or with melamine (made by heating cyanamide),

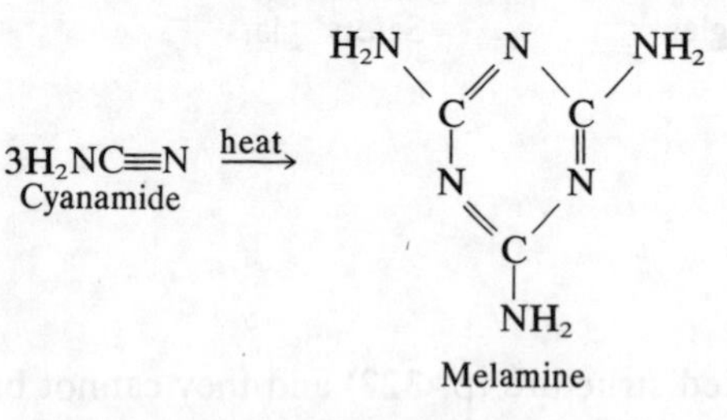

Melamine

Water is eliminated between the carbonyl group of the methanal and two —NH$_2$ groups, e.g.

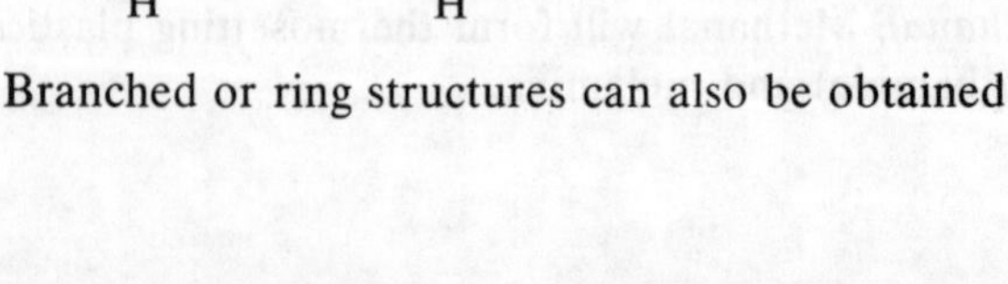

Branched or ring structures can also be obtained.

The resulting thermosetting plastics, e.g. Melaware, Beetle & Formica are harder than Bakelite and have better heat-resisting properties; they can also be made in a wider range of colours.

b. Glyptals. Alkyd resins. These can be made from benzene-1,2-dicarboxylic (phthalic) acid and propane-1,2,3-triol (glycerol). Each of the hydroxyl groups in the triol forms an ester linkage with one of the carboxyl groups of an acid molecule, e.g.

$$
\begin{array}{c}
CH_2-OH \\
| \\
CH-OH \\
| \\
CH_2-OH
\end{array}
+3\
\begin{array}{c}
COOH \\
\bigcirc COOH
\end{array}
\rightarrow
\begin{array}{c}
CH_2-O-CO-\bigcirc COOH \\
| \\
CH-O-CO-\bigcirc COOH \\
| \\
CH_2-O-CO-\bigcirc COOH
\end{array}
$$

The tri-ester contains three unused carboxyl groups per molecule and these can form ester links with other glycerol molecules so that a complex polyester can build up. A wide variety of products is possible for many different triols or diols and dicarboxylic acids can be used.

Glyptals are mainly used as bonding resins and in alkyd paints.

c. Polyurethanes. These are made from diisocyanates and diols or triols, the $-N{=}C{=}O$ and $-OH$ groups linking together to form urethane groups

$$
-N{=}C{=}O + -OH \rightarrow -\overset{H}{\underset{}{N}}-\overset{O}{\underset{}{C}}-O-
$$
Urethane group

Using $OCN-(CH_2)_6-NCO$ and $HO-(CH_2)_4-OH$ gives a product, Perlon U, similar to Nylon (p. 331),

$$
\left[-O-\overset{O}{\overset{\|}{C}}-\overset{H}{\underset{}{N}}-(CH_2)_6-\overset{H}{\underset{}{N}}-\overset{O}{\overset{\|}{C}}-O-(CH_2)_4- \right]_n
$$
Perlon U

By using methyl-2,4-diisocyanobenzene, made from methyl-2,4-dinitrobenzene, and polyesters containing three $-OH$ groups, polyurethane foams can be made. The foaming is caused by treatment with water; this reacts with some of the free —NCO groups resulting in further polymerisation and simultaneous evolution of carbon dioxide,

$$2 - N{=}C{=}O + H_2O \rightarrow \underset{\overset{|}{H}}{-N}\overset{O}{\underset{}{-C-}}\underset{\overset{|}{H}}{N-} + CO_2$$

Polyurethanes are used in floor finishes and hard-wearing paints, as stretch fibres and adhesives, and, in the form of foam, in making pillows, packaging and insulating materials.

d. Epoxy resins. These are polyethers formed by condensation polymerisation between epichlorohydrin (p. 188) and bisphenols (compounds containing two —C₆H₄—OH groups). In the first stage of the polymerisation, carried out in alkaline solution, NaCl is eliminated, and this is followed by the opening of the oxide ring, e.g.

The resulting polymer still contains terminal epoxy groups and can be cross linked by reaction with diamines or acid anhydrides, e.g.

The products vary from viscous liquids to high melting point solids. They are used as surface coatings, as adhesives and in making glass fibre laminates.

e. Natural resins. Natural resins are sticky substances, insoluble in water, which are secreted by plants and animals or which exude from incisions in the barks of trees.

Shellac is exuded by insects living on trees; it comes mainly from India and is used in polishes and varnishes. *Rosin* is a residue from the distillation of turpentine; it is used in sizing paper, in making floor-coverings and in soap-making. *Copal* is the name given to several resins which exude from trees growing in North America, the East Indies and Madagascar; the resins are hard materials used in making copal varnishes and cements.

5 Synthetic fibres

Silk, wool and cotton (p. 319) have been in use as naturally occurring fibres from time immemorial but, in recent years, various synthetic fibres have found increasing uses both by themselves and in conjunction with silk, wool or cotton. Many, e.g. Nylon, Terylene, Orlon, Acrilan, are made from thermoplastic polymers; others, e.g. artificial silks, are made from naturally occurring cellulose.

a. Nylon. Nylon was first synthesised by Carothers in 1935. The original type, now known as Nylon-6.6, is made by condensation of hexanedioic (adipic) acid and hexane-1,6-diamine (hexamethylene diamine), i.e.

$$+ HOOC—(CH_2)_4—COOH + H_2N—(CH_2)_6—NH_2 + HOOC—(CH_2)_4—COOH +$$

$$\downarrow$$

$$—OC—(CH_2)_4—CO—NH—(CH_2)_6—NH—CO—(CH_2)_4—CO—$$

Both raw materials are readily available from a number of sources. The acid, for instance, can be made from cyclohexane by catalytic oxidation

Cyclohexane $\xrightarrow[\text{oxidation}]{\text{catalytic}}$ $\begin{array}{l} CH_2—CH_2—COOH \\ CH_2—CH_2—COOH \end{array}$

Hexanedioic acid

and the diamine from butadiene,

$$CH{=}CH_2 \atop CH{=}CH_2 \quad \xrightarrow{Cl_2} \quad CH{-}CH_2Cl \atop CH{-}CH_2Cl \quad \xrightarrow{NaCN} \quad CH{-}CH_2{-}CN \atop CH{-}CH_2{-}CN$$

Butadiene

$$\Big\downarrow \text{catalytic hydrogenation}$$

$$CH_2{-}CH_2{-}CH_2{-}NH_2 \atop CH_2{-}CH_2{-}CH_2{-}NH_2$$

Nylon-6.6, e.g. Bri-nylon, is an example of a polyamide in which CH_2 chains are linked by peptide linkages (p. 306). The general formula is

$$\cdots C{-}(CH_2)_x{-}\overset{O}{\overset{\|}{C}}{-}\overset{H}{\overset{|}{N}}{-}(CH_2)_y{-}\overset{H}{\overset{|}{N}}{-}\overset{O}{\overset{\|}{C}}{-}(CH_2)_x{-}\overset{O}{\overset{\|}{C}}\cdots$$

Nylon-6 e.g. Enkalon, Celon (made from caprolactam) is very similar

Cyclohexanone   An oxime   Caprolactam   Nylon-6

$$\xrightarrow{NH_2OH} \qquad \xrightarrow[\text{heat}]{\text{c. } H_2SO_4 \text{ and}} \qquad \xrightarrow{\text{heat}} \ [NH{-}(CH_2)_5{-}CO]_n$$

Nylon is a white, horny material which can be used in bulk or converted into fibres by pressing the molten substance through fine holes. The threads are then strengthened by cold-drawing (p. 324). Nylon filament is relatively cheap, it is resistant to chemical attack, and it combines great tensile strength with low density. It is used in various diameters for making such articles as stockings, bristles, surgical sutures, fishing nets and a wide range of fabrics and lace.

b. Terylene. Terylene was first made about 1940 by the condensation of ethane-1,2-diol (glycol) and benzene-1,4-di-carboxylic (terephthalic) acid,

$$\cdots + HOCH_2{-}CH_2OH + HOOC{-}\!\!\bigcirc\!\!{-}COOH + HOCH_2{-}CH_2OH + \cdots$$

$$\Big\downarrow$$

$$\cdots{-}CH_2{-}CH_2{-}O{-}\overset{O}{\overset{\|}{C}}{-}\!\!\bigcirc\!\!{-}\overset{O}{\overset{\|}{C}}{-}O{-}CH_2{-}CH_2{-}$$

Terylene

It is an example of a *polyester.*

Terylene does not crease or stretch, is resistant to moths and mildew, and is very strong. It is woven into many fabrics, e.g. Crimplene. A similar product in America is known as Dacron.

c. Acrylic fibres. These were first made in America by polymerisation of cyanoethene (acrylonitrile or vinyl cyanide), a process resembling the polymerisation of chloroethene (vinyl chloride, p. 120).

$$2n \quad \overset{H}{\underset{H}{\diagdown}} C = C \overset{H}{\underset{CN}{\diagup}} \quad \rightarrow \quad \text{---}(\text{---}CH_2\text{---}\underset{CN}{CH}\text{---}CH_2\text{---}\underset{CN}{CH})_n\text{---}$$

Cyanoethene

The product can be converted into acrylic fibres by passing a solution of it through a spinneret into hot air. The fibres have a soft, wool-like feel and are very resistant to attack by acids or sunlight. They are most widely used in making clothing fabrics. On heating in the absence of air, hydrogen and nitrogen are lost and carbon fibres remain. These can be incorporated into epoxy and other resins to give very stiff and strong materials of low density.

Cyanoethene is also used for making copolymers with, for example, chloroethene or ethenylethanoate; *Acrilan* and *Courtelle* are typical examples.

d. Fibres from cellulose. Fibres have been manufactured from cellulose for about 100 years. As they resemble silk they are known as artificial silks or *Rayons.*

In the *cuprammonium process,* cotton linters are dissolved in an ammoniacal solution of copper(II) hydroxide (Schweitzer's solution). The resulting solution is then forced through a spinneret into an acid bath. As the solution passes into the acid it is converted into cellulose threads which can be dried, spun together and woven.

In the *viscose process,* more common in America and this country than the cuprammonium process, cotton linters or wood pulp are treated with sodium hydroxide solution and carbon disulphide. A viscous solution of cellulose xanthate is obtained and cellulose threads can be obtained from this by passing through a spinneret into an acid bath. The solution can also be made into thin sheets (*cellophane*) by extrusion into an acid bath through a narrow slit.

Acetate silks are made by first treating cotton linters with a mixture of ethanoic anhydride, ethanoic acid and sulphuric(VI) acid. The cellulose is ethanoylated and cellulose acetate (ethanoate) forms as a white flocculent mass when the solution is added to water. The solid can be dissolved in propanone and fibres can be obtained from the solution by passing through a spinneret into hot air. The fibres are woven into cloth under such trade names as *Celanese*. The cellulose acetate solution is also used in varnishes and lacquers and it can be converted into sheets or films by extrusion through slits.

6 Natural fibres

a. Silk. Natural silk is more expensive than artificial silk but it is also more durable. Its composition depends on the larva from which it is obtained but it consists, essentially, of polypeptide chains arranged in sheets through hydrogen bonding. The main amino-acids concerned are glycine, *L*-alanine, *L*-serine and *L*-tyrosine.

b. Wool. Wool, like silk, consists of polypeptide chains with considerable cross linking through —S—S— bonds originating from the cystine content. The fibres are curly and this accounts for the particularly 'fluffy' nature of woven wool. It is difficult to simulate this effect in synthetic fibres; hence the slogan 'there is no substitute for wool'.

If the —S—S— cross links are weakened or broken the fibres become more pliable. Similar cross-links are broken and, later, reformed in chemical hair-curling processes.

c. Cotton, hemp, flax and jute. These are all, basically, made up of cellulose (p. 319).

7 Natural rubber

a. Occurrence and production. Rubber is obtained by the coagulation of the latex (milky juice) which can be tapped off from tropical trees, particularly species of Hevea. A narrow, sloping incision is made in the bark of the tree to provide a channel down which the latex can flow into a collecting cup. The latex is poured into tanks and coagulated into 'curds' by adding weak acids, e.g. methanoic acid. The curds are filtered off, washed and pressed into sheets which are then heated and masticated to soften the rubber. Pigments, fillers, e.g. zinc oxide and carbon black, and sulphur are then added and the mixture is heated by steam under pressure.

The sulphur plays a particularly important part and is said to *vulcanise* the rubber. It does this by improving its elasticity, toughness and strength and by enabling it to be used over a wider range of temperature. Before the discovery of the vulcanisation process, by Goodyear in 1830, rubber had not been widely used as it was soft and sticky in summer and hard and brittle in winter. If a larger proportion of sulphur is used the product is *ebonite* or *vulcanite*, a hard, non-elastic material.

b. Chemical nature. Natural rubber is a polymer made up of an isoprene, C_5H_8, monomer. The polymer chain still contains $C=C$ bonds and can exist in two geometrically isomeric forms, i.e.

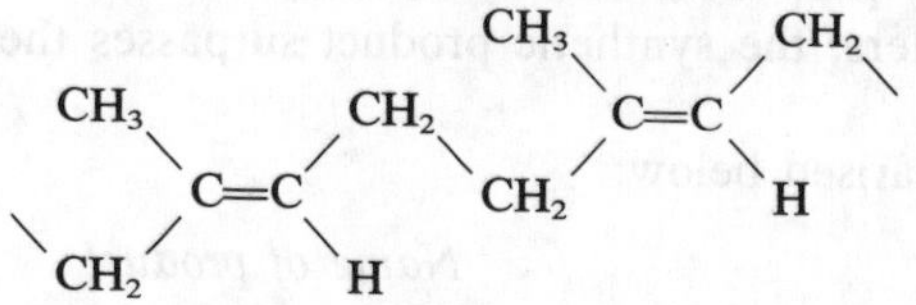

Trans-isomer
(Gutta percha)

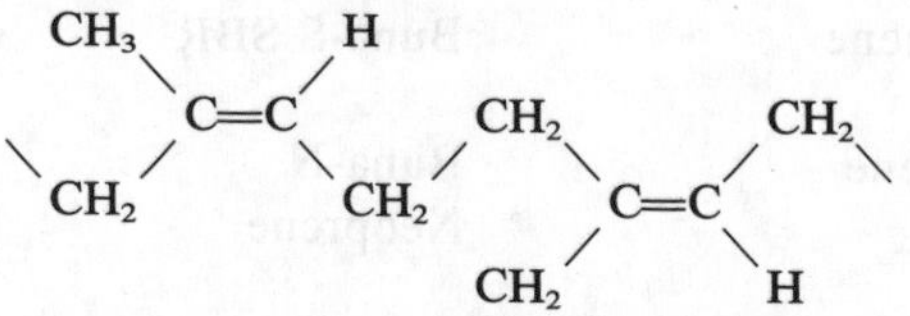

Cis-isomer
(Natural rubber)

The *trans*-form occurs in *gutta-percha*, which is a hard, non-elastic solid. The polymer chains can pack alongside each other giving a semi-crystalline structure. The chain arrangement in the *cis*-isomer, which occurs in natural rubber, does not allow such close packing and the chains are randomly arranged. On stretching, they are straightened and, to some extent, lined up but the stretched state is less stable so that rubber reverts to its original unstretched form.

The precise function of the sulphur in vulcanisation is not known but it involves C—S—C bonds as cross links between the polymer chains.

8 Synthetic rubber

Isoprene (methylbuta-1,3-diene) can be polymerised into a synthetic rubber almost identical with natural rubber using a Ziegler catalyst (p. 326) but the product is expensive because the raw material is, and common synthetic rubbers employ other monomers, e.g.

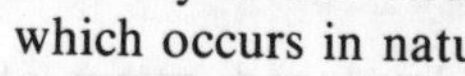

$CH_2=CH-CH=CH_2$

Buta-1,3-diene

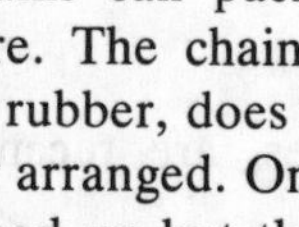

Cyanoethene

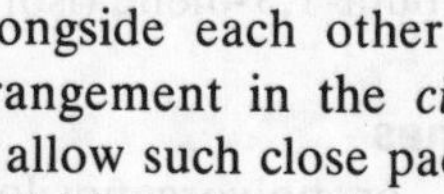

Methylbuta-1,3-diene
(Isoprene)

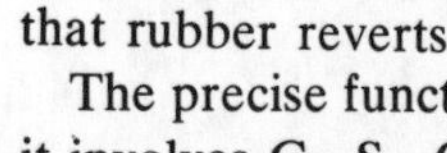

2-chlorobuta-1,3-diene
(Chloroprene)

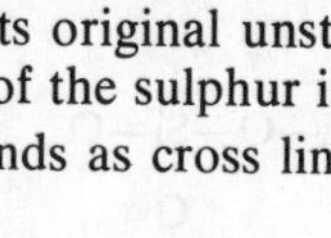

2-methylpropene

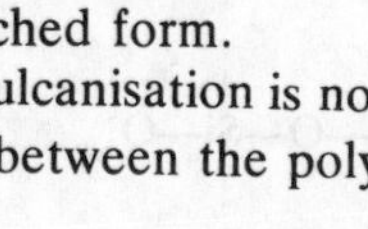

Phenylethene
(Styrene)

These monomers can be polymerised individually or in pairs or threes to give copolymers; lithium, sodium, peroxides or Ziegler catalysts are used, and the conditions can be varied to give rubbers with different properties. For some purposes, natural rubber is better than any synthetic rubber; for others, the synthetic product surpasses the natural one.

Typical products are summarised below:

Monomer(s) used	*Name of product*
Buta-1,3-diene	Buna
Buta-1,3-diene and phenylethene (styrene)	Buna-S SBR
Buta-1,3-diene and cyanoethene	Buna-N
2-chlorobuta-1,3-diene (chloroprene)	Neoprene
2-methylpropene (isobutylene) and methylbuta-1,3-diene (isoprene)	Butyl rubber

9 Silicones

Silicones, or polyorganosiloxanes, are recently developed group of compounds based on

$$-\underset{|}{\overset{|}{Si}}-O-\underset{|}{\overset{|}{Si}}-O-\underset{|}{\overset{|}{Si}}-O \qquad -\underset{|}{\overset{|}{Si}}-O-\underset{|}{\overset{|}{Si}}-O$$

chains with alkyl and/or aryl groups (mainly methyl or phenyl) attached to the silicon atoms. The linear polymers are liquids but some degree of cross linking gives rubber-like solids and more extensive, three-dimensional cross linking leads to resinous solids.

A wide range of products can be made and they have many valuable properties. They are colourless, odourless, non-volatile, insoluble in water and chemically unreactive. They are also good electrical insulators and strongly water-repellant; the liquids have low surface tension so that they are good wetting agents. The silicones maintain these physical properties over a wide range of temperature.

a. Manufacture. Methylchlorosilanes are made by reaction between chloromethane and powdered silicon in the presence of a catalyst of copper at 300 °C. The boiling points of the three products are close together but they are separated by fractional distillation.

CH_3SiCl_3	$(CH_3)_2SiCl_2$	$(CH_3)_3SiCl$
Methyltrichloro-silane (b.p. = 66 °C)	Dimethyldichloro-silane (b.p. = 76 °C)	Trimethylchloro-silane (b.p. = 57 °C)

These substituted silanes give hydroxy compounds on hydrolysis with water and these compounds then condense into linear or cyclic siloxanes, e.g.

$$(CH_3)_2SiCl_2 + 2H_2O \rightarrow (CH_3)_2Si(OH)_2 + 2HCl$$

By using mixtures of the three substituted silanes many different siloxanes can be made and these can, if necessary, be polymerised still further, by elimination of water molecules from adjacent hydroxy groups.

b. Silicone fluids. The simplest silicone fluids are linear polydimethylsiloxanes,

The value of n can vary widely and some of the methyl groups can be replaced by phenyl, or other, groups.

The products are oily liquids useful as lubricants, as hydraulic fluids, as release agents, as water repellants and as spreading agents. They can be converted into greases by adding lithium soaps or other fillers. The products are used as coatings in electrical equipment, and on burette taps.

c. Silicone rubbers. These are based on polydimethylsiloxanes with cross-links between every 100–1 000 silicon atoms, together with silicon(IV) oxide as a filler. They can be made by heating linear polymers with small amounts of organic peroxides to initiate cross-linking, or they are obtainable as two-component products which cross-link and set on mixing at room temperature. They are used in cable insulation, in medical components and spare parts for implant surgery, and as sealants.

d. Silicone resins. These consist of phenyl substituted siloxane units highly cross-linked into three-dimensional structures not unlike that of silica.

The resins are hard, brittle solids generally used in solution. Typical uses include the impregnation of various materials to improve electrical insulation or water repellancy, heat resistant paints and non-stick coatings.

Questions on Chapter 24

1 Write an essay on the preparation and use of plastics. Your account
   should include reasons for choice of monomers, properties of particular plastics which make them useful, and the relationship between
   their structure and their uses. (O and C)

2 'A most important property of unsaturated aliphatic molecules lies
   in their ability to bond to each other to form polymers'. Illustrate
   this quotation by using three commercially important examples. In
   each case give the conditions used for the polymerisation and the
   reasons for them. Show the general mechanisms by which such
   polymerisations may take place. What methods might be used for an
   investigation of the molecular weight of a polymeric product? (W)

3 The deputy head of the research department of a major packaging
   firm recently remarked in conversation that the plastics industry as
   we now know it will be finished within his working lifetime (he is 35)
   because of the shortage of oil. Discuss the implications of his
   remark, assuming it to be true. (L)

4 Give brief discussions, including the statement of one appropriate
   example in each case, of the following topics: (a) The polymerisation
   of alkenes and halogen-alkenes. (b) The formation of polyesters. (c)
   The structure of di- and poly-saccharides. (L)

5 Explain what is meant by the terms 'addition polymerisation' and
   'condensation polymerisation', illustrating your answer in detail by
   reference to one important example of each type. Polymerisation
   processes usually yield long-chain molecules. For a polymer to be
   useful, it must possess adequate 'cross-linkage' between such chains
   so as to assume a 3-dimensional structure. Outline how in certain
   types of polymers cross-linkage is achieved by hydrogen-bonding.
   (L)

6 Write an essay on 'Macromolecules' (polymers). You should base
   your answer on the following points. (a) An explanation of the term
   macromolecule (polymer) with reference to natural and synthetic

polymers, addition and condensation polymerisation, and thermoplastics and thermosetting plastics. (b) Examples of synthetic polymers replacing natural polymers and the advantages and disadvantages of the replacements. (c) Possible future trends and developments in synthetic polymers. Give examples where possible to illustrate your essay. (SUJB)

25

Important Synthetic Reagents

The reagents described in this chapter are all important in the synthesis of many organic compounds. They are, however, generally used on a small scale in the laboratory, rather than industrially.

Grignard Reagents

1 Preparation of Grignard reagents

Grignard reagents are made by refluxing a mixture of clean magnesium filings, ethoxyethane and a bromo- or iodo-alkane or arene, e.g.

$$CH_3—I + Mg \rightarrow CH_3—MgI$$

Methyl magnesium
iodide

$$C_6H_5—I + Mg \rightarrow C_6H_5—MgI$$

Phenyl magnesium
iodide

The apparatus and the reagents must all be dry as Grignard reagents react with water.

The general formula is R—MgX where R stands for an alkyl or an aryl group and X for Cl, Br or I. The reagents are used in etheral solution and attempts to isolate them in the pure state give compounds between them and ethoxyethane.

2 Properties of Grignard reagents

The precise structure of pure Grignard reagents is not known for they have not been isolated. In ethereal solution various species are present in equilibrium, e.g.

$$2R^- + 2Mg^+X \rightleftharpoons 2R—MgX \rightleftharpoons (R—MgX)_2 \rightleftharpoons R_2Mg + MgX_2$$

In most of their reactions they function as nucleophiles due to the polarisation in the $R^{\delta-}$—$Mg^{\delta+}$ bond.

The negatively charged C atom of the R group is attacked by H^+ ions from even the weakest acid, e.g. water, with subsequent elimination of an alkane or arene, e.g.

$$H_3C^{\delta-}—Mg^{\delta+}—I \rightarrow CH_4 + Mg(OH)I$$
$$\overset{\curvearrowleft}{H^+} \quad \overset{\curvearrowleft}{OH^-}$$

Grignard reagents are also able to form addition compounds with other polarised bonds, e.g. $C^{\delta+}{=}O^{\delta-}$ and $C^{\delta+}{\equiv}N^{\delta-}$, resulting in the formation of new C—C links. For example,

$$
\overset{|}{-}C^{\delta+}{=}O^{\delta-} \rightarrow -\overset{|}{C}-O
$$
$$H_3C^{\delta-}—Mg^{\delta+}—I \qquad H_3C \quad Mg—I$$

Iodides are the most reactive Grignard reagents and the most widely used.

a. Conversion into hydrocarbons. Grignard reagents react with water to form alkanes or arenes, e.g.

$$R—MgI + H_2O \rightarrow R—H + Mg(OH)I$$

Other compounds which can provide H^+ ions, e.g. acids, alcohols and amines, react similarly.

b. Conversion into primary alcohols. Grignard reagents react with methanal or epoxyethane to form primary alcohols. The first product is an addition compound but this is then hydrolysed in acid solution, e.g.

The reaction with methanal adds one carbon atom to the Grignard reagent; that with epoxyethane adds two.

c. Conversion into secondary and tertiary alcohols. Reactions, similar to those in (b) with higher aldehydes or with ketones give secondary or tertiary alcohols respectively e.g.

d. Conversion into ketones. Grignard reagents also form addition compounds with nitriles and these can be hydrolysed to ketones, e.g.

e. Conversion into carboxylic acids. Grignard reagents give addition compounds with carbon dioxide and these can be hydrolysed into carboxylic acids. A low temperature is necessary to prevent the initial addition compound reacting with more Grignard reagent and it is convenient to use solid carbon dioxide, e.g.

Other Organometallic Compounds

3 Introduction

Grignard reagents are the best known examples of organometallic compounds (compounds containing C-metal bonds) but other metals also form similar compounds, e.g.

Group 1	Group 2	Group 3	Group 4
C_2H_5Li	$(CH_3)_2Ca$	$[(CH_3)_3Al]_2$	$(C_2H_5)_4Pb$
Ethyllithium	Dimethylcalcium	The dimer of tri-methylaluminium	Tetraethyl lead(IV)
C_2H_5Na	$(C_2H_5)_2Hg$		
Ethylsodium	Diethylmercury		

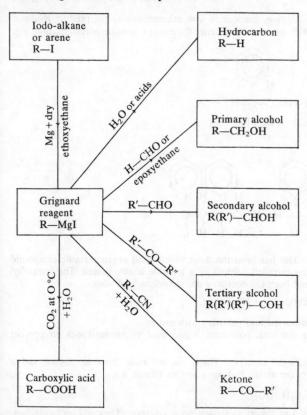

Preparation and properties of Grignard reagent

The bonding between the metal and carbon is substantially ionic when the metal is highly electropositive but mainly covalent with metals of lower electropositivity. It is a matter of the extent of the polarisation in the $C^{\delta-}$—$M^{\delta+}$ bond, and the negative charge on the carbon atom makes the compounds function as nucleophiles.

In general it is the compounds with the highly electropositive metals that are most reactive. Organometallic compounds of sodium and potassium, for example, are generally spontaneously inflammable in air and react with water and carbon dioxide; they are also relatively insoluble in non-polar solvents.

4 Typical examples

a. Organolithium compounds. Lithium which is very similar to magnesium, forms alkyls or aryls by reaction with halogeno-alkanes or arenes in ethoxyethane solution at low temperatures, e.g.

$$CH_3—Br + 2Li \rightarrow CH_3Li + LiBr$$

The lithium compounds are more reactive than Grignard reagents but undergo very similar reactions.

They will also form addition compounds with ethenes and as the product is also a lithium alkyl it reacts with more of the ethene; the result is anionic polymerisation, e.g.

$$H—\underset{H}{\overset{}{C}}=\underset{H}{\overset{}{C}}—H + CH_3Li \rightarrow H—\underset{CH_3}{\overset{H}{C}}—\underset{Li}{\overset{}{C}}—H$$

Phenylethene
(Styrene)

$$H—\underset{H}{\overset{}{C}}=\underset{H}{\overset{}{C}}—H + H—\underset{CH_3}{\overset{H}{C}}—\underset{Li}{\overset{}{C}}—H \rightarrow H—\underset{CH_3}{\overset{H}{C}}—\underset{H}{\overset{}{C}}—\underset{H}{\overset{}{C}}—\underset{Li}{\overset{}{C}}—H$$

b. Tetraethyl lead(IV). This has been the most widely used organometallic compound but its use is now being curtailed because of a possible health hazard. The tetraethyl lead(IV) is manufactured from chloroethane and a sodium/lead alloy,

$$4Na/Pb + 4C_2H_5Cl \rightarrow Pb(C_2H_5)_4 + 3Pb + 4NaCl$$

The excess lead is reconverted into the alloy with sodium.

The lead alkyl is a covalent, poisonous liquid used as an antiknock to prevent pre-ignition in petrol.

c. Aluminium dlkyls. Ziegler catalysts. These can be made from aluminium and a halogenoalkane or from the metal, hydrogen and an ethene, e.g.

$$2Al + 2C_2H_4 + 3H_2 \rightarrow [(C_2H_5)_3Al]_2$$

The alkyls generally exist as dimers as shown in the equation. They add on to ethenes, like lithium alkyls, to form higher alkyls and they can, therefore, initiate anionic polymerisation. Combined with salts such as titanium chloride they form the basis of Ziegler catalysts (p. 325).

d. Organomercury compounds. Many organomercury compounds have been made because they are very toxic, e.g. C_2H_5HgCl, and have possible uses as antiseptics, seed disinfectants and weedkillers.

Diazonium Salts

5. Introduction

Aromatic primary amines react with a mixture of sodium nitrate(III) and dilute acids at temperatures close to 0 °C to form diazonium salts; benzene diazonium chloride is the commonest,

$$NaNO_2(aq) + HCl(aq) \rightarrow HNO_2(aq) + NaCl(aq)$$

$$C_6H_5—NH_2(l) + HNO_2(aq) + HCl(aq) \rightarrow C_6H_5—N_2Cl(aq) + 2H_2O(l)$$
                                        Benzenediazonium
                                             chloride

The process for making these important synthetic reagents, known as *diazotisation*, was discovered by Griess in 1858.

6 Preparation of benzene diazonium chloride

Phenylamine is first dissolved in excess concentrated hydrochloric acid. Enough acid is required to convert the phenylamine into its hydrochloride salt (p. 295), to react with the sodium nitrate(III) to be added later, and to provide an excess so that the reaction is carried out in acid solution.

The acid solution is then cooled by adding ice, and a cooled solution of sodium nitrate(III) containing slightly more than the theoretical requirement of nitrate(III) is added in small portions and with constant stirring. The temperature must be kept close to 0 °C and the addition of the nitrate(III) solution is continued until a drop of the mixture turns starch-iodide paper blue, indicating the presence of free nitric(III) acid.

The resulting solution contains benzene diazonium chloride and is normally used without any further treatment.

7 Properties of diazonium salts

Most diazonium salts are very soluble and so unstable in the dry, solid state that they are liable to decompose explosively. The salts of strong acids are the most stable and the tetrafluoroborate is also insoluble; it can be isolated as a stable, crystalline solid.

The diazonium salts of strong acids are ionic with general structure, $Ar-N^+\equiv NX^-$. They may be regarded as salts of diazonium hydroxides but these hydroxides cannot be isolated.

Diazonium salts undergo three types of reaction. The $-N_2X$ group can be substituted with the elimination of nitrogen gas; they can be reduced; and they will couple with other organic compounds retaining the diazo, $-N=N-$, group.

8 Substitution reactions of benzene diazonium chloride

These reactions involve the replacement of the $-N_2Cl$ group by some other monovalent group either via an S_N reaction involving $C_6H_5^+$ ions e.g.

$$C_6H_5-N^+\equiv N \xrightarrow[(-N_2)]{slow} C_6H_5^+ \xrightarrow[X^-]{quick} C_6H_5-X$$

or via a free radical mechanism involving $C_6H_5\cdot$ radicals.

a. Substitution of $-N_2Cl$ *by* $-OH$. Benzene diazonium chloride is converted into phenol on warming in its aqueous solution; that is why the diazonium salt has to be prepared at low temperatures. The diazonium ion decomposes into a carbonium ion and nitrogen,

$$C_6H_5-N^+\equiv N \xrightarrow{-N_2} C_6H_5^+ \xrightarrow{H_2O} C_6H_5-OH + H^+$$

b. *Substitution of* —N_2Cl *by* —H. This is brought about by reaction with hypophosphorus (phosphinic) acid, H_3PO_2,

$$C_6H_5 - N_2Cl(aq) + H_3PO_2(aq) + H_2O(l)$$
$$\rightarrow C_6H_6(l) + N_2(g) + HCl(aq) + H_3PO_3(aq)$$

c. *Substitution of* —N_2Cl *by* —I. Iodobenzene is formed on warming with potassium iodide solution,

$$C_6H_5-N_2Cl(aq) + KI(aq) \rightarrow C_6H_5-I(l) + N_2(g) + KCl(aq)$$

d. *Substitution of* —N_2Cl *by* —Cl *or* —Br. *Sandmeyer's reaction.* The reagents are hydrochloric acid and copper(I) chloride, or hydrobromic acid and copper(I) bromide, the copper(I) halides acting catalytically. The reactions involve a free radical mechanism, e.g.

$$CuCl(aq) + Cl^-(aq) \rightarrow CuCl_2^-(aq)$$
$$C_6H_5-N_2^+ + CuCl_2^-(aq) \rightarrow C_6H_5\cdot + N_2(g) + CuCl_2(aq)$$
$$C_6H_5\cdot + CuCl_2(aq) \rightarrow C_6H_5-Cl + CuCl(aq)$$

e. *Substitution of* —N_2Cl *by* —F. Addition of potassium tetrafluoroborate, KBF_4, solution to a solution of benzene diazonium chloride gives a precipitate of insoluble benzene diazonium tetrafluoroborate which, when dried, gives fluorobenzene on heating (the Schiemann reaction),

$$C_6H_5-N_2BF_4(s) \xrightarrow{heat} C_6H_5-F(l) + N_2(g) + BF_3(g)$$

f. *Substitution of* —N_2Cl *by* —CN. This involves a reaction similar to the Sandmeyer reaction, the reagent being a solution of copper(I) cyanide in potassium cyanide solution, i.e. $Cu(CN)_4^{3-}$ ions,

$$C_6H_5-N_2Cl \xrightarrow{Cu(CN)_4^{3-}} C_6H_5-CN$$

g. *Substitution of* —N_2Cl *by* —NO_2. This is brought about by treating benzene diazonium chloride with sodium nitrate(III) solution in the presence of Cu^+ ions as catalysts,

$$C_6H_5-N_2Cl + NO_2^-(aq) \xrightarrow{Cu^+} C_6H_5-NO_2 + N_2(g) + Cl^-(aq)$$

9 Reduction of benzene diazonium chloride. Phenylhydrazine

This reduction is brought about by tin(II) chloride, sodium sulphate(IV) or electrolytically. In acid solution the hydrochloride of phenylhydrazine is formed but the free base can be released by warming with sodium hydroxide solution, e.g.

$$C_6H_5-N_2Cl(aq) + 2SnCl_2(aq) + 4HCl(aq) \rightarrow C_6H_5-NH-NH_2 \cdot HCl(aq)$$
$$+ 2SnCl_4(aq)$$

$$C_6H_5-NH-NH_2 \cdot HCl + NaOH(aq) \rightarrow \underset{\text{Phenylhydrazine}}{C_6H_5-NH-NH_2} + NaCl(aq) + H_2O(l)$$

Phenylhydrazine is a pale yellow crystalline solid which is used in preparing phenylhydrazones from aldehydes and ketones (p. 221). It was first made by Emil Fischer (1875) and he used it to investigate the structures of carbohydrates.

10 Coupling reactions of benzene diazonium chloride

Benzene diazonium chloride will react with many substituted arenes with the elimination of HCl between the two molecules. The reactions are often known as coupling reactions but they are really electrophilic substitution reactions not unlike nitration of benzene (p. 129). The diazonium ion, $C_6H_5-N_2^+$, acts as the electrophile, e.g.

$$C_6H_5-N_2^+ + \underset{}{\bigcirc}-X \rightarrow C_6H_5-N=N-\underset{}{\bigcirc}-X + H^+$$

The reactions will only take place when X is a highly activating *o–p* directing group (p. 146) such as —OH, —O⁻ or —N(CH₃)₂. The reactions with phenol and with napthalen-2-ol (β-naphthol), for example, are both carried out in alkaline solution so that it is really the O⁻ group acting as —X and not —OH, the former being more activating than the latter,

$$C_6H_5-N_2Cl + \underset{}{\bigcirc}-OH \xrightarrow{NaOH} C_6H_5-N=N-\underset{}{\bigcirc}-ONa$$

$$C_6H_5-N_2Cl + \underset{}{\bigcirc\bigcirc}-OH \xrightarrow{NaOH} \underset{}{\bigcirc\bigcirc} \overset{N=N-C_6H_5}{\underset{ONa}{}}$$

The coupling generally takes place in the para-position relative to the —X group but if this position is already occupied, as in napthalen-2-ol, the coupling will be in the *ortho*-position.

11 Azo dyes

Compounds containing azo, —N=N—, groups are highly coloured and the group is known as a *chromophore*. Azo dyes are the commonest class of coloured materials but before an azo compound can be useful as a dye other groups, e.g. —NH_2, —$N(CH_3)_2$, —OH, —COOH and —SO_3H, have to be incorporated; they are known as *auxochromes*. Typical examples of commercial dyes are shown below:

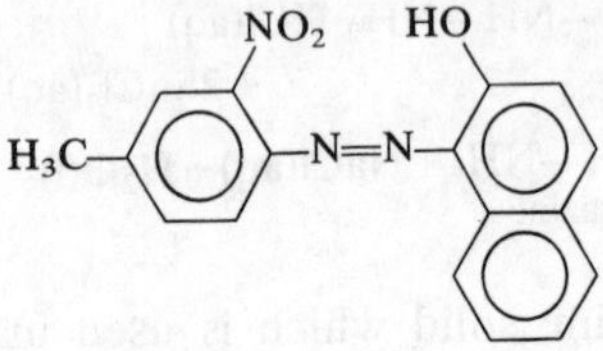

Toluidine red 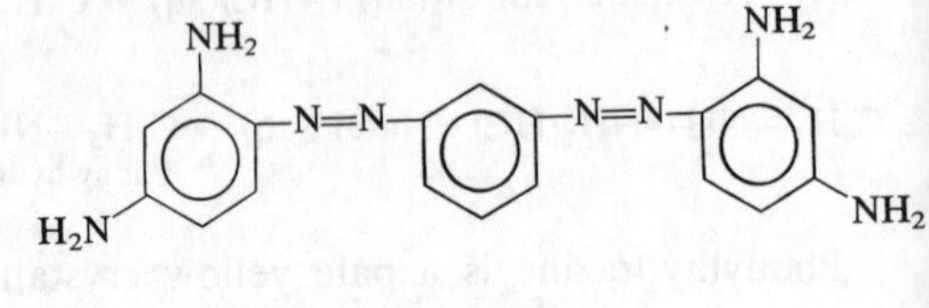   Bismarck Brown

Methyl orange is also an azo dye but it is better known as an indicator,

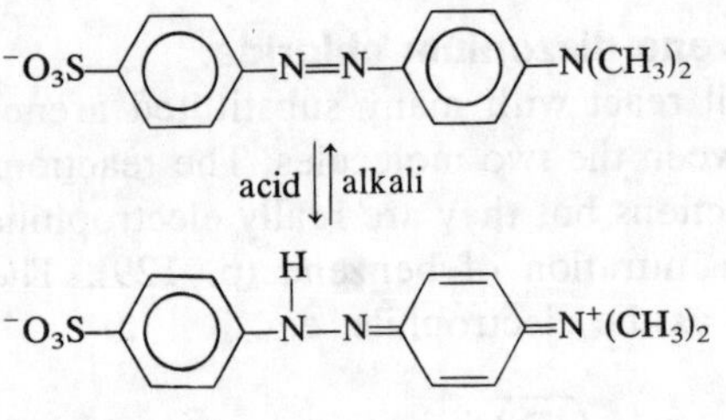

Complex Hydrides

Lithium tetrahydridoaluminate, $LiAlH_4$, and sodium tetrahydridoborate, $NaBH_4$, are the two commonest complex hydrides which have been used in recent years as selective reducing agents. The former is stronger than the latter and their use is limited to the small scale as they are expensive.

12 Lithium tetrahydridoaluminate, LiAlH₄

This is made by the action of lithium hydride on a solution of aluminium chloride in dry ethoxyethane,

$$4LiH + AlCl_3 \rightarrow LiAlH_4 + 3LiCl$$

The resulting ethereal solution is used as the reagent and the reactions

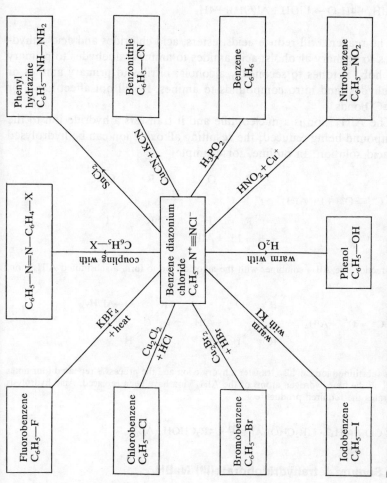

Properties of benzene diazonium chloride

must be carried out in ethereal solution as the lithium tetrahydridoaluminate reacts vigorously with water.

$$LiAlH_4 + 4H_2O \rightarrow LiOH + Al(OH)_3 + 4H_2$$

The reagent will reduce acids, esters, acid chlorides and acid anhydrides to primary alcohols; acid amides to amines; aldehydes to primary
alcohols; ketones to secondary alcohols; nitriles to primary amines or
aldehydes; and nitro compounds to amines. It will not affect C=C or
C≡C bonds.

The AlH_4^- ion is a nucleophile and it transfers a hydride ion to the
compound being reduced; the resulting alkoxide ion can be hydrolysed
in acid solution. In outline, for example,

In practice, the AlH_3 combines with the alkoxide ion to form a substituted AlH_4^- ion,

This substituted ion can also transfer a hydride ion and the process is repeated four times
until all the four hydrogen atoms of the AlH_4^- ion have been replaced. Acid hydrolysis
liberates the required product, e.g.

$$4R_2CO + AlH_4^- \rightarrow (R_2CHO)_4Al^- \xrightarrow{4H^+} 4R_2CHOH + Al^{3+}$$

13 Sodium tetrahydridoborate(III) NaBH₄

This is made by reaction between sodium hydride and methyl bo-
rate(III), at about 250 °C,

$$4NaH + B(OCH_3)_3 \rightarrow NaBH_4 + 3NaOCH_3$$

Sodium tetrahydridoborate is insoluble in ether and as it reacts only
very slowly with water or alcohols its reactions are carried out in
aqueous or alcoholic solution. It is a milder reducing agent than
lithium tetrahydridoaluminate, reducing aldehydes and ketones to alcohols but not reducing acids, esters, anhydrides or amides.

Ethyl-3-oxobutanoate

14 Preparation and structure

Ethyl-3-oxobutanoate is one of the modern systematic names for what has long been called ethyl acetoacetate (ethanoate) or, simply, acetoacetic (ethanoethanoic) ester.

The ester is prepared by warming sodium with ethyl ethanoate in the presence of a little ethanol; a molecule of ethanol is eliminated by condensation between two molecules of the ethyl ethanoate,

$$2CH_3-\overset{\overset{\displaystyle O}{\|}}{C}-O-C_2H_5 \xrightarrow{-C_2H_5OH} \quad H-\overset{\overset{\displaystyle H}{|}}{\underset{\underset{\displaystyle H}{|}}{C}}-\overset{\overset{\displaystyle O}{\|}}{C}-\overset{\overset{\displaystyle H}{|}}{\underset{\underset{\displaystyle H}{|}}{C}}-\overset{\overset{\displaystyle O}{\diagup\|}}{\underset{\underset{\displaystyle O-C_2H_5}{\diagdown}}{C}}$$

Ethyl-3-oxobutanoate

The structural formula given in the above equation indicates the presence of a C=O group. Evidence in favour of this is provided by the reactions of ethyl-3-oxobutanoate with sodium hydrogen-sulphate(IV), hydrogen cyanide, phenylhydrazine and hydroxylamine (pages 217–221). The ester also undergoes, however, a series of reactions with sodium, phosphorus pentachloride and iron(III) chloride solution which indicate the presence of an —OH group. The fact that the ester will decolorise an alcoholic solution of bromine also indicates the presence of a C=C bond.

These facts can only be explained on the assumption that the ester consists of two tautomeric forms (p. 270); one is known as the keto- and the other as the enol-form and they are named accordingly when it is necessary to distinguish between them,

$$H-\overset{\overset{\displaystyle H}{|}}{\underset{\underset{\displaystyle H}{|}}{C}}-\overset{\overset{\displaystyle O}{\|}}{C}-\overset{\overset{\displaystyle H}{|}}{\underset{\underset{\displaystyle H}{|}}{C}}-\overset{\overset{\displaystyle O}{\diagup\|}}{\underset{\underset{\displaystyle O-C_2H_5}{\diagdown}}{C}} \qquad H-\overset{\overset{\displaystyle H}{|}}{\underset{\underset{\displaystyle H}{|}}{C}}-\overset{\overset{\displaystyle OH}{|}}{C}=\overset{\overset{\displaystyle H}{|}}{C}-\overset{\overset{\displaystyle O}{\diagup\|}}{\underset{\underset{\displaystyle O-C_2H_5}{\diagdown}}{C}}$$

The keto form   The enol form
Ethyl-3-oxobutanoate   Ethyl-3-hydroxybut-2-enoate

The truth of this assumption was proved conclusively in 1911 when Knorr succeeded in isolating both forms at very low temperatures. At room temperature the two forms co-exist in dynamic equilibrium with approximately 7 per cent of the enol and 93 per cent of the keto form.

15 Properties

Ethyl-3-oxobutanoate is a colourless, pleasant-smelling liquid sparingly soluble in water. Its importance depends on the facts that it can be hydrolysed in two ways and that it can readily form alkyl derivatives; it can, therefore, be used to synthesise many compounds.

a. Hydrolysis. The ester yields a ketone as the chief product when boiled with dilute aqueous or alcoholic potassium hydroxide solution (*ketonic hydrolysis*); the potassium salt of an acid results on boiling with concentrated alcoholic potassium hydroxide solution (*acid hydrolysis*), e.g.

$$CH_3-\overset{\displaystyle O}{\overset{\|}{C}}-\overset{\displaystyle H}{\underset{\displaystyle H}{\overset{|}{C}}}-\overset{\displaystyle O}{\overset{\|}{C}}-OC_2H_5 \xrightarrow[\text{(dilute KOH)}]{\text{ketonic}\atop\text{hydrolysis}} CH_3-\overset{\displaystyle O}{\overset{\|}{C}}-CH_3 + K_2CO_3 + C_2H_5OH$$

$$CH_3-\overset{\displaystyle O}{\overset{\|}{C}}-\overset{\displaystyle H}{\underset{\displaystyle H}{\overset{|}{C}}}-\overset{\displaystyle O}{\overset{\|}{C}}-OC_2H_5 \xrightarrow[\text{(conc. KOH)}]{\text{acid}\atop\text{hydrolysis}} 2CH_3-COO^-K^+ + C_2H_5OH$$

b. Conversion into alkyl derivatives. The ester forms a sodium salt when treated with sodium ethoxide in ethanol solution and the resulting anion, which is a nucleophile, undergoes a substitution reaction with iodoalkanes to yield an alkyl derivative, e.g.

$$CH_3-\overset{\displaystyle OH}{\overset{|}{C}}=\overset{\displaystyle}{C}-\overset{\displaystyle O}{\overset{\|}{C}}-OC_2H_5 \xrightarrow{C_2H_5O^-Na^+} CH_3-\overset{\displaystyle O^-Na^+}{\overset{|}{C}}=\overset{\displaystyle}{C}-\overset{\displaystyle O}{\overset{\|}{C}}-OC_2H_5$$

$$\Big\downarrow R-I$$

$$CH_3-\overset{\displaystyle O}{\overset{\|}{C}}-\overset{\displaystyle R}{\underset{\displaystyle H}{\overset{|}{C}}}-\overset{\displaystyle O}{\overset{\|}{C}}-OC_2H_5 \rightleftharpoons CH_3-\overset{\displaystyle OR}{\overset{|}{C}}=\overset{\displaystyle}{C}-\overset{\displaystyle O}{\overset{\|}{C}}-OC_2H_5$$

After one alkyl derivative has been formed in this way the process can be repeated to form a dialkyl derivative.

The alkyl derivatives can then be subjected to either ketonic or acid hydrolysis to obtain a variety of ketones or acids.

Diethyl Propanedioate

16 Preparation and structure
Diethyl propanedioate is the modern name for what used to be called ethyl malonate or malonic ester.

It is made by warming an alcoholic solution of potassium cyano-ethanoate with concentrated hydrochloric or sulphuric(VI) acid,

$$H_2C\begin{array}{l} \diagup CN \\ \diagdown COO^-K^+ \end{array} + 2HCl + 2C_2H_5OH \rightarrow H_2C\begin{array}{l} \diagup CO-OC_2H_5 \\ \diagdown CO-OC_2H_5 \end{array} + KCl + NH_4Cl$$

The ester exists in tautomeric forms, like ethyl-3-oxobutanoate, though the proportion of the enol form at room temperature is very small,

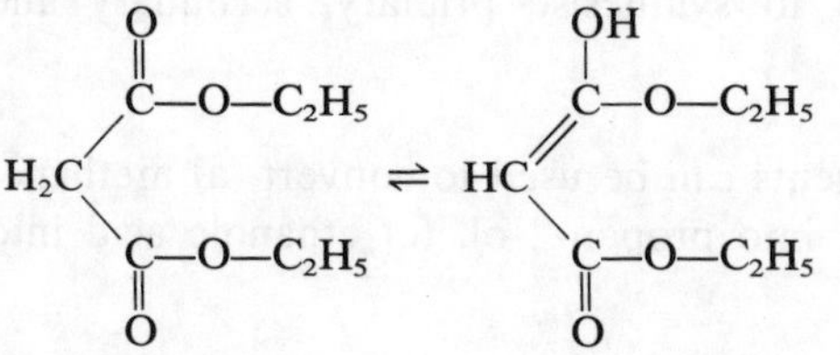

Keto form   Enol form

17 Properties

The ester is a pleasant-smelling liquid useful for synthesising substituted carboxylic acids. Its use for this purpose depends on its conversion into alkyl derivatives which can then be converted into carboxylic acids by hydrolysis and decarboxylation.

a. Formation of alkyl derivatives. Diethyl propanedioate reacts with sodium ethoxide to form a sodium salt in the same way as ethyl-3-oxobutanoate does. The anion of the salt is nucleophilic so that it reacts with iodoalkanes and in this way, one or two alkyl groups can be substituted into the original ester, i.e.

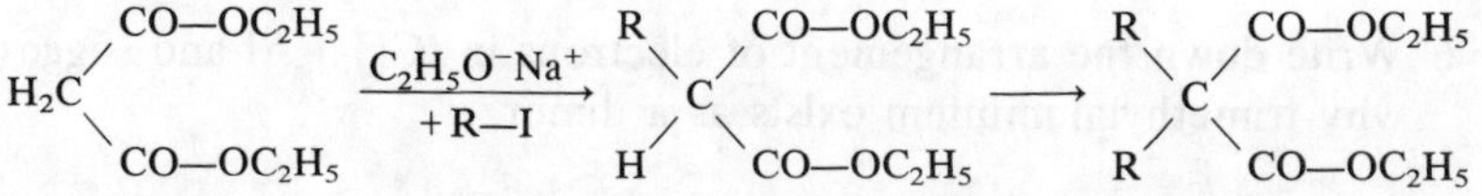

b. Conversion of alkyl derivatives into mono-carboxylic acids. The alkyl derivatives of the ester can be converted into the potassium salts of alkyl substituted propanedioic acids by refluxing with potassium hydroxide solution, and the free dicarboxylic acid can be made by acidification with hydrochloric acid. The dicarboxylic acid loses carbon dioxide (decarboxylation) on heating (p. 243) and is converted into a mono-carboxylic acid, e.g.

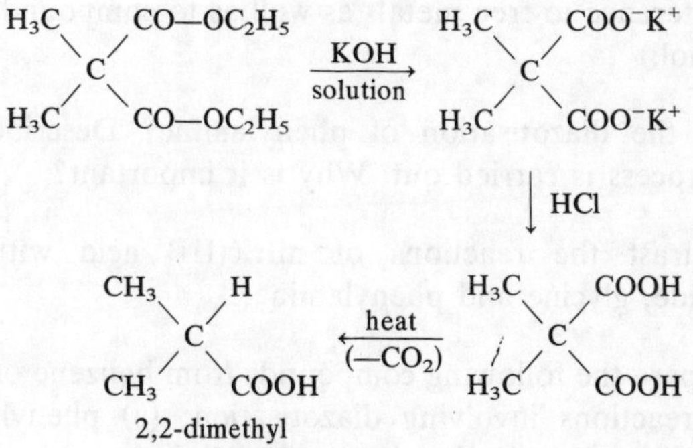

2,2-dimethyl
butanoic acid

Questions on Chapter 25

1 What are Grignard reagents and how are they prepared? Show how they may be used to synthesise primary, secondary and tertiary alcohols.

2 Show how Grignard reagents can be used to convert (a) methanol into ethanol, (b) ethanol into propan-2-ol, (c) ethanoic acid into propanone.

3 Write short notes on (a) Grignard, (b) Zerewitinoff, (c) Frankland, (d) Ziegler.

4 Predict the nature of the reactions between (a) ethylmagnesium iodide and ethyne, (b) $NaC\equiv CH$ and iodomethane.

5 Suggests reasons why (a) the iodide is the most reactive of the Grignard reagents, (b) organomagnesium fluorides do not exist, (c) magnesium and lithium form the commonest organometallic compounds.

6 Write down the arrangement of electrons in $(CH_3)_3Al$ and suggest why trimethylaluminium exists as a dimer.

7 Explain why (a) Grignard reagents add on to $C=O$ bonds much more readily than to $C=C$ bonds, (b) why lithium alkyls are more reactive than Grignard reagents, (c) why ethene does not readily undergo anionic polymerisation.

8 Discuss the importance of metals in organic chemistry. Illustrate your answer with reference to free metals as well as to compounds of metals. (Oxf. Schol.)

9 What is meant by the diazotisation of phenylamine? Describe carefully how the process is carried out. Why is it important?

10 Compare and contrast the reactions of nitric(III) acid with ethylamine, carbamide, glycine and phenylamine.

11 How would you prepare the following compounds from benzene or methylbenzene by reactions involving diazotisation: (a) phenyl-hydrazine, (b) phenol, (c) ethoxybenzene, (d) iodobenzene, (e) methyl orange.

12 Write equations to show the reactions of the following reagents with benzene diazonium chloride: (a) propan-1-ol, (b) an alkaline solution of phenol, (c) copper(I) cyanide and potassium cyanide, (d) hot water.

13 A compound, X, containing 63.7 per cent of bromine and with a relative molecular mass of 251 yields a diazo salt which on reaction with H_3PO_2 gives a solid, B, containing 67.8 per cent of bromine. Only one mono-nitro derivative of B can be made. Give the name and formula of X.

14 An aromatic amine, A, containing 13.09 per cent nitrogen after treatment with nitric(III) acid yielded B which contained 18.13 per cent nitrogen. A, after reacting with an excess of ethanoyl chloride yielded C, 1 g of which required 6.71 cm^3 of M NaOH solution for complete hydrolysis. Deduce the formulae of A, B and C.

15 Explain how you could make a sample of benzaldehyde, labelled with deuterium in the aldehyde group, from methyl benzoate and LiAlD$_4$.

16 Write an account of reduction reactions in organic chemistry.

17 How can the following be synthesised from ethyl-3-oxobutanoate? (a) propanoic acid, (b) 3-methylbutan-2-one, (c) C_6H_5—CH_2—CH_2—COOH, (d) CH_3—CO—$CH(CH_3)$—CH_2—CH_3, (e) 2-methylpropanoic acid, (f) butan-2-one.

18 How can the following be synthesised from diethyl propanedioate? (a) butanedioic acid, (b) CH_3—CH_2—$CH(CH_3)$—COOH, (c) butanoic acid.

Revision Questions

1 The following substance is known as serine (2-amino-3-hydroxy-propanoic acid): $HO—CH_2—CH(NH_2)—COOH$.

From your knowledge of organic chemistry predict what reactions you would expect to take place when this compound is treated with: (a) nitrous acid: (b) ethanol; (c) lithium aluminium hydride; (d) potassium dichromate; (e) hydrochloric acid; (f) phosphorus pentachloride. In each case, state the formula of the reaction products and clearly identify the experimental conditions required for the reactions. (L)

2 You are given unlabelled bottles of the following substances: propan-2-ol (iso-propanol), methanoic acid, benzoyl chloride, methanal, phenylamine, ethanonitrite. In each case give a single positive chemical text (six tests in all) which would enable the bottles to be labelled correctly. (O and C)

3 The properties of an organic compound can sometimes be deduced from the functional groups it contains, but when a compound contains two or more functional groups at the same time (a) the characteristics of one of the groups may be profoundly modified or entirely suppressed, or (b) new characteristics may appear which are not attributable to any one group. Illustrate these statements by reference to the chemistry of ethanoic acid, aminoethanoic acid, ethanamide and benzaldehyde. (O and C)

4 From your knowledge of organic chemistry predict what you would expect to happen when this compound

$$H—O—\langle\ \rangle—CH=CH—CO—NH_2$$

is treated with (a) sodium hydroxide solution, (b) nitrous acid, (c) bromine water, (d) ozone, (e) hydrogen and finely divided palladium, at room temperature. (O and C)

5 Explain, with examples, the following terms: (a) homologous series; (b) benzoylation; (c) diazotisation; (d) polymerisation; (e) hydrogenation. (O and C)

6 How, and under what conditions, does bromine react with the following compounds: (a) benzene, (b) carbamide, (c) phenol, (d) propanone, (e) ethanoic acid. (O and C)

7 Give a classified account of the use of oxidation and reduction in the synthesis of organic compounds (L)

8 Give a classified account of the addition of simple molecules to unsaturated linkages in organic compounds. Your account should include, if possible, any suggested mechanisms to explain the additions (L)

9 Give a classified account of the use of hydrolysis in organic chemistry, paying particular attention to the conditions necessary for the hydrolysis of various types of compound. (L)

10 Discuss the main natural sources of organic compounds. Complement your discourse by giving reaction schemes for the current industrial manufacture of several important organic compounds from natural sources. (SUJB)

11 Give examples of addition and substitution reactions which can be classified as nucleophilic or electrophilic.

12 Outline the various methods available for introducing halogens into organic compounds. Illustrate the application of these methods as widely as you can by specific examples selected from different types of organic compound. (JMB)

13 Discuss logically the methods of forming carbon–oxygen bonds in organic compounds (O.Schol.)

14 Discuss the influence of a benzene ring on the functional groups attached to it.

15 Write an essay on reactions which lead to the formation of a new carbon–nitrogen bond, classifying the reactions according to type as far as possible. (O.Schol.)

16 Give (a) two general methods for increasing the length of a carbon chain and (b) two general methods for diminishing the length of a carbon chain. Illustrate by means of specific examples, with accompanying formulae, names and reagents.

How is polyethylene made? Give two reasons for its widespread use as a container material. Show by reference to the respective chemical structures how polyethylene differs from the polyester Terylene. (JMB)

17 Explain carefully the following terms illustrating each by one example of your own choice: (a) photochemical chlorination, (b) ozonolysis, (c) hydrogenation, (d) 'cracking' of alkanes, (e) saponification. Describe very briefly the commercial application of any three of these processes. (JMB)

18 Explain the following phenomena: (a) The bond lengths between adjacent carbon atoms in the benzene ring are all identical. (b) 1,2-dichloroethene ($CHCl=CHCl$) occurs in two different isomeric forms. (c) Nitric acid alone does not nitrate benzene. However, benzene can be nitrated by a mixture of concentrated nitric acid and concentrated sulphuric acid. (d) The boiling point of a highly branched hydrocarbon is generally found to be lower than that of the isomeric unbranched hydrocarbon. (e) Lactic acid [$CH_3—CH(OH)—CO_2H$] when prepared from propionic (propanoic) acid via α-chloropropionic acid (1-chloropropanoic acid) shows no optical activity. (L)

19 What explanations can you offer for any four of the following observations? (a) Ethers do not react with metallic sodium at ordinary temperatures, but dissolve in concentrated acids. (b) The chlorination of alkanes, in the presence of u.v. light, is sometimes found to be an explosive reaction. (c) When ethene (ethylene) is passed into bromine water containing sodium chloride, one of the reaction products formed is 1-chloro-2-bromoethane. (d) The solubility of alcohols in water decreases as their chain length increases. (e) Upon reaction with nitrous acid, aliphatic primary amines yield alcohols whereas aromatic amines do not. (L)

20 Give an account of the ways in which the following have proved useful in the investigation of organic compounds: (a) The mass spectrometer. (b) Ultra-violet or infra-red spectroscopy. (c) The behaviour of polarised light. (L)

21 How may the following be made from propene ($CH_3CH=CH_2$)? (a) CH_3COCH_3. (b) $(CH_3)_2CHCO_2H$. (c) $CH_3CO\cdot OCH(CH_3)_2$. (d) $(CH_3)_2CHCH_2NH_2$. (e) $CH_3CH_2CH_3$. For each step, indicate briefly the reagents needed, the conditions and the equation. You may assume that any inorganic material you need is available, together with potassium cyanide, acetic anhydride, and ethanol as a solvent. (L)

22 State what is meant by five of the following terms as used in organic chemistry: (a) sulphonation, (b) polymerisation, (c) esterification, (d) ozonolysis, (e) decarboxylation, (f) hydrogenation. Illustrate each answer by an example, specifying the reagents and approximate conditions. (L)

23 How would you carry out the following conversions in aliphatic compounds: (a) CH_3CO_2H to CH_3CHO, (b) CH_3CH_2OH to CH_3CN, (c) CH_3CH_2OH to CH_3CH_2CHO, (d) CH_3CH_2OH to CH_4, (e) CH_3OH to $CH_3—CH_3$? State the reagents and conditions in each case. (L)

24 The first member of an *homologous series* often shows properties which are not typical of the series. Explain the term in italics. Illustrate the statement by reference to (a) the aliphatic aldehydes, (b) the aliphatic carboxylic acids, giving for each series (i) two reactions of the first member which are not typical of the series, (ii) two reactions which are typical of the series. (AEB)

25 Identify **A, B, C, D, E, F, G, H, K** as fully as possible and explain the reactions involved.

(a) **A** is a colourless liquid of molecular weight 106 which is readily oxidised by potassium permanganate. On boiling **A** with aqueous sodium hydroxide, followed by acidification, a white solid **B** is formed. The action of heat on a mixture of **B** with sodalime gives **C**. **B** can be reduced to **D**. **D** reacts with **B** to give **E**.

(b) **F** is prepared by the reaction of ethene with chlorine and water. The hydrolysis of **F** with sodium hydrogencarbonate produces **G** which, when oxidised in stages, gives **H**. The addition of limewater to an aqueous solution of **H** gives a precipitate **K**. (AEB)

26 Suggest possible ways of making

$$CH_3-C^*H_2-O-\underset{\underset{OH}{|}}{\overset{}{C}}^*H-CH_3$$

given supplies of radioactive C^* in the form of KC^*N and C^*O_2.

27 Write short explanatory notes on four of the following: (a) The increase in acid strength as the hydrogen atoms in the $-CH_3$ group in ethanoic acid are successively substituted by chlorine atoms. (b) The function of sulphuric(VI) acid in the nitration of aromatic compounds. (c) The mechanism of an addition reaction of your own choice in carbon chemistry. (d) The rates of hydrolysis of halogenoalkanes. (e) The structures of polypeptides and proteins. (L)

28 Using water enriched with ^{18}O as a source of ^{18}O, suggest methods for the synthesis of the following compounds: (a) $CH_3CH_2O^*CH_2CH_3$, (b) $CH_3CH_2COO^*H$, (c) CH_3CHO^*, (d) $C_6H_5O^*H$, (e) $C_6H_5CH_2O^*H$. The asterisk denotes enrichment with ^{18}O.

29 Give an account of methods available for the production of the carbon–nitrogen link in compounds of the types

$$-\overset{\overset{\textstyle |}{}}{\underset{\underset{\textstyle |}{}}{C}}-NH_2 \qquad \overset{}{>}C=N- \qquad and \qquad -C\equiv N$$

Describe also how such carbon–nitrogen links can be ruptured. (Camb. Schol.)

30 'The reactions of an organic compound may be regarded as the reactions of the sum of the parts of the molecule'. Illustrate what is meant by this statement and indicate its limitations by giving examples where the reactivity of one 'part' of the molecule is influenced by another 'part' of the same molecule. (Camb. Schol.)

31 Suggest reagents to effect the following conversions:
   (a) $Et_2CH \cdot C \equiv CH \rightarrow Et_2CH \cdot CH = CH_2 \rightarrow Et_2CH \cdot CH(OH)Me$
   $$\rightarrow Et_2CH \cdot CO_2H$$
   (b) $MeC \equiv CH \rightarrow C_6H_3Me_3 \rightarrow Me_2(C_6H_3)CO_2H$
   (c) $HC \equiv CH \rightarrow HOCH_2 \cdot C \equiv C \cdot CH_2OH \rightarrow ClCH_2 \cdot C \equiv C \cdot CH_2Cl$
   $$\rightarrow HC \equiv C \cdot C \equiv CH$$
   (d) $EtC \equiv CH \rightarrow EtCOMe \rightarrow MeC \equiv CMe$ (Oxf. Schol.)

32 Write short notes on the types of reaction which establish new carbon–carbon bonds. (Camb. Schol.)

33 Write the structures of the various compounds you would expect to get if the following compound

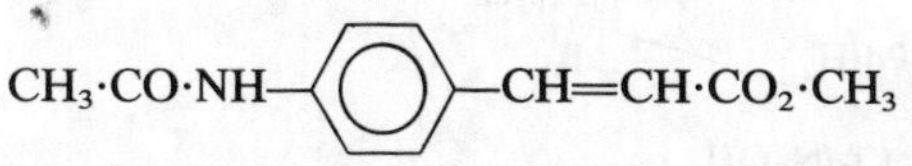

$CH_3 \cdot CO \cdot NH$—⟨O⟩—$CH = CH \cdot CO_2 \cdot CH_3$

was treated with (a) bromine, (b) hot dilute sodium hydroxide solution, (c) H_2/Pt at room temperature, (d) ammonia, (e) $LiAlH_4$, (f) concentrated nitric(V) and sulphuric(VI) acids.

34 Suppose that a row of unlabelled bottles contained water, propanone, ethanol, benzene, trichloromethane, ethanoic acid and ethyl ethanoate. What would you do in a school laboratory to identify each bottle?

35 How would you establish, by chemical methods, that a particular compound had the following structure:
C_6H_5—$CH = CH$—CO—OC_2H_5?

36 Discuss the importance of any five named reactions in organic chemistry.

37 How may hydrogen atoms (a) in straight chain compounds and (b) in benzene be replaced by (i) —Br, (ii) —OH, (iii) —COOH, (iv) —CH$_3$, (v) —CHO?

38 Describe and discuss examples of catalysis in organic chemistry.

39 'The first member of a homologous series often differs in important respects from the others'. Illustrate this statement as widely as possible.

40 Describe any five reactions in which the elements of water are withdrawn from one or more molecular proportions of a single organic compound.

41 It is sometimes said that organic chemistry is much more systematic than inorganic chemistry. What are your views on this?

42 In what respects is it true to say that if an alkyl group is considered as a metallic group then acyl groups are non-metallic.

43 Elucidate the structures of the compounds **A** to **J** and explain the reactions involved.

$$C_7H_6O + CH_3—CO—CH_3 \xrightarrow[\text{Pd/H}_2]{\text{NaOH}} C_{10}H_{10}O$$
$$\quad \textbf{A} \qquad\qquad\qquad\qquad\qquad\qquad \textbf{B}$$

$$C_{10}H_{12}O \xrightarrow[\text{LiAlH}_4]{\text{Cl}_2/\text{NaOH}} C_9H_{10}O_2$$
$$\textbf{C} \qquad\qquad\qquad\qquad \textbf{D}$$

$$C_9H_{12}O \xrightarrow[\text{(2) KCN}]{\text{(1) PBr}_3} C_{10}H_{11}N$$
$$\textbf{E} \qquad \text{HCl}_{aq} \qquad\qquad \textbf{F}$$

$$C_{10}H_{12}O_2 \xrightarrow[\text{(2) AlCl}_3]{\text{(1) SOCl}_2} C_{10}H_{10}O$$
$$\textbf{G} \qquad\qquad\qquad\qquad\qquad \textbf{H}$$

$$\xrightarrow[\substack{\text{KMnO}_4 \\ \text{Boil}}]{} \qquad\qquad \xrightarrow[]{\text{PhNHNH}_2}$$

$$C_8H_6O_4 \qquad\qquad\qquad\qquad C_{16}H_{16}N_2$$
$$\textbf{I} \qquad\qquad\qquad\qquad\qquad\qquad \textbf{J}$$

(O and C)

44 Complete, in as much detail as you can, any or all of the steps in the two schemes outlined below. Insert the formulae of intermediates, isolated or otherwise, which have been omitted. Name the reagents and state briefly the experimental conditions which you consider to be appropriate. Comment on any interesting features of the reactions or compounds involved.

(a)

(b) $CH_3—CH_2—COOH \rightarrow CH_2—CH_2—NH_2 \rightarrow CH_3—CHO$
$\rightarrow CH_3—CH(OH)—COOH$

(Oxf. Schol)

45 Intermediate compounds have been omitted from the following reaction sequences. Write out each sequence *in full*, name each compound, the reagents used and indicate the reaction conditions:

(*a*) 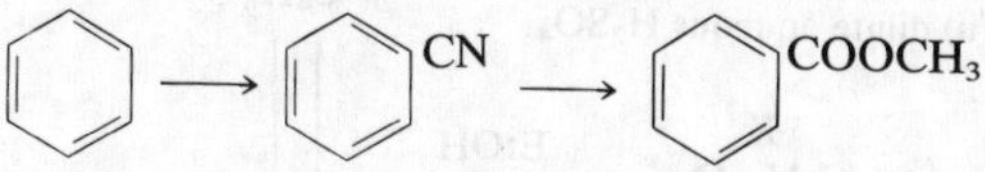

(*b*) $CH_3—CH_2—COOH \rightarrow CH_3—CH_2NH_2 \rightarrow CH_3—CHO \rightarrow$
$CH_3—CH(OH)—CH_2—NH_2$

(*c*) $CH_2{=}CH_2 \longrightarrow CH_2(COOCH_3)—CH_2(COOCH_3)$

(SUJB)

46. Identify **A**, **B**, **C**, **D**, **E** and **F**, giving reagents and reaction conditions.

(a) $CH_3COOH \rightarrow$ **A** $\rightarrow$ **B** $\rightarrow CH_3CN$
(b) $CH_3CH_2OH \rightarrow$ **C** $\rightarrow$ **D** $\rightarrow CH{\equiv}CH$
(c) $C_6H_5NH_2 \rightarrow$ **E** $\rightarrow$ **F** $\rightarrow C_6H_6$.

(AEB)

47 Give the reaction conditions and write the structures of the intermediates **A**, **B**, **C**, **D**, **E**, **F** in the following reaction schemes

(a) 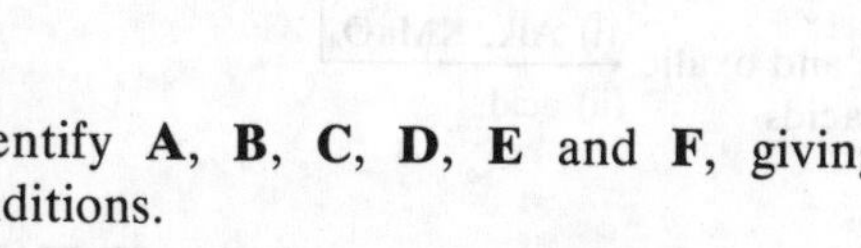

(b) 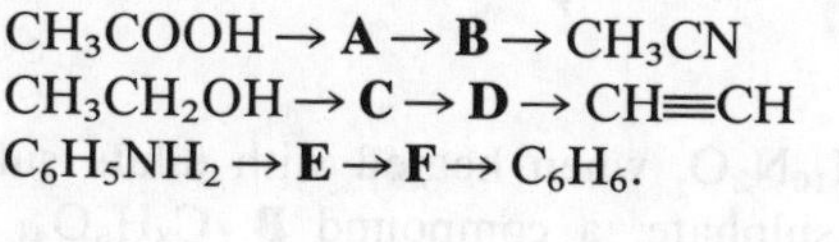

(c) $CH_3OH \longrightarrow$ **E** $\longrightarrow$ **F** $\longrightarrow CH_3COOH$

(AEB)

48 A compound **U**, C_4H_5N, which can be obtained in two stereoisomeric forms, is reduced by hydrogen and a catalyst to give **V**, $C_4H_{11}N$, which exists in one form only. The compound **U**

undergoes the following reactions:

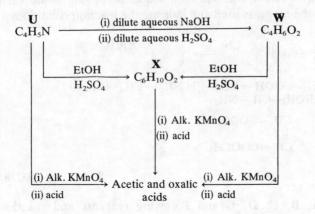

Elucidate the above reactions and give structures for **U**, **V**, **W** and **X**. (Camb. Schol.)

49 A compound **A**, $C_{14}H_{10}N_2O$, when heated with dilute sulphuric acid gave ammonium sulphate, a compound **B**, $C_8H_6O_4$, and a compound **C**, C_6H_7N (as its sulphate). Compounds **B** and **C** behaved as follows:

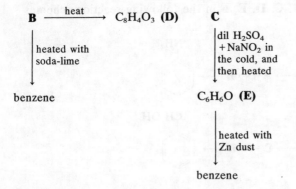

Suggest a formula for **A** and account for the above reactions. (Camb. Schol.

50 Give the structures of the compounds **A**, **B**, **C**, **D**, **E** and **F**, and explain your reasoning.

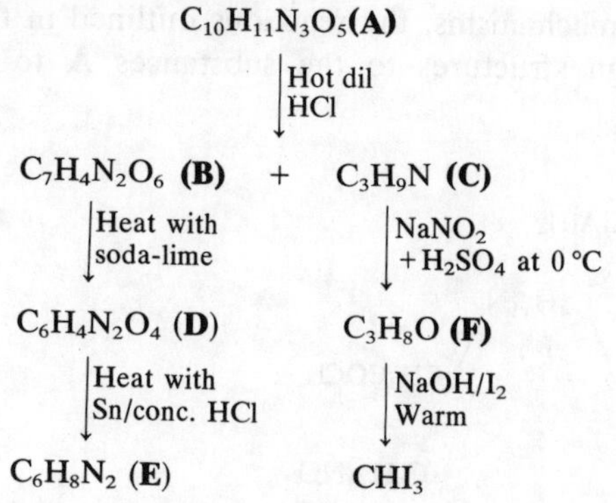

$$C_{10}H_{11}N_3O_5 \text{(A)}$$

$$\downarrow \text{Hot dil HCl}$$

$$C_7H_4N_2O_6 \text{ (B)} \quad + \quad C_3H_9N \text{ (C)}$$

| Heat with soda-lime | NaNO₂ + H₂SO₄ at 0 °C |

$$C_6H_4N_2O_4 \text{ (D)} \qquad\qquad C_3H_8O \text{ (F)}$$

| Heat with Sn/conc. HCl | NaOH/I₂ Warm |

$$C_6H_8N_2 \text{ (E)} \qquad\qquad CHI_3$$

(AEB)

51 In the following reaction schemes, give the reaction conditions and write the structures of the intermediates **A**, **B**, **C**, **D**, **E** and **F**.

(a)

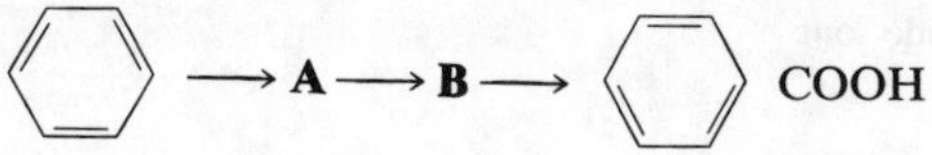

⬡ ⟶ **A** ⟶ **B** ⟶ ⬡COOH

(b)

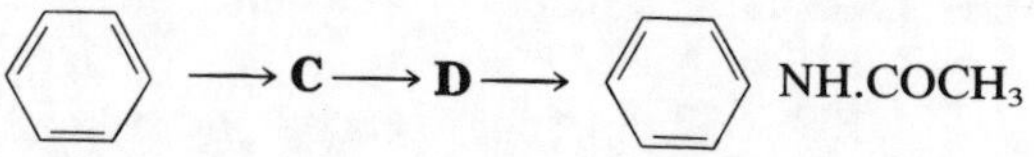

⬡ ⟶ **C** ⟶ **D** ⟶ ⬡NH.COCH₃

(c) CH₃OH ⟶ **E** ⟶ **F** ⟶ CH₃COOH

(AEB)

52 Suggest structures for the compounds shown in the following scheme:

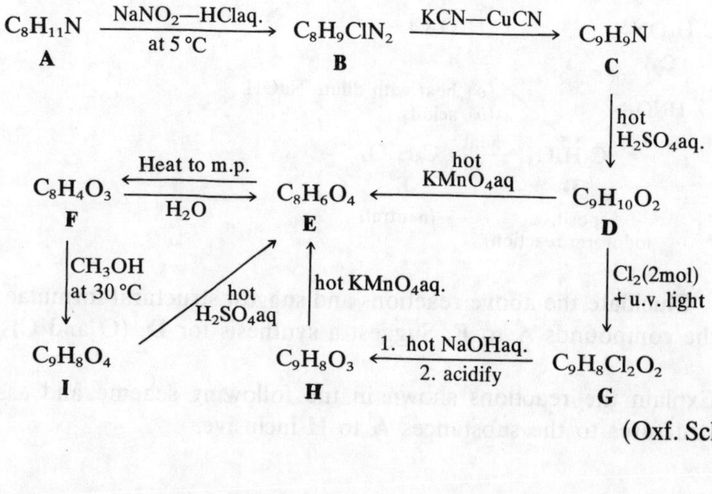

$$C_8H_{11}N \xrightarrow[\text{at 5 °C}]{NaNO_2-HClaq.} C_8H_9ClN_2 \xrightarrow{KCN-CuCN} C_9H_9N$$
$$\text{A} \qquad\qquad\qquad \text{B} \qquad\qquad\qquad \text{C}$$

$$\downarrow \substack{\text{hot} \\ H_2SO_4aq.}$$

$$C_8H_4O_3 \underset{H_2O}{\overset{\text{Heat to m.p.}}{\rightleftharpoons}} C_8H_6O_4 \overset{\text{hot}}{\underset{KMnO_4aq.}{\longleftarrow}} C_9H_{10}O_2$$
$$\text{F} \qquad\qquad\quad \text{E} \qquad\qquad\qquad \text{D}$$

| CH₃OH at 30 °C | hot H₂SO₄aq | hot KMnO₄aq. | Cl₂(2mol) + u.v. light |

$$C_9H_8O_4 \qquad\qquad C_9H_8O_3 \overset{\text{1. hot NaOHaq.}}{\underset{\text{2. acidify}}{\longleftarrow}} C_9H_8Cl_2O_2$$
$$\text{I} \qquad\qquad\qquad \text{H} \qquad\qquad\qquad\quad \text{G}$$

(Oxf. Schol)

53 Explain, without giving mechanisms, the reactions outlined in the following scheme. Assign structures to the substances **A** to **H** inclusive.

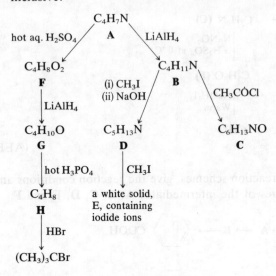

(O and C)

54 A compound **A** undergoes the following reactions:

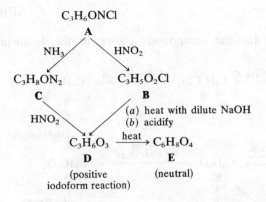

Elucidate the above reactions and suggest structural formulae for the compounds **A** to **E**. Suggest a synthesis for **D**. (O and C)

55 Explain the reactions shown in the following scheme and assign structures to the substances **A** to **H** inclusive.

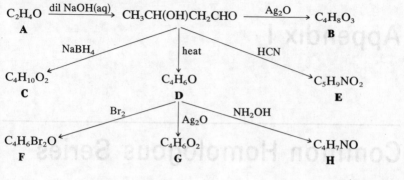

(O and C)

Appendix I

Common Homologous Series

Alkanes	$R-H$	Ketones	$R-\overset{\displaystyle O}{\overset{\|}{C}}-R'$
Alkenes*	$R-\overset{\displaystyle H}{\underset{}{C}}=CH_2$	Carboxylic acids*	$R-C{\overset{\displaystyle O}{\underset{OH}{}}}$
Alkynes*	$R-C\equiv CH$		
Halogenoalkanes (Alkyl halides)	$R-X$	Esters*	$R-C{\overset{\displaystyle O}{\underset{OR'}{}}}$
Primary alcohols*	$R-\overset{\displaystyle H}{\underset{H}{\overset{\|}{\underset{\|}{C}}}}-OH$		
Secondary alcohols	$R-\overset{\displaystyle R'}{\underset{H}{\overset{\|}{\underset{\|}{C}}}}-OH$	Acid anhydrides	$\begin{array}{c} R-C{\overset{\displaystyle O}{\underset{}{}}} \\ O \\ R'-C{\underset{\displaystyle O}{\overset{}{}}} \end{array}$
Tertiary alcohols	$R-\overset{\displaystyle R'}{\underset{R''}{\overset{\|}{\underset{\|}{C}}}}-OH$	Acid chlorides*	$R-C{\overset{\displaystyle O}{\underset{Cl}{}}}$
Ethers	$R-O-R'$		
Aldehydes*	$R-C{\overset{\displaystyle O}{\underset{H}{}}}$	Acid amides*	$R-C{\overset{\displaystyle O}{\underset{NH_2}{}}}$

Nitriles
(Cyanoalkanes)   R—C≡N
(alkyl cyanides)

$$\text{Cyano acids* } R-\overset{\displaystyle H}{\underset{\displaystyle CN}{C}}-COOH$$

Isonitriles
(Isocyanoalkanes)   R—N≝C
(alkyl isocyanides)

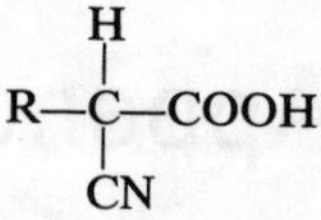

Primary amines   R—NH$_2$

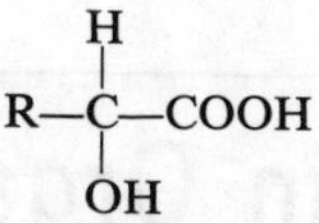

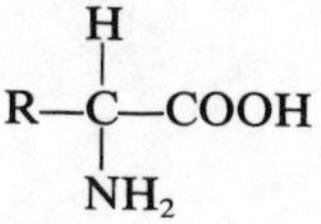

$$\text{Hydroxy acids* } R-\overset{\displaystyle H}{\underset{\displaystyle OH}{C}}-COOH$$

$$\text{Secondary amines } R-\overset{\displaystyle R'}{N}-H$$

$$\text{Tertiary amines } R-\overset{\displaystyle R'}{\underset{\displaystyle R''}{N}}$$

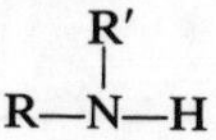

$$\text{Amino acids* } R-\overset{\displaystyle H}{\underset{\displaystyle NH_2}{C}}-COOH$$

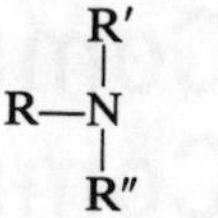

$$\text{Nitroalkanes } R-N\overset{\nearrow O}{\searrow_O}$$

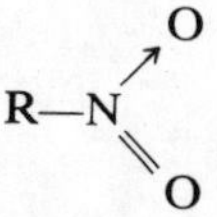

* In the series marked with an asterisk, R in the general formula can stand for an alkyl group or a hydrogen atom. In the other series, R stands only for an alkyl group. R, R' and R'' in the same general formula represent the same or different alkyl groups.

Appendix II

Common Radicals in Organic Compounds

$-CH_3$	Methyl (Me)	
$-C_2H_5$	Ethyl (Et)	
$-C_3H_7$	Propyl (Pr)	
$-C_4H_9$	Butyl (Bu)	
$-C_5H_{11}$	Pentyl	Alkyl groups (R)
$-C_6H_{13}$	Hexyl	
$-C_7H_{15}$	Heptyl	
$-C_8H_{17}$	Octyl	
$-C_9H_{19}$	Nonyl	
$-C_{10}H_{21}$	Decyl	
$-C_6H_5$	Phenyl (Ph)	
$-CH=CH_2$	Ethenyl (Vinyl)	

$-O-N(=O)_2$ (Nitrate V structure)	Nitrate(V)
$-O-N=O$	Nitrate(III) or Nitrite
$-N(=O)_2$ (Nitro structure)	Nitro
$-NH_2$	Primary amine, amino-
$=NH$	Secondary amine (imino)
$\equiv N$	Tertiary amine
$-C\equiv N$	Nitrile, cyanide or cyano-
$-N\overset{\shortmid}{\equiv}C$	Isocyanide or isocyano-

$$\begin{array}{c} O \\ \| \\ -O-S-OH \\ \| \\ O \end{array}$$ Hydrogensulphate(VI)

$$\begin{array}{c} O \\ \| \\ -S-OH \\ \| \\ O \end{array}$$ Sulphonic acid

—OH   Hydroxyl

$$\begin{array}{c} H \\ | \\ -C-OH \\ | \\ H \end{array}$$ Primary alcohol

$$\begin{array}{c} H \\ | \\ -C-OH \\ | \end{array}$$ Secondary alcohol

$$\equiv C-OH$$ Tertiary alcohol

—OCH$_3$ Methoxide }
—OC$_2$H$_5$ Ethoxide } Alkoxide groups (—OR)

$$\begin{array}{c} H \\ | \\ -C \\ \| \\ O \end{array}$$ Aldehyde

$$\begin{array}{c} \diagdown \\ C=O \\ \diagup \end{array}$$ Ketone (keto). Carbonyl group

$$\begin{array}{c} O \\ \| \\ H-C \end{array}$$ Methanoyl ⎫

$$\begin{array}{c} O \\ \| \\ CH_3-C \end{array}$$ Ethanoyl ⎪ Acyl groups

$$\begin{array}{c} O \\ \| \\ C_2H_5-C \end{array}$$ Propanoyl ⎪

$$\begin{array}{c} O \\ \| \\ \bigcirc-C \end{array}$$ Benzoyl ⎭

$$\begin{array}{c} O \\ \| \\ H-C \\ | \\ O- \end{array}$$ Methanoate ⎫

$$\begin{array}{c} O \\ \| \\ CH_3-C \\ | \\ O- \end{array}$$ Ethanoate ⎪ Exist also as anions

$$\begin{array}{c} O \\ \| \\ C_2H_5-C \\ | \\ O- \end{array}$$ Propanoate ⎪

$$\begin{array}{c} O \\ \| \\ \bigcirc-C \\ | \\ O- \end{array}$$ Benzoate ⎭

Answers to Numerical Questions

Chapter	4	(p. 80)		
			9.	$11.2 \, cm^3$
			14	$4 \, cm^3 \, CH_4$: $5.5 \, cm^3 \, H_2$: $0.5 \, cm^3 \, N_2$

	6	(p. 99)		
			1.	61% of CH_4 and 39% of C_3H_8
			5.	CH_4, 37.9 kPa; C_2H_6, 16.16 kPa; C_2H_2, 45.9 kPa
			6.	$209.2 \, kJ \, mol^{-1}$
			18.	(a) 8.04×10^8 (b) 2.6×10^{140}

	7	(p. 116)		
			15.	$12 \, cm^3$ of C_3H_8 and $12 \, cm^3$ of C_3H_6

	8	(p. 124)		
			4.	$44.8 \, cm^3$ of C_2H_6 and $67.2 \, cm^3$ of C_2H
			11.	33.3% each of CH_4, C_2H_4 and C_2H_2
			15.	(a) 5.37×10^{24}, (b) 4.79×10^{17} (c) 2.57×10^{42}

	13	(p. 185)		
			9.	$-278 \, kJ \, mol^{-1}$

	14	(p. 194)		
			3.	86.2
			4.	188.7

	15	(p. 206)		
			8.	1.44 g

	19	(p. 266)		
			17.	74; 1.02 g

	21	(p. 300)		
			19.	28.32%

Index

Notes: For best results in finding the subject, consult the *Contents* list at the beginning of the book as well as this index. Where inclusive page numbers are given, e.g. 259–61, the reference is to the main subject of the text; but where the / sign is used, e.g. 218/20, then the reference is of secondary importance and is not the main subject.